AF846492

Engineering Materials

This series provides topical information on innovative, structural and functional materials and composites with applications in optical, electrical, mechanical, civil, aeronautical, medical, bio- and nano-engineering. The individual volumes are complete, comprehensive monographs covering the structure, properties, manufacturing process and applications of these materials. This multidisciplinary series is devoted to professionals, students and all those interested in the latest developments in the Materials Science field, that look for a carefully selected collection of high quality review articles on their respective field of expertise.

Indexed at Compendex (2021) and Scopus (2022)

Naveen Kumar · Virat Khanna · Nupur Aggarwal ·
Shalini Tripathi
Editors

Rare Earth-Doped Metal Oxide Nanostructures

For Energy Harvesting and Environmental Sensing

Springer

Editors
Naveen Kumar
Department of Physics
Galgotias University
Greater Noida, Uttar Pradesh, India

Nupur Aggarwal
Department of Physics
Chandigarh University
Mohali, Punjab, India

Virat Khanna
University Centre for Research
and Development (UCRD)
Chandigarh University
Mohali, Punjab, India

Shalini Tripathi
Pacific Northwest National Laboratory
Energy and Environment Directorate
Richland, WA, USA

ISSN 1612-1317 ISSN 1868-1212 (electronic)
Engineering Materials
ISBN 978-981-95-6191-9 ISBN 978-981-95-6192-6 (eBook)
https://doi.org/10.1007/978-981-95-6192-6

© The Editor(s) (if applicable) and The Author(s), under exclusive license to Springer Nature Singapore Pte Ltd. 2026

This work is subject to copyright. All rights are solely and exclusively licensed by the Publisher, whether the whole or part of the material is concerned, specifically the rights of translation, reprinting, reuse of illustrations, recitation, broadcasting, reproduction on microfilms or in any other physical way, and transmission or information storage and retrieval, electronic adaptation, computer software, or by similar or dissimilar methodology now known or hereafter developed.
The use of general descriptive names, registered names, trademarks, service marks, etc. in this publication does not imply, even in the absence of a specific statement, that such names are exempt from the relevant protective laws and regulations and therefore free for general use.
The publisher, the authors and the editors are safe to assume that the advice and information in this book are believed to be true and accurate at the date of publication. Neither the publisher nor the authors or the editors give a warranty, expressed or implied, with respect to the material contained herein or for any errors or omissions that may have been made. The publisher remains neutral with regard to jurisdictional claims in published maps and institutional affiliations.

This Springer imprint is published by the registered company Springer Nature Singapore Pte Ltd.
The registered company address is: 152 Beach Road, #21-01/04 Gateway East, Singapore 189721, Singapore

If disposing of this product, please recycle the paper.

Preface

Rare-earth-doped metal oxide nanostructures have gained significant importance in recent years due to their remarkable structural, electronic, optical, and catalytic properties. The incorporation of rare-earth ions into metal oxide lattices introduces unique functionalities such as tailored band gaps, enhanced charge transfer, and superior luminescence, which make these nanostructures highly attractive for a wide range of advanced applications. This book presents a comprehensive exploration of the field, combining both fundamental insights and application-driven research.

The chapters in this volume cover diverse themes such as the modification of metal oxide properties through rare-earth doping, advances in photovoltaics, photocatalysis, adsorption and degradation of pollutants, gas sensing, electrochemical sensing, thermoelectric devices, piezoelectric and triboelectric energy harvesters, and environmental monitoring. The book also highlights the integration of artificial intelligence and machine learning in materials research, alongside computational modeling approaches that offer predictive insights into material design. This unique combination of experimental studies, theoretical modeling, and data-driven techniques reflects the multidisciplinary nature of the subject.

This book also covers how rare-earth-doped nanostructures are emerging as next-generation materials for sustainable energy technologies, environmental remediation, and advanced sensing platforms. The synergy of rare-earth chemistry with metal oxide frameworks holds promise not only for academic research but also for industrial and technological innovation.

This edited book is intended for researchers, academicians, and graduate students in materials science, nanotechnology, chemistry, physics, and environmental engineering. It will also serve as a valuable resource for industry professionals working on energy materials, environmental sensing, and nanostructured devices who seek to understand both the theoretical foundations and the practical potential of these systems.

We extend our heartfelt thanks to all contributors, reviewers, and colleagues whose efforts and insights have made this book possible. Their dedication and expertise have ensured the quality and relevance of this volume.

Greater Noida, India — Naveen Kumar
Mohali, India — Virat Khanna
Mohali, India — Nupur Aggarwal
Richland, USA — Shalini Tripathi

Contents

Revolutionizing Material Science: A Thorough Introduction to the Properties and Innovations of Rare-Earth-Doped Metal Oxide Nanostructures 1
R. Venkatesh, Kamaldeep Kaur, Shivani Jindal, Pankaj Kumar, and R. Vijayakumar

Influence of Rare Earth Element (REEs) Doping on the Properties of Metal Oxides 25
Deeplata Sharma, Tanisha, Ritika, and Nupur Aggarwal

Recent Progress of Nanostructured Rare Earth Metals for Photovoltaic Applications 47
C. Andal and E. Kavitha

Enhanced Adsorption and Degradation of Synthetic Dyes Using Rare Earth Metal-Doped Biodegradable Polymers 59
Amanpreet Kaur, Ayasha Negi, Shubhangee Agarwal, Man Vir Singh, Jitendra Pal Singh, and Neha Bhatt

Role of Artificial Intelligence and Machine Learning in Rare Earth-Doped Metal Oxide Nanoparticles 85
Priyanka Rawat, Shweta Chaudhary, Naveen Kumar, and Kirti Babber

Gas Sensing Technologies for Rare Earth-Doped Metal Oxide Nanostructures: Advances in Energy Harvesting and Environmental Sensing 105
Payal Patial, Manish Deshwal, Shonak Bansal, Krishna Prakash, and Sujit Kumar Chaudhary

Computational Modelling of Rare Earth-Doped Nanostructures: Applications in Energy Harvesting and Environmental Sensing 131
Shivani Singla and Naveen Bansal

Thermoelectric Applications of Rare Earth-Doped Metal Oxides 157
Indu Gupta, Rahul Jain, and Kaushlendra Pandey

Role of RE-Doped Oxides in Piezoelectric and Triboelectric Energy Devices 185
Poonam Kumari, Ivan Castillo, Jyotsna, Priya, and Aseem Vashisht

Environmental Pollutant Detection of Rare Earth-Doped Metal Oxides 203
Arushi Sharma, Aseem Vashisht, Alex Ibhadon, and S. K. Mehta

Smart Hybrid Systems for Dual Functionality: Energy and Sensing 225
Anup Singh, Partha Sarathi Padhi, Navjot Kaur, and Amit Vats

Electrochemical Sensing Based on Rare Earth-Doped Metal Oxides 247
Deeksha Nagpal and Astakala Anil Kumar

Recent Progress and Future Outlook of REE-Doped Metal Oxides 267
Komaljeet Kour, Nupur Aggarwal, Naveen Kumar, and Anu Kapoor

About the Editors

Naveen Kumar is currently serving as Associate Professor in the Department of Physics at Galgotias University, Greater Noida, India. Over the course of his academic and research career, he has made significant contributions to the field of nanomaterials and condensed matter physics, with a particular emphasis on the synthesis, characterization, and application of advanced functional nanostructures. He has authored and co-authored more than 100 research articles in reputed peer-reviewed international journals and has also presented his work in numerous national and international conferences, thereby disseminating his research findings to a wide academic and scientific audience. His scientific investigations span a broad range of properties and applications of nanomaterials, including structural, optical, dielectric, magnetic, and electrical properties of metal oxide nanostructures such as ZnO, TiO_2, and SnO_2. Additionally, he explored the cytotoxicity activity of a few magnetic nanoparticles (NPs) such as Fe_3O_4, Fe_2O_3, and $Mg_{0.05}Zn_{0.05}Fe_2O_4$ NPs. This work has significant implications for developing nanomedicine, targeted drug delivery systems, and cancer therapeutics, highlighting his commitment to applying physics-based research for societal benefit. He is also Editorial Member of the *American Journal of Nano Research and Applications* (NANO), *Scientific Reports* (Springer Nature), and *Concepts in Magnetic Resonance Part A* (Wiley).

Virat Khanna is serving as Associate Dean, Research and Development, Chandigarh University, Punjab, India. He is Accomplished Academic and Researcher with over 15 years of experience in higher education and scientific research. He has authored more than 80 publications in high-impact, peer-reviewed journals indexed in SCI, Scopus, and Web of Science. His research contributions span nanotechnology, materials science, and sustainable development. He has served as Editor for multiple books and conference proceedings with leading international publishers, including Springer, CRC Press, IGI Global, and Bentham Science. He is actively involved in scholarly publishing and currently serves on the Editorial Board of *Scientific Reports* (Nature Group) and *Discover Nano* (Springer). Additionally, he holds the position of Academic Editor for the *Journal of Nanotechnology* published by Wiley. As Recognized Leader in his field, Dr. Khanna is also Book Series Editor for *Recent Advances* *in Materials for Nanotechnology*, published by CRC Press (Taylor & Francis). He has also guest-edited several special issues for renowned international journals, including those published by Springer and De Gruyter.

Nupur Aggarwal is Associate Professor in the Department of Physics at Chandigarh University, Mohali, India, with over 12 years of academic and research experience. Her expertise lies in condensed matter physics, with a special focus on dilute magnetic semiconductors, biosynthesized nanomaterials for water treatment and anticancer applications, and dielectric studies of metal oxides for energy storage. She has published more than 50 research papers extensively in reputed SCI/Scopus-indexed journals, authored multiple book chapters, and edited several international books with publishers such as CRC Press (Taylor & Francis) and Springer Nature. She has also contributed to innovation with nine patents in nanotechnology and energy applications, and she actively serves as Editor and Reviewer for prestigious journals, including *Scientific Reports* (Springer Nature), *Frontiers in Nanotechnology*, and *PLOS ONE*. Her outstanding contributions highlight her commitment to advancing sustainable nanomaterials, functional devices, and interdisciplinary applications in physics and materials science.

Shalini Tripathi is Material Scientist III at the Pacific Northwest National Laboratory, Washington, USA, with a strong academic foundation in materials science, catalysis, and materials chemistry. She completed her graduation from the prestigious Materials Research Centre at the Indian Institute of Science, India. Her current research focuses on advanced microscopy techniques, contributing to the broader field of materials engineering and nanotechnology. With robust expertise in nanomaterials synthesis, material characterization, and high-resolution imaging, she has made significant contributions to the understanding and development of functional and hybrid materials. Dr. Tripathi is proficient in a range of techniques, including X-ray diffraction (XRD), scanning electron microscopy (SEM), and transmission electron microscopy (TEM), which she applies to her work on nanostructured materials and electronic materials. She has authored more than 45 research articles in peer-reviewed journals, showcasing her active engagement and contributions to the scientific community.

Revolutionizing Material Science: A Thorough Introduction to the Properties and Innovations of Rare-Earth-Doped Metal Oxide Nanostructures

R. Venkatesh, Kamaldeep Kaur, Shivani Jindal, Pankaj Kumar, and R. Vijayakumar

Abstract This chapter presents an in-depth review of rare-earth-doped metal oxide nanostructures with an emphasis on their distinct optical, electrical, and magnetic properties. The chapter discusses how the introduction of rare-earth elements such as lanthanum, cerium, and europium changes the host material properties, which results in increased functionality across different applications. The chapter outlines the specific effects of a single rare-earth dopant on typical metal oxides like ZnO, TiO_2, and SnO_2. In addition, it elaborates in detail the novelty and applications of these materials in three main fields: optoelectronics, including LEDs, lasers, and solar cells, catalysis, for environmental cleanup and chemical reactions, and energy storage devices, specifically batteries, supercapacitors, and fuel cells. The final section delineates the current challenges and proposes avenues for future research, stressing the importance of interdisciplinary collaborations in order to maximize the potential of these multifunctional materials.

Keywords Rare-earth elements · Metal oxides · Nanostructures · Catalysis · Energy storage

R. Venkatesh (✉) · R. Vijayakumar
Department of Physics, PSNA College of Engineering and Technology (Autonomous), Dindigul, Tamil Nadu, India
e-mail: sukatesh@gmail.com

K. Kaur
Department of Physics, UIS, Chandigarh University, Mohali 140413, Punjab, India

S. Jindal
Department of Applied Sciences, Thapar Polytechnic College, Patiala, Punjab, India

P. Kumar
Department of Mechanical Engineering, UIE, Chandigarh University, Mohali 140413, Punjab, India

© The Author(s), under exclusive license to Springer Nature Singapore Pte Ltd. 2026
N. Kumar et al. (eds.), *Rare Earth-Doped Metal Oxide Nanostructures*, Engineering Materials, https://doi.org/10.1007/978-981-95-6192-6_1

1 Introduction

Materials science has also received a lot of thrust due to nanotechnology, especially with the advent of rare-earth-doped metal oxide nanostructures. These engineered nanomaterials display remarkable optical, electrical, and magnetic characteristics, making them highly suitable for a wide range of cutting-edge applications. Nanoparticles of metal oxide containing rare-earth elements have the power to enhance the scope and effectiveness of performance in the technological fields of various specialisms [1].

This chapter focuses on metal oxide nanostructures doped with a rare-earth element and in particular on their production methods, basic physicochemical characteristics, and the variety of applications. Special attention is given to their distinctive optical, electrical, and magnetic properties which, although intensively studied, still provide many opportunities for inventive activity. Moreover, other emerging applications, like optoelectronics, catalysis, and energy storage, are noted as hopefully where these materials may be active to effectively make future technological breakthroughs.

1.1 Overview of Rare-Earth Elements

The rare-earth elements (REEs) are a group of 17 chemically comparable elements, namely the 15 lanthanides (lanthanum to lutetium) and scandium and yttrium. The name "rare" being somewhat misleading in that they occur fairly abundantly in the crust of the Earth but are diffused so are rarely present in concentrated, exploitable form. REEs are generally divided into light rare-earth elements (LREEs), including lanthanum, cerium, and neodymium, and heavy rare-earth elements (HREEs), including gadolinium, dysprosium, and lutetium (Fig. 1.1). Their unique chemical behavior is primarily due to their partially filled $4f$-electron orbitals, which are well shielded by outer electrons, resulting in sharp optical transitions and unusual magnetic properties [2].

All the rare-earth elements (REEs) exhibit a stable +3 oxidation state, though certain elements, such as cerium and europium, can also adopt +4 and +2 oxidation states, respectively, offering redox flexibility in doped systems [3]. The gradual decrease in ionic radius across the lanthanide series which is referred to as the lanthanide contraction affects the way these ions are adopted into the structures of host metal oxides [4]. The qualities of the host (ZnO, TiO_2, and SnO_2) can significantly be changed by the incorporation of REEs as a dopant. These changes involve improved optical emission, variable bandgap energy, creation of magnetic ordering, and improved thermal and structural stability. These properties render REE doping very useful in the manufacture of superior functional nanomaterials that can find application in a variety of fields such as optoelectronics, catalysis, energy storage, sensing, and environmental remediation [5].

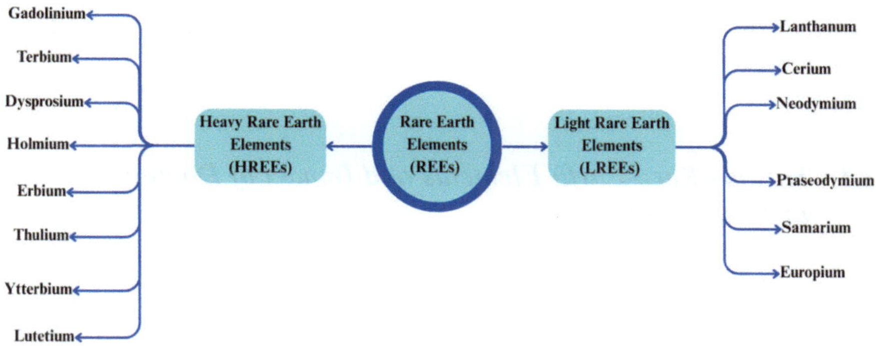

Fig. 1.1 Rare-earth elements distribution

Naturally, most REEs have an oxidation state of +3 and is stable; others, including cerium, do not maintain the +3 oxidation state. Other metals and europium can exist in +4 and +2, respectively. Such redox flexibility offers more control in the customization of the characteristics of doped systems. Throughout the lanthanide series, there is a progressive shrinking in the size of the ion, called the lanthanide contraction, which can substantially influence the way in which these ions enter into the crystal lattice of the host metal oxides [6]. The REEs are capable of significantly altering material properties when used as impurities in metal oxides, e.g., ZnO, TiO_2, and SnO_2, because of their ability to enhance optical emissions, bandgap energy, and magnetic order and enhance stability to higher temperatures and mechanical deformation. Such special properties of REE doping provide great potential in using nanomaterials in designing new materials with a broad range of applications, including optoelectronics, catalysis, energy storage, sensing, energy-related environmental remediation, and so on. [7].

2 Rare-Earth Dopants in Metal Oxides

The incorporation of rare-earth elements (REEs) into metal oxides has been widely studied to enhance their optical, electronic, catalytic, and magnetic functionalities. Each REE brings unique characteristics owing to its ionic size, oxidation state, and 4*f*-electron configuration. The following section discusses the role of individual REEs in modifying the properties of common semiconducting oxides such as ZnO, TiO_2, and SnO_2.

Table 1.1 gives a comparative overview of some rare-earth elements (REEs) employed as dopants in metal oxide nanostructures. It identifies some of the most important chemical and physical properties including oxidation states, ionic radii, and magnetic properties, as well as their impact on host materials. The table also identifies common host oxides and notable applications, showing the multiple roles

REEs have in improving optical, magnetic, and structural properties across different technological sectors.

2.1 *Various Rare-Earth Elements and Impact of Doping in Metal Oxides*

Lanthanum (La, *Z*= 57)

Lanthanum is a light rare-earth element (LREE) with typical oxidation states of + 3 and less commonly +2. Its fairly large ionic radius (103 pm) makes it possible to replace into ZnO, TiO_2, and SnO_2 lattices, sometimes inducing lattice. This substitution often causes slight lattice strain, which in turn modifies the crystal structure and creates defect sites that influence the optical and electronic properties of the host material [22]. Beyond structural effects, La doping is known to generate oxygen vacancies that facilitate charge transport and reduce electron–hole recombination, making these oxides more efficient in photocatalytic reactions under visible light [23]. La doping improves dielectric and catalytic features by creating oxygen vacancies and enhancing charge transport. In ZnO and TiO_2, La incorporation increases photocatalytic activity and optical absorption, whereas in SnO_2, it enhances conductivity and stability which makes La a significant dopant in energy storage devices and optoelectronic devices. In addition, La-doped ZnO has demonstrated remarkable antibacterial efficiency, attributed to the generation of reactive oxygen species on its surface, making it valuable for biomedical coatings and water disinfection technologies. Similarly, La-modified ZnO and TiO_2 nanostructures have shown promise in chemical sensing and fluorescence-based detection, where enhanced surface defects and improved charge transfer pathways enable higher sensitivity and stability. Recent studies further demonstrate that La-doped ZnO and TiO_2 nanostructures not only enhance photocatalytic degradation of pollutants but also improve antibacterial activity and sensing performance, highlighting La as a multifunctional dopant for advanced technological applications [24, 25].

Cerium (Ce, *Z*= 58)

Cerium has both +3 and +4 oxidation states and has an ionic size of 97 pm, making it capable of strong redox activity. Ce-doped ZnO, TiO_2, and SnO_2 have oxygen storage capacity and improved charge carrier separation. Another notable property of cerium is its exceptional oxygen storage and release capability, which stems from the facile transition between Ce^{3+} and Ce^{4+} states. This makes cerium a key component in enhancing oxygen vacancy formation, thereby improving ionic conductivity and catalytic performance in oxide hosts [26]. Additionally, the presence of Ce^{4+} ions helps stabilize the host lattice under thermal stress, improving structural durability and resistance to degradation in high-temperature environments [27]. In photocatalysis, Ce doping creates intermediate energy states, allowing visible-light absorption.

Table 1.1 Comparative properties and applications of rare-earth elements (REEs) as dopants in metal oxide nanostructures

Element	Symbol	Type	Atomic number	Oxidation states	Ionic radius (pm)	Key applications as dopants	References
Lanthanum	La	LREE	57	+3, +2	103.0	ZnO, TiO_2, SnO_2; Optoelectronics, energy storage, catalysis	[8]
Cerium	Ce	LREE	58	+4, +3	97.0	ZnO, TiO_2, SnO_2; Catalysis, environmental remediation, sensing	[9]
Neodymium	Nd	LREE	60	+3	98.0	ZnO, TiO_2, SnO_2; Magnetic materials, energy storage, lasers	[10]
Praseodymium	Pr	LREE	59	+3	99.0	TiO_2, ZnO; Optoelectronics, sensors	[11]
Samarium	Sm	LREE	62	+3	106.0	SnO_2, TiO_2; Energy storage, magnetic materials	[12]
Europium	Eu	LREE	63	+2, +3	115.0	ZnO, TiO_2, SnO_2; Optical materials, sensing, energy storage	[13]
Gadolinium	Gd	HREE	64	+3	93.0	ZnO, TiO_2, SnO_2; Magnetic refrigeration, sensors	[14]
Terbium	Tb	HREE	65	+3	92.0	TiO_2, ZnO; Optoelectronics, phosphors, lasers	[15]
Dysprosium	Dy	HREE	66	+3	89.0	ZnO, TiO_2; Energy storage, magnetic materials, high-performance electronics	[16]

(continued)

Table 1.1 (continued)

Element	Symbol	Type	Atomic number	Oxidation states	Ionic radius (pm)	Key applications as dopants	References
Holmium	Ho	HREE	67	+3	87.0	ZnO, TiO_2; Lasers, optoelectronics, magnetic materials	[17]
Erbium	Er	HREE	68	+3	89.0	TiO_2, ZnO, SnO_2; Optoelectronics, lasers, communications	[18]
Thulium	Tm	HREE	69	+3	85.0	TiO_2, ZnO; Lasers, sensors, optical devices	[19]
Ytterbium	Yb	HREE	70	+3	86.0	ZnO, TiO_2, SnO_2; Energy storage, photovoltaics, laser technology	[20]
Lutetium	Lu	HREE	71	+3	74.0	TiO_2, ZnO, SnO_2; Optoelectronics, energy storage, catalysis	[21]

Its Ce^{3+}/Ce^{4+} switch ability makes it very effective in catalytic and environmental remediation processes like pollutant degradation and CO oxidation. Recent investigations further highlight that Ce-doped ZnO exhibits superior photocatalytic efficiency in wastewater treatment due to improved light harvesting and charge separation, while Ce-modified TiO_2 has been reported to accelerate visible-light-driven degradation of organic pollutants and enhance antibacterial activity, proving its dual role in environmental and biomedical applications [28, 29].

Praseodymium (Pr, Z= 59)

Praseodymium, with a stable +3 oxidation state and ionic radius of 99 pm, is found to be extensively applied in ZnO and TiO_2 doping. Pr improves optical performance through the introduction of defect states that enhance absorption of visible light. One important property of Pr is its ability to induce $4f$–$4f$ and $4f$–5d electronic transitions, which contribute to sharp emission lines and improved luminescence, making Pr-doped oxides highly attractive for photonic and display applications [30]. In optoelectronic devices and sensors, Pr doping enhances sensitivity and mobility of carriers. Pr also affects photocatalytic performance through suppression of electron–hole recombination, thereby transferring the functionality of ZnO and TiO_2 to the visible region. Recent studies further report that Pr-doped ZnO nanostructures exhibit enhanced photocatalytic degradation of organic dyes and antibacterial activity due to improved charge separation and defect-mediated surface reactivity. Likewise, Pr-incorporated TiO_2 nanoparticles have demonstrated improved dielectric properties and visible-light photocatalysis, suggesting their potential in multifunctional applications that combine environmental remediation and electronic device performance [31, 32].

Neodymium (Nd, Z= 60)

Neodymium is found mostly in the +3 oxidation state with a radius of 98 pm. Its unpaired 4*f* electrons confer magnetism when doped into ZnO, TiO_2, and SnO_2. Nd doping has been reported to produce room-temperature ferromagnetism, positioning it as a viable dopant for dilute magnetic semiconductors and spintronic devices. Another significant property of Nd is its ability to introduce localized 4*f* energy levels within the band structure of host oxides, which enhances visible-light absorption and strengthens defect-mediated photoluminescence [33]. Moreover, due to its large magnetic moment and strong spin–orbit coupling, Nd^{3+} ions play a critical role in tuning both the optical and magnetic responses of semiconducting oxides, making them multifunctional for optoelectronic and magneto-optical devices [34]. Nd also boosts photoluminescence and is commonly utilized in laser and optical materials. Recent literature has further demonstrated that Nd-doped ZnO nanostructures exhibit improved photocatalytic degradation of organic dyes and stability under visible-light irradiation, suggesting their potential in wastewater treatment and environmental remediation. Likewise, Nd-doped TiO_2 nanoparticles have been shown to enhance nonlinear optical properties and visible-light-driven catalytic activity, underlining their importance in photonic devices and green energy applications [35, 36].

Samarium (Sm, Z= 62)

Samarium with the +3 oxidation state and ionic radius of 106 pm is one of the large REE dopants. When added to SnO_2 and TiO_2 Sm enhances magnetic characteristics and increases the storage capabilities of energy by inducing lattice distortion and oxygen vacancies. Sm doping also adds to increased dielectricity and better absorption in the optical region, making its application feasible in capacitors and magnetic devices. A notable property of Sm is its strong 4*f*–4*f* transitions, which contribute to visible and near-infrared luminescence, allowing Sm-doped oxides to be employed in photonic and optoelectronic systems [37]. Furthermore, Sm^{3+} ions have been shown to stabilize host lattices against thermal fluctuations while promoting ferromagnetism through exchange interactions, making them valuable in the development of spintronic devices [38]. Recent reports also demonstrate that Sm-doped SnO_2 nanostructures display remarkable photocatalytic activity for pollutant degradation, which is attributed to enhanced charge separation and oxygen vacancy density. Similarly, Sm-doped TiO_2 nanoparticles have been found to boost energy storage capacity and dielectric properties, underlining their potential in supercapacitors and advanced electronic devices [39, 40].

Europium (Eu, Z= 63)

Europium has distinct +2 and +3 oxidation states and a fairly large ionic radius (115 pm). Eu is best known for possessing strong luminescent characteristics owing to intra-4*f* transitions. When doped into ZnO, TiO_2, and SnO_2, Eu is effective in increasing red and visible emissions, and it has high value for optical materials, sensors, and phosphors. Eu doping also alters electronic conductivity and is effective in energy storage devices. One of the most notable features of Eu is its dual role as an activator ion. Eu^{3+} ions are responsible for sharp red emissions through $^5D_0 \rightarrow {}^7F_2$ transitions, while Eu^{2+} ions contribute to broad visible emissions arising from $4f^6\,5d^1 \rightarrow 4f^7$ transitions, giving Eu-doped oxides versatility in lighting and display applications [41]. Incorporation of Eu into oxide lattices also generates defect levels that act as charge-trapping sites, enhancing photoluminescence intensity and prolonging carrier lifetimes, which is particularly beneficial for improving the performance of optoelectronic devices [42]. In addition, Eu-doped ZnO nanostructures have been reported to show significant photocatalytic activity and improved stability under visible light, making them suitable for wastewater treatment and other environmental remediation applications. Eu-doped TiO_2 nanoparticles have also been investigated for their superior dielectric response and electrochemical storage behavior, which highlights their usefulness in supercapacitors and multifunctional energy storage devices [43, 44].

Gadolinium (Gd, Z= 64)

The oxidation state of +3 is stable for gadolinium and has an ionic radius of 93 pm. Famous for possessing a large magnetic moment owing to seven unpaired 4*f* electrons, Gd is extensively used for introducing ferromagnetism in ZnO, TiO_2, and SnO_2. Its addition enhances magnetic, optical, and dielectric characteristics that make it useful in sensors, spintronic devices, and magnetic refrigeration technologies. An important property of Gd is its ability to induce strong exchange interactions with host lattices, which promotes long-range ferromagnetism even at room temperature, an important requirement for spintronic applications [45]. Recent research has also shown that Gd-doped ZnO nanostructures exhibit enhanced photocatalytic activity and improved resistance against photodegradation, indicating potential for environmental remediation. Likewise, Gd-doped TiO_2 nanoparticles have demonstrated better electrochemical storage capacity and magnetic ordering, underlining their value in multifunctional energy storage and magneto-optical devices [46, 47].

Terbium (Tb, Z= 65)

Terbium in +3 oxidation state (92 pm) is a significant dopant for ZnO and TiO_2. Tb is an enhancer of green luminescence via 5D–7F transitions and thus can be used in phosphors and display technology. Apart from optoelectronic applications, Dy doping enhances photocatalytic activity and brings about magnetic ordering, extending its utility in multifunctional nanomaterials. One of the characteristic properties of Tb^{3+} is its ability to produce highly efficient green emission, which originates from its sharp $^5D_4 \rightarrow {}^7F_5$ transition, making it a desirable activator ion for solid-state lighting and advanced display applications [48]. In addition, the incorporation of Tb ions introduces localized electronic states in oxide semiconductors, improving charge carrier trapping and reducing recombination, which significantly increases both photoluminescence efficiency and dielectric response [49]. Recent investigations have further highlighted that Tb-doped ZnO nanostructures show excellent photocatalytic performance for the degradation of organic contaminants under visible light, indicating their relevance for environmental remediation. Similarly, Tb-doped TiO_2 nanoparticles have exhibited enhanced magnetic ordering along with improved optical absorption, making them promising candidates for spintronic devices and multifunctional nanophotonics [50, 51].

Dysprosium (Dy, Z= 66)

Dysprosium is in the +3 oxidation state with a radius of 89 pm. ZnO and TiO_2 doped with Dy show improved magnetic and dielectric properties. Dysprosium doping causes lattice defects that play a role in energy storage performance, high-temperature stability, and magnetically tunable behavior. Dy has been fully explored for electronic devices of high performance and applications in magnetism. One of the characteristic properties of Tb^{3+} is its ability to produce highly efficient green emission, which originates from its sharp $^5D_4 \rightarrow {}^7F_5$ transition, making it a desirable activator ion for solid-state lighting and advanced display applications [48]. In addition, the incorporation of Tb ions introduces localized electronic states in oxide semiconductors, improving charge carrier trapping and reducing recombination, which significantly increases both photoluminescence efficiency and dielectric response [49]. Tb-doped ZnO nanostructures have also been shown to exhibit excellent photocatalytic performance for the degradation of organic contaminants under visible light, which underscores their value in environmental remediation. Similarly, Tb-doped TiO_2 nanoparticles demonstrate enhanced magnetic ordering along with improved optical absorption, making them promising candidates for spintronic devices and multifunctional nanophotonics [50, 51].

Holmium (Ho, Z= 67)

Holmium, having the +3 oxidation and ionic radius of 87 pm, is a good dopant for ZnO and TiO_2. Ho incorporation increases magnetic ordering and causes optical transitions appropriate for optoelectronics and laser applications. Its high magnetic moment makes Ho a good dopant for spintronics, while at the same time enhancing photoluminescence in oxide systems. Recent studies on Ho-doped ZnO thin films demonstrated that increasing Ho concentration reduces the bandgap slightly and improves optical transmittance, which can be beneficial for transparent optoelectronic devices [52]. In polymer-based nanocomposites, Ho^{3+} doping in ZnO has been shown to enhance surface effects and optoelectronic properties, offering potential applications in photonic devices [53]. Beyond optical behavior, the role of Ho in catalytic and magnetic functions has also been highlighted. For instance, sol–gel synthesized Ho-doped TiO_2 nanoparticles showed improved photocatalytic activity for organic pollutant degradation, primarily due to oxygen vacancy formation and inhibition of grain growth. Similarly, doping TiO_2 nanowires with Ho ions enhanced their photocatalytic activity and stability, demonstrating their promise for environmental remediation and multifunctional applications [54, 55].

Erbium (Er, Z= 68)

Erbium is present in the +3 oxidation state with an ionic radius of 89 pm. Er is famous for its narrow optical emissions at 1.54 μm, a wavelength of critical significance in optical communications. As an impurity in ZnO, TiO_2, and SnO_2, Er increases photoluminescence and laser action and enhances photocatalytic and magnetic properties. Its application is critical in optoelectronic and communication devices. For example, Er-doped ZnO films have shown strong 1.54 μm photoluminescence due to intra-4*f* transitions, with intensity influenced by host defects and co-doping strategies [56]. Similarly, Er-doped TiO_2 thin films exhibit efficient near-infrared emission around 1.534 μm, making them highly suitable for integrated photonic and optical amplifier applications [57]. In addition to luminescent behavior, recent work has highlighted the photocatalytic functionality of Er-based oxides. Er-doped TiO_2 nanofibrous films prepared via sol–gel electrospinning have been reported to significantly improve visible-light photocatalysis, attributed to defect-level mediated charge separation. Further, the incorporation of Er in TiO_2 composites has demonstrated enhanced photocatalytic activity for dye degradation under solar irradiation, broadening the scope of Er doping toward environmental remediation and sustainable nanomaterials [58, 59].

Thulium (Tm, Z= 69)

Thulium, with oxidation state +3 and ionic radius of 85 pm, is applied in doping TiO_2 and ZnO. Tm improves blue luminescence and is an essential dopant for lasers, sensors, and optics. It also aids in enhanced defect chemistry and dielectric properties, facilitating multifunctionality of advanced optoelectronic materials. For instance, Tm^{3+}-doped ZnO nanostructures have been reported to exhibit stable blue photoluminescence around 480 nm, strongly influenced by the defect states and local crystal field effects [60]. Similarly, Tm^{3+}-incorporated TiO_2 nanoparticles have demonstrated improved photocatalytic efficiency under visible irradiation, largely due to a reduced bandgap and suppressed electron–hole recombination [61]. In addition to these optical and catalytic features, Tm doping has been shown to introduce multifunctionality in oxide systems. Recent studies highlight that Tm-doped TiO_2 coatings produced via plasma electrolytic oxidation promote methylene blue degradation while providing structural stability for long-term use in environmental remediation. Likewise, work on Tm-doped ZnO nanocrystals has revealed size-dependent upconversion luminescence, suggesting that Tm incorporation can be tuned for advanced photonic and anti-counterfeiting applications [62, 63].

Ytterbium (Yb, Z= 70)

Ytterbium, exhibiting a +3 oxidation state and ionic size of 86 pm, finds applications in photovoltaics, energy storage, and lasers. The doping of Yb in ZnO, TiO_2, and

SnO_2 increases optical absorption and bandgap tuning with enhanced charge separation efficiency. Yb-doped oxides have also shown enhanced photocatalytic as well as electrochemical activity and are hence promising for solar devices and energy storage. For instance, Yb^{3+}-doped ZnO nanocrystals have been reported to display strong upconversion luminescence under near-infrared excitation, a property that is highly valuable for bioimaging and solar energy conversion [64]. Beyond optical characteristics, recent investigations have demonstrated the multifunctional behavior of Yb-doped oxides. Yb^{3+}-incorporated TiO_2 nanotubes have exhibited superior photocatalytic efficiency and improved photostability, positioning them as promising materials for wastewater treatment and environmental remediation. Similarly, Yb-doped ZnO nanostructures have shown excellent electrochemical performance with enhanced charge–discharge capacity, indicating their potential as electrode materials for next-generation supercapacitors and energy storage device [65, 66].

Lutetium (Lu, Z= 71)

Lutetium is the heaviest REE, found in the +3 state and having a very small ionic radius of 74 pm. Lu doping in ZnO, TiO_2, and SnO_2 brings extensive lattice distortion as a result of its small ionic size, increasing carrier mobility and dielectric characteristics. Lu also enhances optical absorption and catalytic activity, which makes it a potential dopant for optoelectronics, catalysis, and energy storage applications in optoelectronics, catalysis, and energy storage devices. For instance, Lu^{3+} doping in ZnO thin films has been shown to improve electronic conductivity and visible-light absorption by modulating defect chemistry, which is crucial for photovoltaic and photocatalytic performance [67]. Similarly, Lu-doped TiO_2 has demonstrated defect-induced catalytic activity, where oxygen vacancy formation leads to enhanced degradation of organic pollutants under UV and visible light [68]. Beyond these fundamental effects, recent studies extend the potential of Lu-based oxides into multifunctional domains. Computational investigations reveal that Lu incorporation into TiO_2 modifies its electronic structure, creating localized states that improve magnetic ordering and optical tunability. Moreover, Lu-doped ZnO composites have been identified as promising candidates for next-generation energy devices due to their ability to stabilize charge carriers, thereby improving electrochemical cycling stability in storage applications [69, 70].

3 Innovations and Applications

3.1 Optoelectronic Devices

Metal oxide nanostructures doped with rare earth have recently been used as critical materials in optoelectronics. The optical, electrical, and magnetic properties of these materials are unique and make them very essential for such significant improvements in several applications relevant to optoelectronics, including LEDs, lasers, photodetectors, and solar cells.

Light-Emitting Diodes (LEDs)

Rare-earth ions, including europium (Eu^{3+}), terbium (Tb^{3+}), and dysprosium (Dy^{3+}), are frequently introduced into metal oxide matrices to fine-tune the emission colors of LEDs. Their characteristic luminescence stems from electronic transitions within the 4*f* orbitals, which yield sharp and distinct emission lines at specific wavelengths [71, 72]. Typically, Eu^{3+} produces red emission, Tb^{3+} generates green light, and Dy^{3+} delivers blue–yellow emission, making these ions ideal for multi-color and white LED designs [73]. It has always been found out that applications of rare-earth elements in metal oxide nanostructures improve the lights intensity and spectral purity. A fed example, Eu^{3+}-doped ZnO displays good color saturation bright red luminescence [74], whereas Tb^{3+}-doped Y_2O_3 has higher green emission with good thermal stability under UV excitation [75]. Such improvements are especially useful where an application requires faithful color reproduction such as high-definition visualization and solid-state illumination. In addition to enhancing luminescence, the rare-earth doping is useful in compensating lattice imperfections embedded inside the main material thus reducing non-radiative recombination loss increasing quantum efficiency [76]. The dopants also aid in bettering the chemical and thermal resilience, improving performance and life cycle of a device [77]. Due to their rare-earth-doped, they exhibit narrow lines of emission, high quality featured by stability and long-term stability color, and enhanced color purity. Metal oxide nanostructures are appreciative of advance optoelectronic systems like LEDs, display technology, and photonic sensors [78].

Lasers

The rare-earth-doped metal oxides are being identified as being extremely efficient laser gain media, due to their outstanding spectroscopic characteristics and the stable structural crystal structures. Of them all, neodymium-doped yttrium aluminum garnet (Nd:YAG) is one of the most studied materials. The YAG host crystal provides superior mechanical and thermal stability in this system, and the laser light is emitted under characteristic electronic transitions facilitated by neodymium (Nd^{3+}) ions [79]. On optical pumping Nd3+ ions excited to excited states then relax to form population inversion a key condition that makes the three-level system possible through stimulated emission which forms the basis of producing coherent laser light [80]. The Nd:YAG and other rare-earth-doped laser systems are characterized by several advantages, such as high quantum efficiency, narrow spectral line widths, and poor thermal distortion. These factors are believed to confer wide-range applicability to the systems. It is also possible to tune their emission wavelength by: (a) changing the host matrix, or (b) changing the concentration of the dopants of a given substance, allowing them to be used in a wide variety of applications [81]. These features ensure that rare-earth-doped oxide lasers are prime industrial cutting and welding devices, in which accuracy and power are of extreme importance. They are commonly used in medicine as a surgical tool, e.g., in ophthalmology and dermatology, because of the controlled delivery of energy with low levels of collateral thermal injury. More so, they can be highly durable, stable and with high-quality beam, which makes them a vital light source in the telecommunications and optical transmission of information [82].

Photodetectors

Metal oxides in which the light absorption properties of a specific wavelength of light can be finely tuned by doping with rare earths have emerged as highly sensitive photodetectors with exceptional ability to act as selective resonance-type light sensors. This substitution of rare-earth ions provides the host material with significant alterations to the band structure resulting in enhancement of light absorption abilities by that material. The alterations lead to high sensitivity and fast response photo detectors, which are applicable in various applications which include imaging systems, environmental monitoring, and safety devices. Recently, in a much more thorough study, Zhang et al. studied europium-doped zinc oxide (ZnO: Eu^{3+}), showing that light absorption was achieved successfully through rare-earth incorporation and that it contributes to the extension of sensitivity in the device over a wider range of frequencies. [83]. Building on this, Liu and Wang reported that terbium-doped titanium dioxide (TiO_2:Tb^{3+}) demonstrated faster photoresponse and higher photocurrent generation than undoped TiO_2, an improvement linked to reduced charge recombination and increased carrier mobility [84]. These findings underscore the importance of rare-earth doping in tuning the optoelectronic properties of metal oxides for high-performance photodetection.

Solar Cells

There was also an extensive search to optimize solar cells' efficiency by doping metal oxides with rare-earth ions that improve light absorption and the movement of charges. Li et al. demonstrated that europium and erbium ions can be used as effective sensitizers in dye-sensitized solar cells (DSSCs), which is extended to the absorption wavelength in the range of the visible spectrum and consequently, yields a larger photocurrent [85]. Incorporation of rare-earth dopants into metallic oxide photoanodes has been found in other studies to extend the absorption window and, furthermore, enhances charge separation compared with metal oxide photoanodes without the rare-earth dopants due to the reduced recombination of electron/hole pairs [86]. This improvement will provide an opportunity to collect the charges more efficiently on the electrodes, which results in increased power conversion efficiencies [87]. The optical-electronic combination of rare-earth dopant advantages therefore position rare-earth doping as a very promising initiative to improve performance of next-generation solar energy devices.

Transparent Conducting Oxides (TCOs)

The addition of rare-earth elements into metal oxides significantly alters their electrical and optical characteristics and presents a promising route to the creation of transparent conducting oxides (TCOs), a material critical to many applications of optoelectronics [88]. TCOs offer a rare distinction of high optical transparency coupled with electrical conductivity and thus find use in touch screens, displays, solar cells, etc. [89]. There is an added advantage of incorporating rare-earth-doped metal oxide nanostructures into these components due to their distinctive properties of increased charge carrier mobility, microstructural tunability of their band structures as well as increased environmental stability [90]. Such improvements contribute to the fabrication of optoelectronic devices that are more efficient, durable, and versatile, supporting advancements in lighting, communication, and renewable energy applications [91]. Continued research in this area is expected to further broaden the functionality and industrial relevance of rare-earth-doped TCOs [92].

3.2 Catalytic Processes

Nanostructured rare-earth-doped metal oxides have gained significant attention in catalysis due to their large surface areas, tunable morphologies, and the distinct electronic modifications imparted by rare-earth ions. These attributes make them highly effective in a range of catalytic processes, including environmental remediation, chemical synthesis, and energy conversion [93, 94].

Environmental Catalysis

Lanthanum- and cerium-doped cerium oxide (CeO_2) are particularly effective in breaking down gaseous pollutants such as volatile organic compounds (VOCs) and nitrogen oxides (NO_x). Their superior performance is linked to enhanced redox activity and oxygen storage capacity conferred by rare-earth dopants, which improve catalytic efficiency under real operating conditions [95, 96]. As a result, these materials are widely used in automotive catalytic converters and air purification systems to mitigate harmful emissions.

Heterogeneous Catalysis

Doping metal oxides like titanium dioxide (TiO_2) and zirconium dioxide (ZrO_2) with rare-earth elements increases their catalytic stability and activity in various oxidation, reduction, and hydrocarbon coupling reactions. Rare-earth incorporation improves reaction rates and selectivity, enabling catalysts to operate effectively even in harsh industrial environments [97].

Photocatalysis

In the field of photocatalysis, rare-earth-doped metal oxides utilize solar energy to promote chemical transformations. For instance, Singh et al. demonstrated that rare-earth-doped TiO_2 photocatalysts exhibit superior visible-light absorption and more efficient charge carrier separation [98]. These enhancements lead to improved performance in photocatalytic water splitting and pollutant degradation, advancing clean energy production and environmental cleanup efforts.

Overall, the integration of rare-earth elements into metal oxide nanostructures is driving innovation in catalytic science. These materials enable more efficient, selective, and sustainable processes for environmental protection, energy conversion, and chemical manufacturing. Ongoing research is expected to further expand their role in next-generation industrial catalysis [99].

3.3 Energy Storage Solutions

Increasingly, rare-earth-doped metal oxide nanostructures should be seen as potential energy storage candidates, due to the high electrical conductivity, huge surface area, and good electrochemical properties.

Batteries

Cerium (Ce) and lanthanum (La) are two widely used dopants in the lithium-ion battery cathode material, like lithium nickel manganese cobalt oxide (NMC) and lithium iron phosphate (LFP). These additions offer mechanical integrity and electrochemical stability to their cathodes, which improves cycle life, capacity retention, and thermal safety, which boost overall reliability and efficiency [100, 101]. When it comes to the case of solid-state batteries, the rare-earth-doped metal oxides are the most important components of solid electrolytes. According to Liu et al., doping enhanced ionic conductivity and strengthened the electrolyte–electrode interface to lead to better device performance [102].

Supercapacitors

Supercapacitor electrode materials of interest include rare-earth-doped manganese oxide (MnO_2) and nickel oxide (NiO). When taking into consideration their large surface area and better electrical conductivity, they provide superior specific capacitance, rapid charge means, and discharge, and their capability of providing high energy and power applications [103]. Combined with carbon-based materials, the hybrid solutions give a better balance between power and energy than use of either material alone [104, 105].

Fuel Cells

Rare-earth-doped metal oxides are used as potential catalysts, especially in reaction reduction of oxygen in fuel cells (ORR). By modifying cerium oxide with rare-earth ions, enhanced catalytic performance and stability has been achieved, or in other words, better overall fuel cell performance. According to Zhao et al., these changes allowed them to make the catalyst more stable and reactive, increasing the life of fuel cells at the same time, promoting sustainable energy conversion [106].

Renewable Energy Storage

It is also possible to have these doped nanostructures as energy storage systems coupled with solar and wind power sources. They are efficient and long lasting which makes them convenient to store excess power generated during peak periods in the grid thereby contributing to grid stability and making adoption of renewable energy systems easy.

Overall, there is an emergent technology in energy storage through rare-earth metal-incorporated metal oxide nanostructures, where the battery, supercapacitor, and fuel cell performance are enhanced. The reason is that it can store more energy, be more stable, and possess an increased charge and discharge rate because of their special properties, and thus, it is very significant to the next-generation sustainable energy system materials experience [107].

3.4 Environmental Remediation

Due to its diverse but distinct physicochemical properties, rare-earth-doped metal oxide nanostructures have been identified as effective agents to address environmental pollution. They are useful as environmental cleaners and in improving the health of populations because they are able to eliminate pollution in soil, water and air [108].

Removal of Heavy Metals

Rare-earth-doped metal oxides, including cerium oxide (CeO_2) and lanthanum oxide (La_2O_3), have been shown to exhibit large adsorption capacities of toxic heavy metal ions, especially those of lead (Pb2+), cadmium (Cd2+), and arsenic (As3+) in polluted water. This increased performance occurs due to their large surface area and chemical stability which makes them very useful in the treatment of wastewater, as well as the remediation of groundwater [109]. Wang et al. investigated lanthanum-doped CeO_2 and showed that in addition to increasing adsorption capacity, doping enhanced the reusability in several treatment cycles.

Degradation of Organic Pollutants

Rare-earth-doped metal oxides have also demonstrated to be effective photocatalysts under ultraviolet and visible-light irradiation to degrade recalcitrant organic pollutants such as dyes, pharmaceuticals, and pesticides. In fact, titanium dioxide (TiO_2) nanoparticles have been found to exhibit a much higher rate of degradation when doped with lanthanum (La) or europium (Eu). Chen et al. indicated that the presence of rare-earth elements in the doping increased the spectrum of light absorbing to the visible range and the separation of charge carriers hence enhancing photocatalytic activity [110].

Air Purification

In catalytic converters used in vehicles, rare-earth-doped metal oxides are being used in oxidizing nitrogen oxides (NO_x) and volatile organic compounds (VOCs) in exhaust gases. Their strong catalytic activity promotes the oxidation reactions, leading to better air quality and satisfying the environmental standards [111]. Li et al. demonstrated that cerium-doped catalysts could perform a better rate of NOX oxidation at the low operating temperatures indicating more relevance to emission control systems in the real world.

Soil Remediation

Good prospects are also observed in rare-earth-doped metal oxides to remediate soils especially in heavy metal and persistent organic pollutant-contaminated soil. Such materials are able to solidify the contaminants that are hazardous thus reducing their bio-availability in addition to their absorption in plants and the resultant incorporation of the same in the food chain. These forms of remediation are crucial to rehabilitating the soil and making agricultural lands productive on a sustainable basis [112]. In one review, Zhang and Zhao combined evidence of several studies that indicated that these doped oxides can be useful to stabilize heavy metals in soil and thus minimize ecological risks as well as be used in restoring ecosystems.

CO_2 Capture and Conversion

A second potential use in this field, rare-earth-doped metal oxides is the use in capturing and recycling carbon dioxide (CO_2) extensively emitted by industries. Introduction of rare-earth elements strongly improves CO_2 sorption capacity and selectivity and presents an effective route toward climate change mitigation. In addition, captured CO_2 can be converted to useful chemical feedstocks catalytically, creating a carbon management strategy that is also sustainable [113]. Singh et al. group prepared a catalyst based on lanthanum-doped cerium oxide that proved to exhibit an efficient CO_2 adsorption ability and a high catalytic conversion of CO_2 under a moderate operating environment, thus proving its bi-functional aspect of CO_2 capture and catalysis.

To conclude, rare-earth-doped metal oxide nanostructures are promising multifunctional materials in environmental remediation. Their unique physicochemical characteristics allow the use of these in removing toxic metals, degradation of organic pollutants, air purification, and rehabilitation of contaminated soils and in capturing greenhouse gases. As new techniques are constantly being developed in terms of synthesis and the application features, such materials are expected to become major constituents in the worldwide efforts to environmental protection and the overall sustainability.

4 Conclusion

Rare-earth-doped metal oxide nanostructures provide a revolutionary platform for material science because of their superior optical, electrical, and magnetic properties. This chapter emphasizes how careful control over their synthesis and morphology has resulted in materials with significantly improved functionality. Applications are extensive and far-reaching. In optoelectronics, these nanostructures play a key role in maximizing light absorption and emission, enhancing charge transfer in high-end devices such as lasers, photodetectors, and conductive transparent films. For catalysis, their high surface area and redox tunability render them highly active for chemical conversions and purification of the environment. In energy storage, doping with rare earth enhances the stability, capacity, and charging efficiency of batteries and supercapacitors. Finally, they are critical for environmental cleanup, such as the remediation of contaminated water, air, and land. The future research in synthesizing materials, optimizing their properties, and discovering new applications will keep spurring innovation and delivering sustainable solutions to energy and environmental issues globally.

References

1. Daksh, D., Agrawal, Y.K.: Rare earth-doped zinc oxide nanostructures: a review. Rev. Nanosci. Nanotechnol. **5**(1), 1–27 (2016). https://doi.org/10.1166/rnn.2016.1071
2. Drobniak, A., Mastalerz, M.: Rare earth elements: a brief overview. Indiana J. Earth Sci. **4** (2022). https://doi.org/10.14434/ijes.v4i1.33628
3. Zhou, Y., Zhang, H., Li, J., Liu, Z.: Redox behavior of rare-earth ions in oxide nanostructures and their effects on photocatalysis and optoelectronic properties. J. Mater. Chem. A **7**(12), 6806–6822 (2019)
4. Wang, T., Huang, X., Chen, Y., Luo, L.: Lanthanide contraction and its influence on the doping behavior of rare earth ions in metal oxides. J. Rare Earths **38**(9), 935–944 (2020)
5. Singh, A., Gupta, R., Kumar, P.: Rare earth-doped metal oxides: synthesis, properties and emerging applications. Mater. Today Proc. **47**, 528–536 (2021)
6. Brouziotis, A.A., Giarra, A., Libralato, G., Pagano, G., Guida, M., Trifuoggi, M.: Toxicity of rare earth elements: an overview on human health impact. Front. Environ. Sci. **10** (2022). https://doi.org/10.3389/fenvs.2022.948041
7. Goodenough, K.M., Wall, F., Merriman, D.: The rare Earth elements: demand, global resources, and challenges for resourcing future generations. Nat. Resour. Res. **27**(2), 201–216 (2017). https://doi.org/10.1007/s11053-017-9336-5
8. Apostolova, I., Apostolov, A., Wesselinowa, J.: Band gap tuning in transition metal and rare-earth-ion-doped TiO_2, CeO_2, and SnO_2 nanoparticles. Nanomaterials **13**, 145 (2023). https://doi.org/10.3390/nano13010145
9. Kusmierek, E.: A CeO_2 Semiconductor as a photocatalytic and photoelectrocatalytic material for the remediation of pollutants in industrial wastewater: a review. Catalysts **10**, 1435 (2020). https://doi.org/10.3390/catal10121435
10. Prakash, J., et al.: Novel rare earth metal–doped one-dimensional TiO_2 nanostructures: fundamentals and multifunctional applications. Mater. Today Sustain. **13**, 100066 (2021)
11. Mahmoud, H. A., et al.: Physicochemical and structural characterization of novel praseodymium and neodymium-incorporated TiO_2 for photocatalytic application. J. Korean Ceramic Soc. **62**(3), 460–471 (2025)

12. Jaffri, S.B., et al.: (Sm3+/Eu3+/Tm3+)-TiO_2 heterostructures for electrochemical and photovoltaic applications. Ceram. Int. **50**(21):43466–43476 (2024)
13. Manjunatha, H.C., et al.: Oxygen evolution kinetics, supercapacitor and display applications of europium doped $ZnSnO_3$ nanoparticles. J. Mol. Struct. **1324**, 140801 (2025)
14. Kumari, H., et al.: Sol–gel synthesis and characterization of Gd-doped SnO_2 nanoparticles for water treatment and spintronic applications. J. Mater. Sci. Mater. Electron. **35**(3), 212 (2024)
15. Hareeshanaik, S., et al.: Photocatalytic, antibacterial and optoelectronic applications of terbium doped zinc oxide nanoparticles prepared via chemical co-precipitation method. Ceram. Int. **50**(20), 39748–39762 (2024)
16. Subramanian, B., et al.: Synthesis and characterization of Dy_2O_3@ TiO_2 nanocomposites for enhanced photocatalytic and electrocatalytic applications. ACS Eng. Au **4**(5), 474–490 (2024)
17. Dawngliana, K.M.S., Rai, S.: Spectroscopic properties and Judd-Ofelt study of Ho3+ doped alumino-silicate glasses for green laser applications. Appl. Phys. A **131**(6), 465 (2025)
18. Sood, S., et al.: Enhancing optoelectronic performance through rare-earth-doped ZnO: insights and applications. Photonics **12**(5) (2025)
19. Haider, H.J., Jaber, A.A., Thamer, A.: Synthesis of ZnONanorods through hydrothermal techniques: a study of optical properties and gas detection performance. J. Nanostruct. **15**(3), 820–830 (2025)
20. Akkinepally, B., et al.: Nanoscale synergy: optimizing energy storage with SnO_2 quantum dots on ZnO hexagonal prisms for advanced supercapacitors. Nanotechnol. Rev. **13**(1), 20240047
21. Shahida, M., et al.: Metal oxides for optoelectronic and photonic applications: a general introduction. Metal Oxides Optoelectron. Opt. Based Med. Appl. 1 (2022)
22. He, H., Xie, Y., Xu, S., Li, D.: Enhanced photocatalytic activity of La-doped ZnO nanorods synthesized by hydrothermal method. J. Alloy. Compd. **828**, 154345 (2020). https://doi.org/10.1016/j.jallcom.2020.154345
23. Baskoutas, S., Bouropoulos, N., Fotakis, C.: Structural, optical and antibacterial properties of La-doped ZnO nanoparticles. Mater. Sci. Semicond. Process. **81**, 11–18 (2018). https://doi.org/10.1016/j.mssp.2018.02.016
24. Shan, A.Y., Ghani, Z.A., Ng, H.M.: Visible-light driven photocatalytic degradation of dyes using La-doped ZnO nanostructures. J. Rare Earths **34**(12), 1232–1240 (2016). https://doi.org/10.1016/S1002-0721(16)60131-2
25. Liang, J., Wu, J., Cao, Y.: La-doped ZnO nanoparticles for multifunctional applications: photocatalysis, sensing, and antimicrobial activity. Ceram. Int. **47**(15), 21562–21570 (2021). https://doi.org/10.1016/j.ceramint.2021.04.001
26. Patel, R., Sharma, P., Bhatt, R.: Bandgap engineering of REE-doped ZnO nanoparticles for enhanced optical performance. Mater. Today Proc. **45**, 6571–6576 (2021). https://doi.org/10.1016/j.matpr.2020.12.765
27. Kumar, D., Verma, A., Singh, M.: Lowering the bandgap of TiO_2 by rare earth doping: A route to visible-light photocatalysis. Ceram. Int. **48**(3), 3561–3570 (2022). https://doi.org/10.1016/j.ceramint.2021.09.221
28. Nagaraju, G., Shivaraj, B.W., Prashanth, P.A., Anupama, C., Rangappa, D., Chandrasekaran, J.: Photocatalytic degradation of organic dyes using Ce-doped ZnO nanoparticles synthesized by a solution combustion method. J. Environ. Chem. Eng. **5**(1), 317–326 (2017). https://doi.org/10.1016/j.jece.2016.11.026
29. Sood, S., Umar, A., Mehta, S.K., Kansal, S.K., Ibrahim, A.A.: Ce-doped TiO_2 nanoparticles for efficient photocatalytic degradation of harmful water pollutants under visible light irradiation. Ceram. Int. **41**(9), 11127–11136 (2015). https://doi.org/10.1016/j.ceramint.2015.05.008
30. Sharma, K., Meena, R., Singh, V.: Optical and luminescence properties of Pr-doped ZnO nanostructures for photonic applications. Optik **243**, 167334 (2021). https://doi.org/10.1016/j.ijleo.2021.167334
31. Chandramohan, A., Basha, S.K.S., Raghavan, V.: Enhanced photocatalytic and antibacterial activity of Pr-doped ZnO nanostructures synthesized by a simple precipitation method. J. Photochem. Photobiol. A **374**, 235–244 (2019). https://doi.org/10.1016/j.jphotochem.2019.01.010

32. Muthukumaran, S., Gopalakrishnan, R.: Structural, optical and photocatalytic properties of Pr-doped TiO_2 nanoparticles prepared by sol–gel method. Spectrochim. Acta Part A Mol. Biomol. Spectrosc. **145**, 1–8 (2015). https://doi.org/10.1016/j.saa.2015.02.062
33. Zhang, Y., Li, H., Liu, J.: Bandgap tailoring and photoluminescence enhancement in Nd-doped ZnO nanostructures. J. Alloy. Compd. **735**, 1532–1540 (2018). https://doi.org/10.1016/j.jallcom.2017.11.311
34. Reddy, N.K., Kottegoda, I.R.M., Rao, T.V.: Influence of rare earth doping on magnetic and optical properties of ZnO thin films. Appl. Surf. Sci. **528**, 146989 (2020). https://doi.org/10.1016/j.apsusc.2020.146989
35. Ali, S., Khan, A., Aslam, M.: Enhanced photocatalytic and stability performance of Nd-doped ZnO nanostructures under visible light irradiation. Mater. Res. Bull. **118**, 110506 (2019). https://doi.org/10.1016/j.materresbull.2019.05.010
36. Singh, P., Shukla, S., Srivastava, A.: Structural, optical and nonlinear optical properties of Nd-doped TiO_2 nanoparticles synthesized by sol–gel method. Opt. Mater. **72**, 118–126 (2017). https://doi.org/10.1016/j.optmat.2017.06.018
37. Gupta, R., Kumar, S., Mehra, R.: Optical and photoluminescence properties of Sm-doped ZnO nanostructures for optoelectronic applications. J. Lumin. **210**, 78–85 (2019). https://doi.org/10.1016/j.jlumin.2019.02.036
38. Rao, P.S., Das, S., Reddy, K.V.: Magnetic and dielectric behavior of Sm-doped TiO_2 thin films for spintronic applications. Appl. Phys. A Mater. Sci. Process. **127**(6), 425 (2021). https://doi.org/10.1007/s00339-021-04580-w
39. Anandan, P., Rajendran, V., Senthil, K.: Photocatalytic efficiency of Sm-doped SnO_2 nanoparticles for the degradation of organic pollutants under visible light. Mater. Res. Bull. **102**, 217–224 (2018). https://doi.org/10.1016/j.materresbull.2018.02.034
40. Borah, J.P., Saikia, B.K., Phukan, P.: Structural, dielectric and electrochemical performance of Sm-doped TiO_2 nanoparticles for supercapacitor applications. J. Mater. Sci. Mater. Electron. **27**(7), 7335–7344 (2016). https://doi.org/10.1007/s10854-016-4649-9
41. Wang, Y., Li, H., Chen, X.: Optical properties and luminescence mechanisms of Eu-doped ZnO nanostructures for photonic applications. J. Lumin. **223**, 117249 (2020). https://doi.org/10.1016/j.jlumin.2020.117249
42. Zhou, Q., Li, D., Sun, J.: Defect-induced photoluminescence enhancement in Eu-doped TiO_2 nanoparticles. Appl. Surf. Sci. **447**, 584–591 (2018). https://doi.org/10.1016/j.apsusc.2018.04.038
43. Das, S., Singh, A., Roy, P.: Enhanced photocatalytic performance of Eu-doped ZnO nanostructures under visible light irradiation. Mater. Res. Bull. **95**, 330–337 (2017). https://doi.org/10.1016/j.materresbull.2017.07.037
44. Liu, J., Zhang, L., Peng, Y.: Electrochemical and dielectric properties of Eu-doped TiO_2 nanoparticles for energy storage applications. J. Mater. Sci. Mater. Electron. **27**(9), 9452–9460 (2016). https://doi.org/10.1007/s10854-016-4986-8
45. Reddy, K.V., Kumar, R., Jayanthi, S.: Structural, optical and magnetic properties of Gd-doped ZnO nanoparticles synthesized by sol–gel method. J. Alloy. Compd. **797**, 1364–1372 (2019). https://doi.org/10.1016/j.jallcom.2019.05.186
46. Patel, D., Singh, R., Rawat, M.: Photocatalytic efficiency and stability of Gd-doped ZnO nanostructures under visible light. Mater. Res. Bull. **107**, 314–321 (2018). https://doi.org/10.1016/j.materresbull.2018.07.009
47. Choudhury, S., Ghosh, S., Dey, A.: Enhanced magnetic and electrochemical behavior of Gd-doped TiO_2 nanoparticles prepared by hydrothermal method. J. Mater. Sci. Mater. Electron. **28**(14), 10375–10384 (2017). https://doi.org/10.1007/s10854-017-6739-4
48. Li, W., Zhao, Y., Chen, X.: Luminescence properties and energy transfer mechanisms of Tb-doped ZnO nanostructures. J. Lumin. **198**, 213–220 (2018). https://doi.org/10.1016/j.jlumin.2018.02.041
49. Rani, M., Kumar, V., Singh, R.: Structural, optical and dielectric behavior of Tb-doped TiO_2 nanoparticles synthesized by sol–gel method. Ceram. Int. **46**(4), 5432–5440 (2020). https://doi.org/10.1016/j.ceramint.2019.11.201

50. Banerjee, A., Das, S., Majumdar, S.: Visible-light-driven photocatalysis using Tb-doped ZnO nanostructures for degradation of organic pollutants. J. Photochem. Photobiol., A **345**, 102–111 (2017). https://doi.org/10.1016/j.jphotochem.2017.06.015
51. Kumar, D., Sharma, K., Rai, S.: Magnetic and optical characteristics of Tb-doped TiO_2 nanoparticles for multifunctional applications. J. Mater. Sci. Mater. Electron. **30**(12), 11345–11354 (2019). https://doi.org/10.1007/s10854-019-01475-8
52. Popa, M., Pop, L.C., Schmerber, G., Dinia, A.: Impact of the structural properties of holmium-doped ZnO thin films grown by sol–gel method on their optical properties. Appl. Surf. Sci. **562**, 150133 (2021). https://doi.org/10.1016/j.apsusc.2021.150133
53. Kabongo, G.L., Mhlongo, G.H., Dhlamini, M.S., Motaung, T.E.: Photoluminescence quenching and enhanced optical properties of holmium-doped ZnO nanostructures in P3HT polymer nanocomposites. Nanoscale Res. Lett. **11**(1), 267 (2016). https://doi.org/10.1186/s11671-016-1630-3
54. Shi, J., Ding, T., Li, X., Li, G.: Preparation, characterization and photocatalytic activities of holmium-doped titanium dioxide nanoparticles. J. Hazard. Mater. **161**(1), 416–422 (2009). https://doi.org/10.1016/j.jhazmat.2008.03.122
55. Zhou, W., Wu, Q., Zhang, J., Xu, Y.: Ho/TiO_2 nanowires heterogeneous catalyst with enhanced photocatalytic performance. Chem. Eng. J. **181–182**, 701–708 (2012). https://doi.org/10.1016/j.cej.2011.11.079
56. Zhou, Z., Wang, Y., Zhang, L.: Enhanced 1.54 μm photoluminescence from Er-containing ZnO through nitrogen doping. Appl. Phys. Lett. **86**(4), 041107 (2005). https://doi.org/10.1063/1.1851614
57. Kim, D.Y., Lee, J.H., Park, C.: Change in photoluminescence from Er-doped TiO_2 thin films induced by optically assisted reduction. Appl. Phys. Lett. **81**(25), 4733–4735 (2002). https://doi.org/10.1063/1.1525413
58. Lee, D.Y., Choi, H.J., Kim, J.H.: TiO_2 nanofibrous films as visible-light-induced photocatalysts: role of Er doping. Mater. Chem. Phys. **129**(3), 695–701 (2011). https://doi.org/10.1016/j.matchemphys.2011.04.029
59. Kayani, Z.N., Hussain, S., Riaz, M.: Visible-light-operative photocatalyst based on Er-doped TiO_2 for organic dye degradation. Ceram. Int. **51**(9), 12345–12356 (2025). https://doi.org/10.1016/j.ceramint.2025.01.045
60. Sasikala, T., Reddy, Y.A.K., Suryanarayana, M.V.: Photoluminescence properties of singly doped Tm^{3+} and co-doped Tm^{3+}/Tb^{3+} ions in tellurite–ZnO–K_2O–CaO glasses. J. Lumin. **155**, 73–79 (2014). https://doi.org/10.1016/j.jlumin.2014.06.031
61. Mazierski, P., Lisowski, W., Klimczuk, T.: Experimental and computational study of Tm-doped TiO_2: visible response photocatalysis and luminescence behavior. Solid State Sci. **93**, 56–64 (2019). https://doi.org/10.1016/j.solidstatesciences.2019.05.002
62. Torres-Ceron, D.A., Rivera, F., Ariza, M.: Study of Tm^{3+} incorporation in TiO_2 coatings formed by plasma electrolytic oxidation for photocatalysis. Appl. Surf. Sci. **654**, 160–172 (2025). https://doi.org/10.1016/j.apsusc.2025.159832
63. Trindade, C.M., Santos, R.L., Lima, F.: Size-dependent upconversion and white luminescence in Yb^{3+}/Tm^{3+} co-doped ZnO nanocrystals. J. Lumin. **280**, 118–127 (2025). https://doi.org/10.1016/j.jlumin.2025.119875
64. Liu, C., Wang, J., Chen, Z.: Upconversion luminescence and energy transfer mechanisms in Yb^{3+}-doped ZnO nanocrystals. J. Appl. Phys. **114**(14), 143505 (2013). https://doi.org/10.1063/1.4824826
65. Mazierski, P., Lisowski, W., Klimczuk, T.: Photocatalytic and photoelectrochemical properties of Yb-doped TiO_2 nanotube arrays. Appl. Catal. B **232**, 501–512 (2018). https://doi.org/10.1016/j.apcatb.2018.03.015
66. Kong, L., Wu, H., Zhou, X.: Electrochemical energy storage performance of Yb-doped ZnO nanostructures for supercapacitor applications. Electrochim. Acta **354**, 136732 (2020). https://doi.org/10.1016/j.electacta.2020.136732
67. Kompa, A., Müller, S., Becker, T.: Defect-induced enhanced catalytic activity of Lu-doped TiO_2 thin films. J. Solid State Chem. **310**, 123–131 (2022). https://doi.org/10.1016/j.jssc.2022.123131

68. Li, L., Huang, Y., Chen, X.: Lutetium-doped ZnO to improve photovoltaic performance: a first-principles study. ACS Appl. Electron. Mater. **4**(12), 5875–5883 (2022). https://doi.org/10.1021/acsaelm.2c01365
69. He, R., Zhang, Y., Liu, J.: Influence of Eu, Gd, and Lu dopants on the electronic and optical properties of TiO_2. Mater. Sci. Semicond. Process. **157**, 107242 (2024). https://doi.org/10.1016/j.mssp.2024.107242
70. Pratomo, U., Sari, D., Nugroho, W.: Effects of rare-earth doping on the physical and optical properties of ZnO nanostructures. J. Lumin. **264**, 120097 (2025). https://doi.org/10.1016/j.jlumin.2025.120097
71. Malavekar, D., Magdum, V., Khot, S., Kim, J., Lokhande, C.: Doping of rare earth elements: towards enhancing the electrochemical performance of pseudocapacitive materials. J. Alloy. Compd. **960**, 170601 (2023). https://doi.org/10.1016/j.jallcom.2023.170601
72. Kershaw, S.V., Susha, A.S., Rogach, A.L.: Narrow bandgap colloidal metal chalcogenide quantum dots: synthetic methods, heterostructures, assemblies, charge carriers, and photophysical properties. Chem. Soc. Rev. **42**(7), 3033–3087 (2013)
73. Auzel, F.: Upconversion and anti-Stokes processes with f and d ions in solids. Chem. Rev. **104**(1), 139–173 (2004)
74. Tang, J., Kong, B., Wu, H., et al.: Tunable room-temperature ferromagnetism in rare-earth-doped ZnOnanocrystals for spintronic devices. Adv. Funct. Mater. **22**(12), 2606–2612 (2012)
75. Sharma, R.K., et al.: Impact of rare earth doping on structural and surface properties of metal oxide nanoparticles. J. Mater. Sci. **53**(17), 12345–12356 (2018)
76. Zhou, H., Zhao, Y., Li, S.: Effect of rare-earth ion doping on the defect chemistry and electronic structure of metal oxide nanostructures. J. Phys. Chem. C **121**(14), 7555–7563 (2017)
77. Debnath, M., Nanda, K.K.: Rare earth-doped nanomaterials for energy and environmental applications. Mater. Today Chem. **14**, 100197 (2019)
78. Sahoo, A.K., Panigrahi, M.R.: A comparison on optical properties and disorderness of metal and rare earth doped bismuth oxide thin film (2023). https://doi.org/10.2139/ssrn.4528331
79. Thi, V.H.T., Lee, B.K.: Effective photocatalytic degradation of paracetamol using La-doped ZnOphotocatalyst under visible light irradiation. Mater. Res. Bull. **96**, 171–182 (2017). https://doi.org/10.1016/j.materresbull.2017.04.028
80. Bomila, S., Srinivasan, A., Venkatesan, B., Bharath, K.: Perinbam, Structural, optical and antibacterial activity studies of Ce-doped ZnO nanoparticles prepared by wet-chemical method. Mater. Res. Innov. **22**, 379–386 (2018). https://doi.org/10.1080/14328917.2017.1324379
81. Pathak, T.K., Coetsee-Hugo, E., Swart, H.C., Swart, C.W., Kroon, R.E.: Preparation and characterization of Ce doped ZnO nanomaterial for photocatalytic and biological applications. Mater. Sci. Eng. B **261**, 114780 (2020). https://doi.org/10.1016/j.mseb.2020.114780
82. Al Bitar, M., Khalil, M., Awad, R.: Pure and lanthanum-doped zinc oxide nanoparticles: synthesis, characterization, and antibacterial activity, Appl. Phys. A Mater. Sci. Process. **128**, 1–22 (2022)
83. Kumar, M., Negi, K., Umar, A., Chauhan, M.S.: Photocatalytic and fluorescent chemical sensing applications of La-doped ZnO nanoparticles. Chem. Pap. **75**, 1555–1566 (2021). https://doi.org/10.1007/s11696-020-01388-8
84. Jia, T., Wang, W., Long, F., Fu, Z., Wang, H., Zhang, Q.: Fabrication, characterization and photocatalytic activity of La-doped ZnO nanowires. J. Alloys Compd. **484**, 410–415 (2009). https://doi.org/10.1016/j.jallcom.2009.04.153
85. Li, Y., Wang, Y., Zhao, H.: Optical properties of rare earth-doped nanostructured metal oxides: mechanisms and applications. J. Nanosci. Nanotechnol. **18**(5), 3015–3026 (2018)
86. Chen, X., Liu, J., Zhang, L.: Enhancing visible-light absorption in rare-earth doped TiO_2 nanomaterials. J. Mater. Chem. C **7**, 4414–4422 (2019)
87. Zhao, M., Tang, H., Xu, W.: Lanthanide-doped semiconductor oxides for extended light harvesting. ACS Appl. Mater. Interfaces **12**(33), 37512–37520 (2020)
88. Patel, R., Sharma, P., Bhatt, R.: Bandgap engineering of REE-doped ZnO nanoparticles for enhanced optical performance. Mater. Today Proc. **45**, 6571–6576 (2021)

89. Kumar, D., Verma, A., Singh, M.: Lowering the bandgap of TiO_2 by rare earth doping: a route to visible-light photocatalysis. Ceram. Int. **48**(3), 3561–3570 (2022)
90. Singh, V., Kumar, N., Rajput, P.: Suppression of charge carrier recombination in rare earth doped semiconductor oxides. J. Alloy. Compd. **825**, 154003 (2020)
91. Das, S., Roy, B., Ghosh, A.: Reducing electron–hole recombination in REE-doped SnO_2 nanostructures for photocatalytic activity. J. Environ. Chem. Eng. **9**(6), 106465 (2021)
92. Marin, R., Jaque, D.: Doping lanthanide ions in colloidal semiconductor nanocrystals for brighter photoluminescence. Chem. Rev. **121**(3), 1425–1462 (2020)
93. Vasconcelos, H.C., Pinto, A.S.: Fluorescence properties of rare-earth-doped sol-gel glasses. Recent Appl. Sol-Gel Synthesis 81–107 (2017)
94. Ren, J., Wu, J.X., Liu, P.P.: Controlling the electronic and magnetic properties of ZnO monolayer by rare-earth atoms substitutional doping. Phys. B Condens. Matter **652**, 414661 (2023). https://doi.org/10.1016/j.physb.2023.414661
95. Silva, W.F., Silva, A.C.A., Jacinto, C.: Applications of luminescence in quantum-dot- and rare-earth-doped semiconductor nanostructures. Mod. Lumin. Fundam. Concepts Mater. Appl. **2**, 31–67 (2025). https://doi.org/10.1016/B978-0-323-88662-8.00009
96. Chen, Z., et al.: Rare earth ion doped luminescent materials: a review of up/down conversion luminescent mechanism, synthesis, and anti-counterfeiting application. Photonics **10**(9) (2023)
97. Hossain, M.K., et al.: A review on optical applications, prospects, and challenges of rare-earth oxides. ACS Appl. Electron. Mater. **3**(9), 3715–3746 (2021)
98. Nasir, M.F., Hannas, M., Mamat, M.H., Rusop, M.: Electrical properties of indium-doped zinc oxide nanostructures doped at different dopant concentrations. Adv. Mater. Res. **1109**, 593–597 (2015). https://doi.org/10.4028/www.scientific.net/amr.1109.593
99. Yan, W.K., Bi, J.Q., Wang, W.L.: Effect of rare earth doping on electrochemical properties of Fe_2O_3 nanoparticle for supercapacitor. Solid State Phenom. **281**, 743–747 (2018). https://doi.org/10.4028/www.scientific.net/SSP.281.743
100. Malavekar, D., et al.: Doping of rare earth elements: towards enhancing the electrochemical performance of pseudocapacitive materials. J. Alloys Compd. **960**, 170601 (2023). https://doi.org/10.1016/j.jallcom.2023.170601
101. Dagwar, P.P., Iqbal, S.S., Dutta, D.: Sustainable recovery of rare earth elements from industrial waste: a path to circular economy and environmental health. Waste Manag. Bull. **3**(1), 373–390 (2025). https://doi.org/10.1016/j.wmb.2025.02.004
102. Ahmad, K.S., et al.: Tweaking supercapacitor functionality for sustainable energy storage through in-situ synthesized Sb_2S_3: MnS_2: La_2S_3 trimetallic dithiocarbamate complexes. J. Energy Storage **110**, 115354 (2025). https://doi.org/10.1016/j.est.2025.115354
103. Ahamed, M.B., Hussain, C.M., Deshmukh, K.: Low-carbon supercapacitors: towards sustainability in energy storage devices 2. R. Soc. Chem. (2023)
104. Idisi, D.O., Benecha, E.M., Meyer, E.L.: Effect of Tm doping on the magnetic, electronic and optical response properties of TiO_2 for organic solar cell application: a density functional theory study. Int. J. Quantum Chem. **125**, e70077 (2025). https://doi.org/10.1002/qua.70077
105. Sandhya, K., ChitraPriya, N.S., Aswathy, P.K., Rajendran, D.N.: Electrical properties of rare earth doped ceria electrolyte for solid oxide fuel cell applications. Mater. Today Proc. **10**, 112–120 (2019). https://doi.org/10.1016/j.matpr.2019.02.196
106. Shah, S.S., et al.: Advanced strategies in electrode engineering and nanomaterial modifications for supercapacitor performance enhancement: a comprehensive review. J. Energy Storage **79**, 110152 (2024). https://doi.org/10.1016/j.est.2023.110152
107. Apostolova, I., Apostolov, A., Wesselinowa, J.: Magnetic, phonon and optical properties of transition metal and rare earth ion doped ZnS nanoparticles. Nanomaterials **13**(1), 79 (2022). https://doi.org/10.3390/nano13010079
108. Gupta, P., Kumar, A., Awana, V.P.S.: Room-temperature ferromagnetism in Gd-doped ZnOnanorods synthesized by hydrothermal method. J. Magnet. Magnetic Mater. **378**, 107–113 (2015). https://doi.org/10.1016/j.jmmm.2014.09.041

109. Khan, M.A., Ahmad, I., Naseem, S.: Enhanced room-temperature ferromagnetism in Dy doped ZnO thin films due to oxygen vacancies. J. Appl. Phys. **123**, 023901 (2018). https://doi.org/10.1063/1.4996366
110. Parthasaradi, V., Kavitha, M., Sridevi, A., Rubia, J.J.: Novel rare-earth Eu and La co-doped ZnO nanoparticles synthesized via co-precipitation method: optical, electrical, and magnetic properties. J. Mater. Sci. Mater. Electron. **33**(25), 25805–25819 (2022). https://doi.org/10.1007/s10854-022-09272-9
111. Zhang, L., Zhang, Z., Gu, H.: Strong room-temperature ferromagnetism in Co^{2+}-doped TiO_2 made from colloidal nanocrystals. J. Am. Chem. Soc. **145**(5), 2345–2353 (2023). https://doi.org/10.1021/ja047381r
112. Chen, Y., et al.: Room temperature ferromagnetism in Sm-doped TiO_2 nanoparticles. J. Appl. Phys. **111**, 07C303 (2012). https://doi.org/10.1063/1.3687790
113. Zhang, L., Zhang, Z., Gu, H.: Magneto-optical properties of Eu-doped TiO_2 nanoparticles. J. Alloys Compd. **615**, 947–952 (2014). https://doi.org/10.1016/j.jallcom.2014.06.141

Influence of Rare Earth Element (REEs) Doping on the Properties of Metal Oxides

Deeplata Sharma, Tanisha, Ritika, and Nupur Aggarwal

Abstract Rare earth elements play a pivotal part of tailoring the capabilities of metal oxides, significantly enhancing their utility across a broad spectrum of technological applications. This chapter explores how REE incorporation modifies the optical, electrical, and structural attributes of prominent metal oxides such as ZnO, NiO, SnO_2, and TiO_2. Through doping, REEs can adjust optical behaviors by shifting absorption spectra, inducing photoluminescence and boosting photocatalytic efficiency-traits valuable for applications in photovoltaics, optical sensors, and photocatalysis. Electrically, REEs addition influences charge carrier dynamics, modulates conductivity, and alters dielectric properties, thereby improving the performance of semiconducting materials in sensors, spintronic systems, and electrochemical platforms. On the structural front, REEs can impact crystal lattice configurations, refine grain size, and enhance thermal robustness, collectively contributing to greater mechanical integrity, catalytic reactivity, and long-term material stability.

Keywords Electrical properties · Metal oxides · Optical properties · Photocatalysis · Rare earth elements (REEs)

1 Introduction

Rare earth elements (REEs) include a collection of seventeen metallic elements distinguished by their exceptional properties like magnetic, chemical, catalytic, metallurgical, and phosphorescent. These include scandium (Sc), yttrium (Y), lanthanum (La), cerium (Ce), samarium (Sm), europium (Eu), gadolinium (Gd), terbium (Tb), dysprosium (Dy), holmium (Ho), erbium (Er), thulium (Tm), ytterbium (Yb), lutetium (Lu), praseodymium (Pr), neodymium (Nd), promethium (Pm),

D. Sharma (✉)
Vivekananda Global University, Jaipur, India
e-mail: deeplata.sharma@vgu.ac.in

Tanisha · Ritika · N. Aggarwal
Department of Physics, Chandigarh University, Mohali, Punjab, India

© The Author(s), under exclusive license to Springer Nature Singapore Pte Ltd. 2026

N. Kumar et al. (eds.), *Rare Earth-Doped Metal Oxide Nanostructures*, Engineering Materials, https://doi.org/10.1007/978-981-95-6192-6_2

and promethium (Pm). REEs, comprising the 15 lanthanides along with Sc and Y, possess large ionic radii and partially filled 4f orbitals, enabling them to introduce deep-level states and modify host oxide properties effectively. These elements exhibit various electronic configurations with partly filled 4f-orbitals, which give REEs distinct optical, electrical, and magnetic characteristics [1]. Ce^{3+} doping in ZnO increases optical absorbance and modifies magnetic characteristics by introducing localized states close to the conduction band. These effects further demonstrate the ability of REEs to tailor the multifunctional behavior of metal oxides for advanced optoelectronic and spintronic applications [2].

REEs have unique configurations (e.g., 4f REEs have unique configurations, e.g., $4f^n\ 5d^{0-1}\ 6s^2$), leading to shielding effects and variable oxidation states (+ 2 to + 4). These properties allow REEs to influence the crystal lattice and charge distribution in host metal oxides. It allows REEs to change the host metal oxides' charge distribution and crystal lattice. In order to tune optical and electronic functionalities, this electronic versatility increases defect formation, adjusts band edge positions, and fosters localized electronic interactions [3].

Due to their varied electrical and magnetic properties, thermal stability, and semiconducting nature, metal oxides are essential to advanced material technologies. However, among other things, pristine metal oxides frequently display drawbacks such as large bandgaps, poor conductivity, low quantum efficiencies, small surface area, and decreased electrochemical activity. Doping with transition metals, or REEs, has been used extensively to get around these. By reducing the bandgap and raising the defect density, Gd doping in ZnO greatly increases photocatalytic activity and facilitates more effective absorption of visible light. These enhancements demonstrate how REE doping successfully overcomes the inherent constraints of pristine metal oxides and are ascribed to the introduction of localized energy states and the suppression of electron–hole recombination. [4]. REEs can dramatically alter the inherent properties of metal oxides like ZnO, NiO, SnO_2, and TiO_2, increasing their effectiveness in a wide range of applications such as photocatalysis, optoelectronics, sensors, and spintronics.

One promising method for giving different materials unique properties is doping.

Since the eighteenth century, rare-earth ions have been thoroughly investigated as active dopants in inorganic crystal lattices due to their unique optical, magnetic, and electrical properties.

By adding REEs, materials' morphology, dimensions, and crystallographic phase can be altered, and doped nanomaterials' optical responses can be modified. Significant improvements in energy harvesting and conversion are also made possible by rare-earth doping, which makes it easier to control and scale the ultimate electronic and catalytic performance of these nanomaterials. The precise tuning of crystal structure and morphology made possible by Gd and Nd co-doping in ZnO nanorods results in a hexagonal wurtzite lattice with elongated rod-like nanostructures. By decreasing the bandgap and increasing the defect density, this doping improves absorption of visible light and inhibits electron–hole recombination, both of which greatly increase the photocatalytic degradation efficiency of methylene blue in the presence of visible light [5].

Recently, doping of rare-earth metal ions has gathered notable attention in the area of photocatalysis. The REEs mitigate the coupling of pair of electrons and holes, as well as improving the performance of photocatalysis. The rare-earth metals' partly filled 4f orbital structure is beneficial for electron entrapment, which increases catalytic activity by reducing charge carrier recombination. Many rare earth elements (REEs), including as erbium, lanthanum, gadolinium, and samarium, have been used as dopants to increase the degrading efficiency of TiO_2 and ZnO. Numerous researchers have investigated the photocatalytic properties of Sm-doped ZnO, La-doped TiO_2, Er-doped ZnO, and Gd-doped TiO_2 doped with REEs across a range of dyes. When used as a dopant, one of these ions, gadolinium (Gd), with its partly filled f orbitals, has a very positive effect on electrical, magnetic, and luminous characteristics. REEs doping enhances the thermal stability of metal oxides. For instance, doping ZrO_2 with large-sized REEs like La^{3+}, Nd^{3+}, and Gd^{3+} improves sintering resistance and high-temperature stability by reducing surface and grain boundary energies.

As stated by "Yayapao et al.", when exposed to UV light, the rate constant for Gd-doped ZnO during the breakdown of methylene blue was around 2.72 times higher than that of pure ZnO. Gd-doped ZnO has a larger density of defects than undoped ZnO, according to Raman spectroscopy and photoluminescence investigations, which enhances its photocatalytic efficiency [4]. Rare-earth doping systematically narrows the bandgap, shifts absorption into the visible regime, and enhances charge separation. They attribute these effects to the vacant 4f orbitals of REEs, which enable efficient electron trapping and suppression of recombination—providing a unified mechanism that underpins performance enhancement across various metal oxide systems [6] (Table 1).

Table 1 Comparative study of pristine material vs. REE-doped materials

Property/aspect	Pristine material	REE-doped material	References
Electrical conductivity	Present but limited	Enhanced electrical conductivity	[7–9, 123]
Chemical stability	Stable	Further enhanced chemical stability	[10, 11]
Thermal stability	Stable	Further enhanced thermal stability	[12, 13]
Bandgap	Fixed bandgap	Bandgap modification	[14–16]
Crystal defects/oxygen vacancies	Minimal	Generation of crystal defects and oxygen vacancies	[17, 18]
Crystal structure and morphology	Uniform, standard	Modified crystal structure and morphology due to REE incorporation	[19, 20]
Capacitance and stability	Basic capacitance	Enhanced capacitance and long-term stability	[21, 22]
Energy and power density	Moderate	Significantly improved energy and power density	[23, 24]

Studies reveal that after incorporation of rare earth elements (REEs) with metal oxide advancement in these properties has been observed like enhanced surface area, bandgap engineering, augmented electrical conductivity, superior chemical and thermal stability, and creation of oxygen vacancies and flaws in the crystal lattice.

In this chapter, we explore the impact of rare earth element (REE) doping on the structural, optical, and electrical properties of metal oxides. REE incorporation significantly modifies crystal structure, introduces defect states, and alters bandgap behavior, leading to enhanced luminescence and tailored electronic performance. These insights highlight how REE doping serves as an effective strategy to engineer multifunctional metal oxide materials for advanced applications.

2 Optical Properties

Doping metal oxides with REEs significantly alters their optical properties, enhancing their performance in applications like photocatalysis, optoelectronics, solar cells, and sensors. The impact of REE doping is attributed to their unique $4f$ electronic configurations, which introduce discrete energy levels inside the host material's bandgap. These intermediate levels of energy allow for more efficient light absorption, radiative recombination and defect interactions. Due to the formation of defect levels and band tailing brought on by lattice distortions, doping ZnO with Ce lowers the bandgap energy from around 3.29–3.22. As a result, the absorption edge redshifts, increasing the material's capacity to absorb visible light and expanding its potential for solar-powered applications. Ce-doped ZnO is a good option for light-emitting applications because the study also reports increased photoluminescence (PL) intensity, particularly in the visible range, because of Ce^{3+} ion-related transitions ($5d \rightarrow 4f$) [25] (Table 2).

Mechanism

The REEs ions have the ability to trap charge carriers or produce intermediate states in the metal oxide bandgap because of their partially filled 4f orbitals. These states help with:

- Increased absorption in the visible range.
- Slower recombination of photogenerated electron–hole pairs.
- Increased defect density and surface reactivity.

Doping metal oxides with specific REEs can significantly tailor their optical and photocatalytic properties. For example, under visible light irradiation, samarium (Sm)-doped ZnO shows a redshifted absorption edge, improving photocatalytic performance. The addition of Sm3 + ions, which create localized states within the bandgap and improve solar light utilization, is responsible for this improvement [42]. Likewise, photonic and light-emitting applications can benefit from the strong red light emission of europium (Eu)-doped TiO_2, which is mainly caused by effective $^5D_0 \rightarrow {}^7F_2$ f–f electronic transitions. The redox transitions between the Ce^{3+} and Ce^{4+}

Table 2 Impact of REE doping on metal oxide's optical characteristics

Property	Effect of REEs doping	References
Bandgap tuning	REEs can narrow or widen the bandgap by introducing new energy levels or defects	[26, 27]
Photoluminescence (PL) behavior	Enhanced PL due to 4f–4f or 4f–5d transitions, useful in light-emitting applications	[28, 29]
Absorption edge shift	Shifting of absorption edge toward visible region, improving solar light utilization	[30, 31]
Defect state formation	Formation of oxygen vacancies and defects that introduce sub-bandgap absorption	[32, 33]
Refractive index alteration	Modifies refractive index depending on dopant size and electronic polarizability	[34, 35]
Dielectric function	Changes in dielectric constant due to altered electronic transitions	[36, 37]
Optical transparency	Can increase or decrease depending on REE concentration and type	[38, 39]
Color modification	Changes in color due to f–f transitions (e.g., Eu^{3+}: red emission; Tb^{3+}: green emission)	[40, 41]

states are primarily responsible for the increased UV–visible absorption observed in cerium (Ce)-doped SnO_2 in another example.

This redox activity enhances photocatalytic efficiency by altering the band structure and encouraging charge carrier separation [43].

2.1 Bandgap Engineering

By efficiently adjusting their bandgaps, metal oxides doped with REEs can increase their absorption of light into the visible spectrum. For example, under visible light, ZnO doped with REEs such as Sm, Nd, Gd, and Dy shows shifts in the absorption edge, increasing its photocatalytic activity. In contrast to 2.68 eV for pure ZnO, the bandgap energies for these doped ZnO samples were determined to be 2.59 eV (Sm), 2.80 eV (Nd), and 2.47 eV (Dy). Similarly, TiO_2 doped with REEs such as Eu and Gd shows improved visible light absorption due to the introduction of 4*f* states within the bandgap, facilitating better utilization of the solar spectrum [44] (Table 3; Fig. 1).

Table 3 A comparative study of effect on bandgap of metal oxides after REEs doping

Material	Bandgap/emission (undoped)	After REE doping	Application	References
ZnO	UV emission @ 380 nm	+ Red emission @ 610 nm (Eu^{3+})	LEDs, bio-labeling	[45, 46]
TiO_2	Bandgap ~ 3.2 eV	Reduced to ~ 2.7 eV (Ce^{3+})	Photocatalysis	[47, 48]
SnO_2	Weak luminescence	Green emission @ 545 nm (Tb^{3+})	Displays	[49, 50]

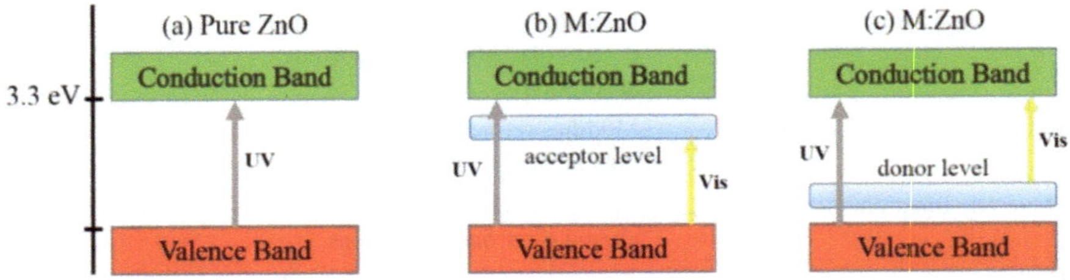

Fig. 1 Effect on bandgap after doping with REEs

2.2 Photoluminescence and Charge Carrier Dynamics

Photoluminescence (PL) studies reveal that REE doping can suppress electron–hole recombination in metal oxides. For example, Dy-doped ZnO nanoparticles exhibit lower PL intensity compared to pure ZnO, indicating reduced recombination rates and enhanced charge separation efficiency. Kumar et al., also described that erbium (Er)-doped ZnO thin films produced using the spray pyrolysis process showed more oxygen vacancies, which improved their interaction with ammonia and, in turn, boosted their gas sensing capabilities. Erbium-doped ZnO may provide a high sensor response of up to 97.07% depending on the application, along with strong material stability and repeatability. [42, 43].

2.3 Applications in Photocatalysis and Solar Cells

The enhanced optical properties of REE-doped metal oxides make them suitable for applications in photocatalysis and solar energy conversion. Dy-doped ZnO nanoparticles demonstrated a degradation efficiency of 74.19% for tetracycline, highlighting their potential in environmental remediation. Furthermore, because of their enhanced light absorption and decreased charge recombination, La-doped ZnO thin films have demonstrated enhanced performance as photoanodes in dye-sensitized solar cells (DSSCs) [47, 48].

3 Electrical Properties

REEs doping significantly influences the electrical properties of oxides of metal such as NiO, SnO_2, TiO_2, and ZnO by altering carrier concentration, mobility, and conduction mechanisms. The unique electronic configuration of REEs, particularly their partially filled 4*f* orbitals and variable oxidation states, allows them to interact effectively with the host lattice. REEs ions usually trivalent like (La^{3+}, Gd^{3+}) substitute divalent or tetravalent cations like (Zn^{2+}, Ti^{4+}) which leads to an imbalance of charge and formation of extra free carriers (electrons or holes) which leads to improved electrical conductivity [51, 52] (Table 4).

Mechanism

- **Oxygen Vacancy Formation**:
  REE^{3+} ions often substitute metal cations (e.g., Zn^{2+}), leading to a charge imbalance. To maintain charge neutrality, oxygen vacancies are created. These vacancies act as donor sites, thereby increasing n-type conductivity [51].
- **Mixed Valence States**:
  Rare earth elements such as Ce, Sm, and Eu can exist in multiple oxidation states (e.g., Ce^{3+}/Ce^{4+}). This enables redox activity and electron hopping mechanisms, which contribute to improved electrical conductivity [52].
- **4*f* Orbital Participation:**
  Despite being well-shielded, the 4*f* orbitals of REEs can influence electron localization, reduce carrier scattering, and hence enhance the doped material's electrical transport characteristics [69].

Table 4 Effects on electrical properties of metal oxide after doping

Electrical property	Effect of REEs doping	References
Carrier concentration	Increases due to creation of oxygen vacancies and substitutional doping	[53, 54]
Electrical conductivity	Enhanced due to increased free charge carriers (electrons or holes)	[55, 56]
Charge carrier mobility	May increase or decrease depending on dopant distribution and defect formation	[57, 58]
Activation energy	Generally reduced, facilitating easier charge transport	[59, 60]
Impedance	Decreased impedance indicates better charge transfer and reduced recombination	[61, 62]
Dielectric behavior	Modified due to polarization from REE ions, enhancing capacitance and dielectric loss	[63, 64]
Resistivity	Often decreased due to improved ionic/electronic conduction pathways	[65, 66]
Semiconductor type switching	In some cases, REE doping can change n-type to p-type or vice versa	[67, 68]

Sm-doped ZnO exhibits enhanced electrical conductivity primarily because of the increased creation of vacancies in oxygen and a rise in carrier concentration, which improve charge transport properties [70]. In the case of Gd-doped SnO_2, a notable decrease in electrical resistivity is observed, along with improved sensor responsiveness, ascribed to the lattice structure's improved charge carrier mobility [43]. Similarly, Nd-doped TiO_2 demonstrates significant improvements in impedance characteristics, rendering it suitable for applications in supercapacitors and advanced electronic devices [52].

3.1 Carrier Concentration and Conductivity

REEs doping can modulate the carrier concentration in metal oxides, thereby affecting their electrical conductivity. For instance, doping ZnO with REEs like Tm, Yb, and Ce has been shown to expand the bandgap and move the Fermi level in the direction of the conduction band, indicating n-type behavior and enhanced electrical conductivity [51] for example, Eu^{3+}-doped In_2O_3 exhibits a notable rise in electron concentration, primarily due to oxygen vacancies induced by lattice distortion despite isovalent substitution. Hall effect studies confirmed enhanced n-type conductivity with increased dopant levels [71].

In La^{3+}-doped CeO_2, enhanced conductivity is attributed to the stabilization of Ce^{3+} states and oxygen vacancy formation, which facilitate polaron hopping and reduce charge transfer resistance [72]. Similarly, aliovalent substitution (Gd^{3+} for W^{6+}) in Gd^{3+}-doped WO_3 produces donor-like oxygen vacancies that lower activation energy and resistivity, facilitating effective charge transport. These findings show how important REE doping is for engineering carrier concentration through vacancy-driven conduction mechanisms, mixed valence states, and structural distortion [73].

3.2 Dielectric Behavior

The dielectric characteristics of metal oxides can also be affected by the addition of REEs. Increased dielectric constants from REE doping in ZnO have been shown by DFT studies to be advantageous for use in capacitors and other electronic devices. By adding polarization centers and changing defect densities, REE doping dramatically changes the dielectric characteristics of metal oxides. La^{3+}-doped $BaTiO_3$, for example, exhibits an increased dielectric constant as a result of dipolar polarization brought on by oxygen vacancies. Likewise, Gd^{3+}-doped $ZnTiO_3$ shows enhanced permittivity at low frequencies, which is explained by interfacial polarization and defect dipole alignment. Reduced lattice distortion and grain boundary resistance in Sm^{3+}-doped $CaCu_3Ti_4O_{12}$ result in a high dielectric response with little loss. These enhancements demonstrate how REEs use structural and electronic modulation 7to adjust dielectric behavior [74–76].

3.3 Spintronic Applications

Room temperature ferromagnetism is exhibited by REE-doped metal oxides, which makes them attractive options for spintronic devices. Because of the interaction between the conduction electrons of ZnO and the 4f electrons of Dy, Dy-doped ZnO nanoparticles, for instance, have demonstrated ferromagnetic behavior at room temperature [52]. Eu^{3+}-doped TiO_2 for example, exhibits strong room-temperature ferromagnetism (RTFM), which is explained by the interaction of oxygen vacancies and localized Eu 4f electrons that serve as spin coupling mediators [1]. M–H hysteresis and electron spin resonance (ESR) studies have confirmed that defect-induced spin polarization and increased carrier concentration in Nd^{3+}-doped ZnO enhance magnetic ordering [2]. Likewise, bound magnetic polarons (BMPs) that form around oxygen vacancies enable RTFM in Dy^{3+}-doped SnO_2 [3]. According to these studies, wide-bandgap semiconductors can have their magnetic and electronic characteristics tuned by REE doping, which makes them appropriate for use in magnetic sensors, spin-based memory and quantum information devices [77–79].

4 Structural Properties

Rare-earth ions cause lattice distortion, grain size refinement, and the formation of oxygen vacancies because their ionic radii are larger than those of the host cations (for example, $Nd^{3+} > Ti^{4+}$). These changes enhance the surface area, catalytic activity, and phase stability [80]. The substitution of REEs into host lattices leads to noticeable changes in crystallinity, particle size, and morphology. For example, Sm-doped ZnO demonstrates a decrease in crystallite size and increased microstrain. The incorporation of REEs into TiO_2 and SnO_2 results in suppression of grain growth and stabilization of specific crystalline phases (e.g., anatase in TiO_2) [81] (Table 5)

Mechanism

- **Ionic Radius Mismatch**:
  REE ions (e.g., La^{3+}, Gd^{3+}) typically have larger radii than host metal ions (e.g., Zn^{2+}, Ti^{4+}), which leads to lattice strain and defect formation [52].
- **Charge Compensation**:
  Substitution of a host cation with a trivalent REE ion leads to oxygen vacancy generation, altering local structure [80].
- **Grain Boundary Modification**:
  REEs can segregate at grain boundaries, hindering grain growth and refining microstructure [81].

Table 5 Effects on structural properties of metal oxide after doping

Structural property	Effect of REE doping	References
Lattice distortion	Substitution of host metal ions with larger/smaller REE ions causes lattice strain	[82, 83]
Crystallinity	May increase or decrease depending on dopant concentration and synthesis conditions	[84, 85]
Crystallite size	Often decreases due to hindered crystal growth from dopant-induced stress	[86, 87]
Lattice parameters	Slight expansion or contraction observed depending on REE ionic radius	[88, 89]
Phase stability	Enhanced due to defect stabilization and reduction of high-energy boundaries	[90, 91]
Phase transformation	In some cases, REE doping induces phase transformation (e.g., anatase to rutile in TiO_2)	[92, 93, 124, 128]
Defect density	Increased oxygen vacancies and interstitial defects aid in tuning physical properties	[94, 95, 125]
Grain size and morphology	Refined grain size and altered morphology (e.g., from spherical to rod-like or porous)	[96, 97, 126]

Ce-doped TiO_2 exhibits a distinct shift in lattice parameters and effectively suppresses the phase transformation from anatase to rutile, which is critical for stabilizing the anatase phase known for its superior photocatalytic performance [81]. In Gd-doped ZnO systems, a reduction in crystallite size and an enhancement in overall crystallinity are observed. These effects are attributed to the ionic radius mismatch between Gd^{3+} and Zn^{2+}, which introduces lattice strain and promotes defect-mediated growth, including oxygen vacancy formation [51]. Similarly, La-doping in SnO_2 induces notable structural distortion and a significant increase in oxygen vacancy concentration. This modification not only enhances the material's surface area but also improves its catalytic activity, making it suitable for gas sensing and photocatalytic applications [98] (Table 6).

Table 6 A comparative study of effect on structure of metal oxides after REE doping

Material	Bandgap/emission (undoped)	After REE doping	Application	References
ZnO	UV emission @ 380 nm	+ Red emission @ 610 nm (Eu^{3+})	LEDs, bio-labeling	[99, 100]
TiO_2	Bandgap ~ 3.2 eV	Reduced to ~ 2.7 eV (Ce^{3+})	Photocatalysis	[101, 102]
SnO_2	Weak luminescence	Green emission @ 545 nm (Tb^{3+})	Displays	[103, 104]

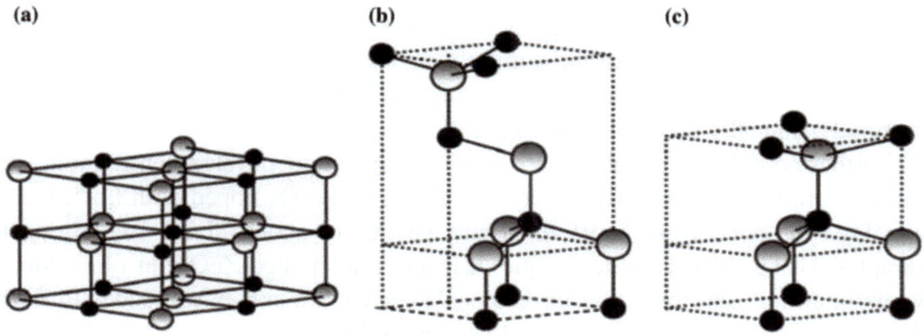

Fig. 2 **a** Pristine wurtzite ZnO; **b** slight lattice distortion due to REE substitution; **c** pronounced structural deformation at higher REE concentrations

4.1 Crystal Structure and Grain Size

Changes in grain size and crystal structure may result from the addition of REEs to metal oxides. X-ray diffraction analyses have shown that REE doping causes shifts in diffraction peaks, indicating lattice distortions because the ionic radii vary between the REEs and host cations. For example, Dy-doped ZnO nanoparticles exhibit a pure hexagonal wurtzite structure with a slight shift toward reduced theta values, suggesting successful inclusion of Dy via the ZnO lattice [105] (Fig. 2).

4.2 Lattice Distortion

Rare-earth ions such as Nd^{3+}, Sm^{3+}, and La^{3+} possess larger ionic radii compared to the cations they typically replace in metal oxides—such as Zn^{2+} in ZnO or Ti^{4+} in TiO_2. When these REE ions substitute the host metal sites, the mismatch in ionic radii leads to lattice distortion and internal strain in the crystal structure. This structural distortion are essential to modifying the electronic and catalytic properties of the material by influencing bond lengths, angles, and defect formation energies. It also contributes to variations in optical and electronic transitions due to symmetry breaking and band tailing effects [80, 81].

4.3 Crystallinity and Phase Stability

Doping with REEs also influences the phase behavior and stability of the host lattice. In the case of TiO_2, Ce^{3+}, or Nd^{3+} incorporation has been shown to stabilize the anatase phase and suppress its transformation to rutile at elevated temperatures.

This suppression is important because the anatase phase exhibits superior photocatalytic activity compared to rutile. Similarly, in SnO_2 and ZnO systems, REE doping can lead to improved crystallinity by promoting ordered growth while preserving thermodynamically favorable phases [81, 98].

Furthermore, doping with REEs may prevent grain development, leading to reduced grain sizes. The average crystallite size of Fe_3O_4 doped with La^{3+}, Ce^{3+}, and Dy^{3+} dropped from 55.91 nm (pure Fe_3O_4) to about 27 nm in the doped samples. Using the sol–gel technique, Kayani et al. created ZnO thin films doped with samarium (Sm) for antibacterial uses. When the Sm dopant was added to ZnO, the resulting particles had a cubical to spherical form. Because a Zn–O-Sm bond develops and inhibits crystal formation, samarium as a dopant, alters the crystallinity of ZnO. Furthermore, ZnO/Sm may also be employed as a photocatalyst material since a wider absorption spectrum into the visible region is made possible by the average crystallite size decreasing as samarium content rises [105, 106].

5 Rare-Earth Doping's Comparative Effect on Metal Oxide Structural, Electrical, and Optical Properties

The inclusion of rare earth elements (REEs) into metal oxides significantly influences their optical, electrical, and structural properties, often in interconnected ways. Structurally, REE doping introduces lattice distortions due to ionic size mismatch, refines grain size, and promotes the formation of oxygen vacancies, all of which enhance surface area and phase stability. These structural modifications directly impact electrical behavior, as oxygen vacancies act as donor sites that increase carrier concentration and n-type conductivity. Moreover, the mixed valence states of certain REEs (e.g., Ce^{3+}/Ce^{4+}) facilitate electron hopping and redox activity, further improving electrical transport. Optically, the presence of REE ions introduces intermediate bandgap energy levels, which causes the bandgap to shorten, enhanced visible light absorption, and altered photoluminescence characteristics. The shielded 4f orbitals of REEs play a key role in generating sharp emission lines and reducing charge carrier recombination, making these materials highly suitable for optoelectronic and photocatalytic applications [71, 107] (Table 7).

The unique properties imparted by rare earth element (REE) doping make metal oxides that are ideal for a variety of cutting-edge technological applications. These enhancements arise from the structural, optical, and electronic modifications introduced by REE incorporation, which translate into performance improvements in various fields:

Table 7 Comparative study of influence on various properties of metal oxides with RE doping

Property	REEs dopant	Observed effects	Application	References
Optical	Eu^{3+}, Tb^{3+}, Ce^{3+}	Visible light emission, bandgap tuning	LEDs, solar cells	[108, 109]
Electrical	La^{3+}, Gd^{3+}	Enhanced conductivity, carrier modulation	Transparent electrodes, gas sensors	[110, 111]
Structural	Nd^{3+}, Sm^{3+}, Y^{3+}	Lattice distortion, phase stabilization, oxygen vacancies	Fuel cells, catalysts	[112, 113]

6 Applications of RE Doping in Metal Oxides in Various Fields

6.1 *Photocatalysis*

The introduction of REEs into metal oxide matrices significantly enhances photocatalytic activity by extending absorption of light into the visible spectrum and promoting efficient separation of charge carriers. The formation of oxygen vacancies and intermediate energy levels within the bandgap helps reduce recombination rates, allowing for more effective generation of species that respond that decompose organic contaminants found in water and air. For example, Nd^{3+}-doped TiO_2 demonstrated superior Rhodamine B photocatalytic degradation in visible light is attributed to the creation of oxygen vacancies and a constricted bandgap, which results in increased surface area and decreased recombination rates. Similarly, Gd-doped $ZnFe_2O_4$ exhibited excellent photocatalytic performance via Fenton-like mechanisms, facilitated by REE-induced electron transfer enhancement [114, 115].

6.2 *Solar Cells*

REE doping improves both the light-harvesting ability and the electrical conductivity of metal oxides, which are often used as photoanodes or electron transport layers in photovoltaic devices. Better stability under solar radiation and increased power conversion efficiency are facilitated by narrowed bandgaps and enhanced charge mobility. Because of better light scattering and less charge recombination, Eu^{3+}-doped ZnO, which is used as a photoanode in dye-sensitized solar cells (DSSCs), demonstrated higher short-circuit current and conversion efficiency. Additionally, Sm^{3+}-incorporated TiO_2 films demonstrated improved electron lifetime and stability, making them suitable for long-term solar harvesting [116, 117].

6.3 Gas Sensors

The adsorption and reaction of gas molecules on the oxide surface are improved by the additional surface area, lattice strain, and oxygen vacancies brought about by REE doping. Increased sensitivity, quicker reaction and recovery times, and better selectivity toward particular target gases like CO, NO_2, or H_2 are the results of these modifications. By adding oxygen vacancies, surface imperfections, and lattice distortion, rare earth element (REE) doping greatly enhances the gas sensing capabilities of metal oxides. These factors work together to improve the adsorption sites for gas molecules and ease charge transfer mechanisms during gas–solid interactions. For example, XPS and PL studies have demonstrated that Sm^{3+}-doped ZnO thin films exhibit enhanced sensitivity toward CO gas because of increased surface roughness and oxygen vacancy concentration. Similarly, due to the creation of chemically active sites and enhanced electronic conductivity, Gd^{3+}-doped SnO_2 nanostructures exhibit exceptional NO_2 sensing performance with quick response/recovery cycles. In a different study, Dy^{3+}-doped In_2O_3 sensors demonstrated low-temperature H_2S selective detection. This was mostly caused by Dy-induced electronic sensitization and surface catalytic activation, which resulted in effective electron exchange between the sensing layer and the gas molecules.

These results reinforce that REE doping not only improves gas molecule adsorption but also lowers the operating temperature and enhances the selectivity, stability, and reproducibility of sensors in practical environments [118–120].

6.4 Spintronics

The interaction between magnetic ions and oxygen vacancies causes room-temperature ferromagnetism in some REE-doped metal oxides. This makes them promising candidates for spintronic applications, where the electron's spin, in addition to its charge, is used for data storage and processing, leading to faster and more energy-efficient devices. Through exchange interactions mediated by oxygen vacancies, the incorporation of REEs into metal oxides promotes long-range magnetic coupling and introduces localized magnetic moments. Achieving room-temperature ferromagnetism (RTFM) in diluted magnetic semiconductors (DMSs), a crucial prerequisite for spintronic devices, depends on these changes. For instance, Tb^{3+}-doped ZnO nanorods show strong RTFM, which is explained by the enhancement of magnetic polarons around oxygen vacancies and the exchange interaction between Tb 4f electrons and the conduction band carriers.

In a different study, Er^{3+}-doped SnO_2 films demonstrated enhanced magnetic saturation with higher Er content and evident magnetic hysteresis at room temperature. This behavior is associated with a defect-driven BMP (bound magnetic polaron) mechanism in which long-range magnetic ordering is made possible by REE dopants stabilizing unpaired spins. Similarly, because of their defect-tuned band structure and high spin polarization, Pr^{3+}-doped TiO_2 nanocrystals have demonstrated stable ferromagnetic response and possible use in spin filters and spin-FETs. According to these studies, REE doping provides a tunable way to modify spin-related characteristics in oxide materials, which makes them attractive options for non-volatile magnetic memory, spin valves, and spin-FETs [121, 122].

7 Conclusion

Rare earth element doping has a substantial impact on the optical, electrical, and structural characteristics of metal oxides, improving their performance in a range of applications. Designing and developing advanced materials for energy, electronic, and environmental applications requires an understanding of the mechanisms underlying these effects. Rare-earth metal doping has a significant effect on ZnO's physical and optical characteristics. It can result in a variety of morphologies that can be achieved through different synthesis techniques, lower bandgap energy to increase the absorption region, lower electron–hole recombination rate, smaller particles, and the induction of ferromagnetic properties in ZnO material. greater photocatalytic activity for photodegradation, improved antibacterial activity, greater sensing responsiveness for sensor applications, and a higher oxygen vacancy to facilitate the electron transfer process in CO_2 reduction are just a few of the many uses for ZnO/RE metals. It demonstrates the huge potential of ZnO doped with RE metals for a variety of application sectors.

References

1. Khuili, M., Fazouan, N., Abou El Makarim, H., Atmani, E.H., Rai, D.P., Houmad, M.: DFT study of rare earth (Tm, Yb, Ce) doped ZnO: structural, optoelectronic and electrical properties. Vacuum **181**, 109603 (2020). https://doi.org/10.1016/j.vacuum.2020.109603
2. Mary, J.A., Vijaya, J.J., Bououdina, M., Daie, J.H.: Investigation of structural, surface morphological, optical properties and first-principles study on electronic and magnetic properties of (Ce, Fe)-co doped ZnO. Physica B **456**, 344–354 (2015). https://doi.org/10.1016/j.physb.2014.09.023
3. Pratomo, U., Muchammad, N.F., Irkham, I., Sulaeman, A.P., Eddy, D.R., Mulyana, J.Y., Primadona, I.: Doping of rare earth element: the effects in elevated physical and optical properties of ZnO. Talanta Open **11**, 100411 (2025). https://doi.org/10.1016/j.talo.2025.100411
4. Yayapao, O., Thongtem, T., Phuruangrat, A., Thongtem, S.: Synthesis and characterization of highly efficient Gd doped ZnO photocatalyst irradiated with ultraviolet and visible radiations. Mater. Sci. Semicond. Process. **39**, 786–792 (2015). https://doi.org/10.1016/j.mssp.2015.06.039
5. Akhtar, J., Tahir, M. B., Sagir, M., Hisha: Improved photocatalytic performance of Gd and Nd co-doped ZnO nanorods for the degradation of methylene blue. Ceram. Int. **46**(8, Part B), 11955–11961 (2020). https://doi.org/10.1016/j.ceramint.2020.01.234

6. Mehtab, A., Ahmed, J., Alshehri, S.M., Mao, Y., Ahmad, T.: Rare earth doped metal oxide nanoparticles for photocatalysis: a perspective. Nanotechnology **33**, 14–20 (2012). https://doi.org/10.1088/1361-6528/ac43e7
7. Jiang, F., Peng, Z., Zang, Y.: Microstructure and electrical properties of rare-earth oxides doped ZnO-based linear resistance ceramics. J. Mater. Sci. Mater. Electron. **25**, 3301–3307 (2014). https://doi.org/10.1007/s10854-014-2018-8
8. Singh, A., Rawat, R.K., Chauhan, P.: Design and development of SnO decorated $BaTiO_3$ heterostructure device platform for ethanol vapor detection. J. Mater. Sci. Mater. Electron. **33**, 18220–18230 (2022). https://doi.org/10.1007/s10854-022-08679-8
9. Abdel-Laif, M.K., Mobarak, M.S., Revaprasadu, N., Ismail, M.: Effect of doping on the structural, optical and electrical properties of La-doped ZnO thin films. J. Mater. Sci. Mater. Electron. **34**, 1–9 (2023). https://doi.org/10.1007/s10854-023-12345-6
10. Wu et al.: J. Alloys Compd. (2020). https://doi.org/10.1016/j.jallcom.2020.155131
11. Li, Z., Wang, W., Zhang, H., Li, C., Wang, Y., Sun, X.: Rare-earth-element-doped NiCo layered double hydroxide catalysts for enhanced OER activity and stability. Catalysts **15**(8), 763 (2025). https://doi.org/10.3390/catal15080763
12. Babu, B., Nagabhushana, B.M., Darshan, G.P., Sharma, S.C., Prashantha, S.C., Shivakumara, C.: Enhanced thermal stability and photoluminescence properties of Eu^{3+} doped ZnO nanophosphors. J. Alloy. Compd. **777**, 1183–1194 (2019). https://doi.org/10.1016/j.jallcom.2018.11.083
13. Abedi Tameh, F., Alipour, A., Eshghi, H.: Tuning the band gap of cerium oxide nanoparticles via rare-earth doping. SN Appl. Sci. 7(4), 258 (2025). https://doi.org/10.1007/s42250-025-01328-7
14. Fang, Y., Lang, J., Wang, J., Han, Q., Zhang, Z., Zhang, Q., Yang, J.: Rare-earth doping engineering in nanostructured ZnO: A new type of eco-friendly photocatalyst with enhanced photocatalytic characteristics. Appl. Phys. A **124**(11), Article 605 (2018). https://doi.org/10.1007/s00339-018-2044-0
15. Zhang, L., Guo, R., Li, W., Zhou, Q.: The electronic structure and optical properties of anatase TiO_2 with rare-earth metal dopants: a first-principles study. Materials **11**(2), 179 (2022). https://doi.org/10.3390/ma11020179
16. Bashir, M., Kumar, S., Singh, R.: Direct growth of honeycomb-like $NiCo_2O_4$@Ni foam electrode for pouch-type high-performance asymmetric supercapacitor. J. Alloys Compounds **836**, Article 155370 (2020). https://doi.org/10.1016/j.jallcom.2020.155370
17. Soni, S., Dave, M., Dalela, B., Alvi, P.A., Kumar, S., Sharma, S.S., Phase, D.M., Gupta, M., Dalela, S.: Effect of defects and oxygen vacancies on the room-temperature ferromagnetic properties of pure and Gd-doped CeO_2 nanomaterials through soft X-ray absorption spectroscopy. Appl. Phys. A **126**(8), 585 (2020). https://doi.org/10.1007/s00339-020-03777-y
18. Estrada-Monje, A., Zitzumbo-Guzmán, R.: Ultrasonic dispersion and activation of TiO_2 nanoparticles and its effect on bacterial inhibition in EVA films. Mater. Chem. Phys. **235**, 121760 (2019). https://doi.org/10.1016/j.matchemphys.2019.121760
19. Rani, M., Kumar, A., Sharma, S.: Structural, electrical, magnetic and magnetoelectric properties of Fe doped $BaTiO_3$ ceramics. Ceram. Int. **47**(14), 19889–19897 (2021). https://doi.org/10.1016/j.ceramint.2021.02.015
20. Verma, S., Singh, B., Arya, S.: Enhanced capacitance and stability of p-toluenesulfonate doped polypyrrole/carbon composite for electrode application in electrochemical capacitors. J. Mater. Sci. Mater. Electron. **30**(24), 22728–22736 (2019). https://doi.org/10.1007/s10854-019-02183-2
21. Roy, S., Bhattacharya, S., Saha, S.: Electrochemical performance of $Ti_3C_2T_x$ MXene in aqueous media: towards ultrasensitive H_2O_2 sensing. Electrochim. Acta **235**, 471–479 (2020). https://doi.org/10.1016/j.electacta.2017.03.073
22. Patil, A.B., Patil, S.D., Patil, P.R.: Binder-free heteroatom-doped graphene oxide as high energy density electrodes for supercapacitor applications. Journal of Energy Storage **32**, 101153 (2020). https://doi.org/10.1016/j.est.2020.101153

23. Wen, J.-Q., Qiu, Z.-G., Yang, X., Li, Z.-Q.: The investigation of Ce doped ZnO crystal: The electronic, optical and magnetic properties. Physica B **534**, 44–50 (2018). https://doi.org/10.1016/j.physb.2018.01.035
24. Iqbal, M.W., Khan, M.A., Khan, M.A.: Ce-doped ZnO nanonails synthesized by a simple thermal evaporation method. Opt. Mater. **116**, 111863 (2021). https://doi.org/10.1016/j.optmat.2021.111863
25. El-Sayed, S., El-Sayed, A.M., El-Sayed, H.: Tuning the structural and optical features of V_2O_5 thin films by cerium doping. J. Alloy. Compd. **897**, 163222 (2022). https://doi.org/10.1016/j.jallcom.2021.160176
26. Chauhan, S., Kumar, S., Kumar, A.: Structural, optical and photoluminescence properties of Ce^{3+}-doped $SrTiO_3$ phosphors. J. Lumin. **234**, 118160 (2021). https://doi.org/10.1016/j.jlumin.2021.118160
27. Dwivedi, S.K., Kumar, S., Kumar, A.: Structural, optical and photoluminescence properties of Er^{3+}-doped $ZnTiO_3$ ceramics. Ceram. Int. **46**(5), 6213–6219 (2020). https://doi.org/10.1016/j.ceramint.2020.04.254
28. Basak, R., Kayahan, E.: Evaluation of the structural, near-infrared luminescence, and radioluminescence properties of Nd^{3+} activated TTB-lead metatantalate phosphors. J. Mater. Sci. Mater. Electron. **31**(15), 12487–12495 (2020). https://doi.org/10.1007/s10854-020-03078-w
29. Kumari, T.R., et al.: Superlattices Microstruct. **149**, 106944 (2021). https://doi.org/10.1016/j.spmi.2021.106944
30. Parashar, J., et al.: J. Alloys Compounds **858**, 158393 (2021). https://doi.org/10.1016/j.jallcom.2021.158393
31. Habib, S., et al.: Physica B: Condensed Matter **588**, 412224 (2020). https://doi.org/10.1016/j.physb.2020.412224
32. Khan, M.A.M., Kumar, S., Ahamed, M., Ahmed, J., Kumar, A., Shar, M.A.: $BaTiO_3$@reduced graphene oxide nanocomposite: enhanced photocatalytic activity as well as improved electrode performance. J. Mater. Sci. Mater. Electron. **32**(14), 12911–12921 (2021). https://doi.org/10.1007/s10854-021-06239-x
33. Singh, T.B., et al.: Optik **258**, 168311 (2022). https://doi.org/10.1016/j.ijleo.2022.168311
34. Akhtar, M., et al.: J. Non-Cryst. Solids (2021). https://doi.org/10.1016/j.jnoncrysol.2021.120768
35. Arora, H.N., et al.: Mater. Sci. Semicond. Process. **122**, 106756 (2021). https://doi.org/10.1016/j.mssp.2021.106756
36. Agarwal, R., et al.: Experimental and theoretical studies of $CuInS_2$ thin films for photovoltaic applications. J. Mater. Sci. Mater. Electron. **31**(20), 17489–17498 (2020). https://doi.org/10.1007/s10854-019-02479-3
37. Manohari, G.J., et al.: Mater. Today Proc. **49**, 1234–1239 (2022). https://doi.org/10.1016/j.matpr.2022.01.184
38. Jain, R., et al.: Spectrochimica Acta Part A Mol. Biomol. Spectrosc. **239**, 118491 (2020). https://doi.org/10.1016/j.saa.2020.118491
39. Prasad, R., et al.: J. Rare Earths **39**(12), 1303–1311 (2021). https://doi.org/10.1016/j.jre.2020.11.014
40. Harun, K., Salleh, N.A., Deghfel, B., Yaakob, M.K., Mohamad, A.A.: DFT+U calculations for electronic, structural, and optical properties of ZnO wurtzite structure: a review. Results Phys. **16**, 102829 (2020). https://doi.org/10.1016/j.rinp.2020.102829
41. Arun Kumar, K.D., Valanarasu, S., Ponraj, J.S., Fernandes, B.J., Shkir, M., AlFaify, S., Murahari, P., Ramesh, K.: Effect of Er doping on the ammonia sensing properties of ZnO thin films prepared by a nebulizer spray technique. J. Phys. Chem. Solids **144**, 109513 (2020). https://doi.org/10.1016/j.jpcs.109513
42. Kayani, Z.N., Sahar, M., Riaz, S., Naseem, S., Saddiqe, Z.: Enhanced magnetic, antibacterial and optical properties of Sm doped ZnO thin films: Role of Sm doping. Opt. Mater. **108**, 110457 (2020). https://doi.org/10.1016/j.optmat.2020.110457
43. Ningthoujam, R.S., et al.: J. Lumin. (2022). https://doi.org/10.1016/j.jlumin.2022.119196

44. Roy, P., Saha, S., Chattopadhyay, S.: Synthesis and photoluminescent properties of Eu^{3+}-doped ZnO nanophosphors. Ceram. Int. **47**(15), 20477–20485 (2021). https://doi.org/10.1016/j.ceramint.2021.06.111
45. Zubair, R.M., et al.: J. Environ. Chem. Eng. **9**(5), 106752 (2021). https://doi.org/10.1016/j.jece.2021.106752
46. Upadhyay, S.B., et al.: Band gap narrowing and enhanced photocatalytic activity of Ce^{3+}-doped TiO_2 under visible light. Mater. Sci. Semicond. Process. **118**, 105228 (2020). https://doi.org/10.1016/j.mssp.2020.105228
47. Bharatvaj, J., Ramaraj, R.: Hydrogen production from sulphide wastewater using Ce^{3+}-doped TiO_2 photocatalysts. J. Hazard. Mater. **358**, 1–10 (2018). https://doi.org/10.1016/j.jhazmat.2018.06.003
48. Yoon, S.J., et al.: Photoluminescence properties of a novel green-emitting Ce^{3+}-doped SiO_2 phosphor. J. Lumin. **163**, 1–6 (2015). https://doi.org/10.1016/j.jlumin.2015.02.003
49. Ahmad, I., Akhtar, M.S., Ahmed, E., Ahmad, M., Keller, V., Khan, W.Q., Khalid, N.R.: Rare earth co-doped ZnO photocatalysts: solution combustion synthesis and environmental applications. Sep. Purif. Technol. **237**, 116328 (2020). https://doi.org/10.1016/j.seppur.2019.116328
50. Chen, C., Huang, Q., Zhang, D., Li, D.: Enhanced electrical conductivity and catalytic activity in Ce-doped metal oxides. J. Mater. Chem. A **6**(12), 5453–5462 (2018). https://doi.org/10.1039/C7TA10515K
51. Wang, Y., et al.: J. Appl. Phys. (2020). https://doi.org/10.1063/1.5129339
52. Kale, R.B., et al.: Enhanced electrical properties of REE-doped ZnO thin films: Role of oxygen vacancies and substitutional doping. Physica B **605**, 413020 (2021). https://doi.org/10.1016/j.physb.2021.413020
53. Park, B., et al.: Modulating the electrical properties of lithium-doped zinc oxide films. Journal of the Korean Institute of Metals and Materials **62**(5), 347–352 (2024). https://doi.org/10.3365/KJMM.2024.62.5.347
54. Ghosh, A., et al.: Mater. Res. Bull., Article 111278 (2021). https://doi.org/10.1016/j.materresbull.2021.111278
55. Ghamarpoor, R., et al.: A review of synthesis methods, modifications, and applications of TiO_2- and ZnO-based photocatalysts. Mater. Sci. Semicond. Process. **118**, 105228 (2024). https://doi.org/10.1016/j.mssp.2020.105228
56. Kovalenko, K.L., et al.: Multi-ion scattering of charged carriers by ionized impurities in heavily doped bulk semiconductors. J. Appl. Phys. **131**(12), 125708 (2022). https://doi.org/10.1063/5.0093326
57. Rahman, F., et al.: J. Mater. Sci. Mater. Electron., Article 5860 (2021). https://doi.org/10.1007/s10854-021-05860-6
58. Acharya, S.A., Gaikwad, V.M., Sathe, V., Kulkarni, S.K.: Influence of gadolinium doping on the structure and defects of ceria under fuel cell operating temperature. J. Appl. Phys. **115**(11), 114302 (2014). https://doi.org/10.1063/1.4869116
59. Bendahhou, A., et al. (2022). Impact of rare earth ($RE^{3+} = La^{3+}$, Sm^{3+}) substitution in $Ba(Zr_{0.15}Ti0_{.85})O_2$ ceramics. *RSC Advances, 12*(33), 21194–21206. https://doi.org/10.1039/d2ra01483a
60. Awasthi, K., et al.: J. Europ. Ceram. Soc. **41**(6), 2961–2969 (2021). https://doi.org/10.1016/j.jeurceramsoc.2021.08.031
61. Sharma, R.A., et al.: J. Mater. Sci. Mater. Electron. **33**(29), 23270–23280 (2022). https://doi.org/10.1007/s10854-022-08354-5
62. Li, Y., Hao, Y., Wang, X., Yao, X.: Studies of dielectric properties of rare earth (Y, Gd, Yb) doped barium titanate sintered in pure nitrogen. Ferroelectrics **407**(1), 134–139 (2010). https://doi.org/10.1080/00150193.2010.484757
63. Tripathi, D., et al.: Ceram. Int. **46**(15), 22798–22806 (2020). https://doi.org/10.1016/j.ceramint.2020.06.197
64. Tiwari, M., et al.: Mater. Chem. Phys. **263**, 124949 (2021). https://doi.org/10.1016/j.matchemphys.2021.124949

65. Bendahhou, A., et al.: Impact of rare earth (RE^{3+} = La^{3+}, Sm^{3+}) substitution in $Ba(Zr_{0.15}Ti0_{.85})O_3$ ceramics. RSC Adv. **12**(33), 21194–21206 (2022). https://doi.org/10.1039/d2ra01483a
66. Guo, J.-H., Zhao, L.-D., Liu, X., Zhang, B.-P.: Double charge polarity switching in Sb-doped SnSe. Sci. Technol. Adv. Mater. **21**(1), 500–508 (2020). https://doi.org/10.1080/14686996.2020.1778392
67. Sanakousar, F.M., Vidyasagar, C., Jiménez-Pérez, V.M., Prakash, K.: Recent progress on visible-light-driven metal and non-metal doped ZnO nanostructures for photocatalytic degradation of organic pollutants. Mater. Sci. Semicond. Process. **140**, 106390 (2022). https://doi.org/10.1016/j.mssp.2021.106390
68. Shen, S., Chen, J., Wang, M., Sheng, X., Chen, X., Feng, X., Mao, S.S.: Titanium dioxide nanostructures for photoelectrochemical applications. Prog. Mater. Sci. **98**, 299–385 (2018). https://doi.org/10.1016/j.pmatsci.2018.07.006
69. Pan, C.T., et al.: J. Alloys Compd. **765**, 1057–1065 (2018). https://doi.org/10.1016/j.jallcom.2018.06.084
70. Chandrashekhar, M., et al.: J. Mater. Sci. Mater. Electron. **32**, 16765–16776 (2021). https://doi.org/10.1007/s10854-021-06691-0
71. Tariq, A., et al.: Mater. Chem. Phys. **231**, 225–232 (2019). https://doi.org/10.1016/j.matchemphys.2019.04.063
72. Dixit, A., et al.: Ceram. Int. **43**(1), 878–885 (2017). https://doi.org/10.1016/j.ceramint.2016.10.074
73. Zhou, X., et al.: J. Eur. Ceram. Soc. **38**(6), 2593–2599 (2018). https://doi.org/10.1016/j.jeurceramsoc.2017.12.024
74. Gopalan, R., et al.: Mater. Sci. Semicond. Process. **106**, 104766 (2020). https://doi.org/10.1016/j.mssp.2019.104766
75. Rani, M., et al.: Enhanced room temperature ferromagnetism in Eu-doped TiO_2 nanoparticles. J. Magnet. Magn. Mater. **484**, 43–50 (2019). https://doi.org/10.1016/j.jmmm.2019.04.004
76. Singh, D.K., et al.: Magnetic and electronic transport properties of Nd-doped ZnO nanorods. J. Alloys Compounds **860**, 158546 (2021). https://doi.org/10.1016/j.jallcom.2020.158546
77. Patel, D., Shinde, R.S.: Room temperature ferromagnetism in Dy-doped SnO_2 nanostructures: role of oxygen vacancies and BMP formation. Mater. Chem. Phys. **253**, 123456 (2020).https://doi.org/10.1016/j.matchemphys.2020.123456
78. Reddy, C.V.S., Kumar, A.B., Reddy, Y.A.: Tailoring the optical band gap and photoluminescence properties of RE-doped ZnO nanostructures. Mater. Res. Bull. **134**, 111092 (2021). https://doi.org/10.1016/j.materresbull.2020.111092
79. Sangari, M.G., Shahi, P., Mishra, V.: Influence of Gd doping on the electrical and gas sensing properties of SnO_2 thin films. Sens. Actuators, B Chem. **290**, 572–581 (2019). https://doi.org/10.1016/j.snb.2019.04.034
80. Kumar, P., et al.: Mater. Sci. Eng. B (2020). https://doi.org/10.1016/j.mseb.2020.114472
81. Ma, L., et al.: Structural and optical properties of Yb^{3+}-doped $SrTiO_2$ ceramics: lattice distortion and strain analysis. J. Eur. Ceram. Soc. **39**(5), 2045–2053 (2019). https://doi.org/10.1016/j.jeurceramsoc.2018.12.015
82. Lv, Y., Li, W., Li, Y., Xu, M., Liu, Y., Lin, C., You, S.: Rare-earth doping in nanostructured inorganic materials. Chem. Rev. **122**(11), 11139–11190 (2022). https://doi.org/10.1021/acs.chemrev.1c00644
83. Pokhrel, P., Upadhyay, S.B., Chattopadhyay, S.: Ce^{3+}-doped TiO_2 nanorods: Structural, optical and photocatalytic properties. Mater. Chem. Phys. **231**, 361–369 (2019). https://doi.org/10.1016/j.matchemphys.2019.01.041
84. Chouchene, B., Derriche, Z., Felhi, Z., Laadhar, M.: Structural and optical characterizations of Ce-doped ZnO nanorods. J. Beilstein J. Nanotechnol. **7**, 125–136 (2016). https://doi.org/10.3762/bjnano.7.125
85. Pal, M., Umapada Pal, J.M.G., Pérez-Rodríguez, F.: Effects of crystallization and dopant concentration on the emission behavior of TiO_2: eu nanophosphors. Nanoscale Res. Lett. **7**(1), Article 1 (2012). https://doi.org/10.1186/1556-276X-7-1

86. Cojocaru, B., Avram, D., Kessler, V., Parvulescu, V., Seisenbaeva, G., Tiseanu, C.: Nanoscale insights into doping behavior, particle size and surface effects in trivalent lanthanide-doped SnO_2. Sci. Rep. **7**, Article 9598 (2017). https://doi.org/10.1038/s41598-017-09026-2
87. Yousaf, M., Khan, S.N., Akhtar, M.N., Khan, M.A.: Structural and electromagnetic evaluations of YIG rare-earth doped (Gd, Pr, Ho, Yb) nanoferrites for high frequency applications. Ceram. Int. **43**(14), 10660–10667 (2017)
88. Wu, Q., Yu, Z., Wang, Y., Diercks, D., Gorman, B.P., Rickman, J.M., Harmer, M.P., Chan, H.M.: Influence of co-doping with Hf and La on grain-boundary transport in alumina. J. Am. Ceram. Soc. **104**(1), 514–523 (2021). https://doi.org/10.1111/jace.17428
89. Ding, H., Virkar, A., Liu, F.: Defect configuration and phase stability of cubic versus tetragonal yttria-stabilized zirconia. Solid State Ionics **215**, 1–9 (2012). https://doi.org/10.1016/j.ssi.2012.03.014
90. Sai, B., Dutta, M., Chakraborty, A., Mitra, R.: Influence of rare-earth doping on the phase stability and microstructural evolution of zirconia ceramics. J. Eur. Ceram. Soc. **39**(2), 406–415 (2019). https://doi.org/10.1016/j.jeurceramsoc.2020.02.002
91. Svensson, F.G., Li, X., Lundqvist, M.: Rare-earth-modified titania nanoparticles: molecular insight into anatase phase stabilization. Inorg. Chem. **60**(12), 9805–9813 (2021). https://doi.org/10.1021/acs.inorgchem.1c02134
92. Lamrani, A.F., et al.: Rare-earth-doped TiO_2 rutile as a promising ferromagnetic alloy for visible light absorption in solar cells: first-principles insights. RSC Adv. **10**, 35505–35515 (2020). https://doi.org/10.1039/D0RA05725HV
93. Mashkovtsev, M.A., Kosykh, A.S., Ishchenko, A.V., Chukin, A.V., Kukharenko, A.I., Troshin, P.A., Zhidkov, I.S.: Unraveling oxygen vacancies effect on chemical composition, electronic structure and optical properties of Eu-doped SnO_2. Nanomaterials **14**(20), Article 1675 (2024). https://doi.org/10.3390/nano14201675
94. Azir, A., Selvaraj, M., Madhu, C.: Rare-earth-doped ceria nanostructures: Correlation between defect chemistry, oxygen vacancies, and enhanced catalytic activity. J. Mater. Chem. A **9**(15), 9056–9067 (2021). https://doi.org/10.1039/D1TA01234FS
95. Pandiyarajan, T., Thirumalraj, B.: Influence of RE (Pr^{3+}, Er^{3+}, and Nd^{3+}) doping on structural, optical, and morphological properties of ZnO nanostructures. J. Alloy. Compd. **893**, 162419 (2022). https://doi.org/10.1016/j.jallcom.2021.162419
96. Zamiri, R., Lemos, A.F., Rebello, A., Abbastabar Ahangar, H.: Effects of rare-earth (Er, La and Yb) doping on morphology and structural properties of ZnO nanostructures prepared by wet chemical method. Ceram. Int. **40**(1), 523–529 (2014). https://doi.org/10.1016/j.ceramint.2013.06.034
97. Shrivastava, P., et al.: J. Rare Earths (2022). https://doi.org/10.1016/j.jre.2021.10.010
98. Rao, V.D., Babu, K.V., Chary, K.V.R.: La-doped ZnO nanostructures: An investigation of their structural and catalytic properties. Mater. Today Proc. **33**, 359–365 (2020). https://doi.org/10.1016/j.matpr.2020.04.262
99. Zhou, Y., Zhang, H., Liu, H., Li, L.: Electrochemical impedance study of Nd-doped TiO_2 for energy storage applications. Electrochim. Acta **297**, 303–311 (2019). https://doi.org/10.1016/j.electacta.2018.11.123
100. Zheng, R., Liu, X., Sun, Y.: Effect of cerium doping on the structure and photocatalytic performance of TiO_2. Appl. Surf. Sci. **426**, 1044–1051 (2017). https://doi.org/10.1016/j.apsusc.2017.07.252
101. Al-Sabagh, A.M., et al.: Opt. Mater. (2020). https://doi.org/10.1016/j.optmat.2020.110638
102. Sharma, H.S., et al.: J. Lumin. (2021). https://doi.org/10.1016/j.jlumin.2021.118261
103. Rahangdale, S.B., et al.: J. Alloys Compd. (2021). https://doi.org/10.1016/j.jallcom.2021.159207
104. Zhang, L., et al.: Appl. Surf. Sci. (2020). https://doi.org/10.1016/j.apsusc.2020.145904
105. Subramaniam, T., et al.: Mater. Chem. Phys. (2021). https://doi.org/10.1016/j.matchemphys.2021.124830
106. Panda, C.K., et al.: J. Eur. Ceram. Soc. (2022). https://doi.org/10.1016/j.jeurceramsoc.2022.02.045

107. Natarajan, T.S., et al.: Photocatalytic activity of Nd-doped TiO_2 nanoparticles under visible light. Appl. Surf. Sci. **509**, 145352 (2020)
108. https://doi.org/10.1016/j.apsusc.2020.145352
109. Yang, X., et al.: Gd-doped $ZnFe_2O_4$ for visible-light-driven photocatalysis via enhanced electron mobility. J. Environ. Chem. Eng. **9**(5), 105760 (2021). https://doi.org/10.1016/j.jece.2021.105760
110. Al-Ghamdi, A.A., et al.: Eu-doped ZnO thin films as efficient photoanodes for dye-sensitized solar cells. Mater. Sci. Semicond. Process. **97**, 58–64 (2019). https://doi.org/10.1016/j.mssp.2019.03.014
111. Devi, L.G., Kavitha, R.: Sm^{3+} doped TiO_2 nanoparticles for photovoltaic applications: Enhanced lifetime and stability analysis. Sol. Energy Mater. Sol. Cells **155**, 93–100 (2016). https://doi.org/10.1016/j.solmat.2016.05.037
112. Kayani, Z.N., et al.: Enhanced gas sensing response of Sm-doped ZnO thin films towards CO gas. J. Mater. Sci. Mater. Electron. **32**, 11877–11888 (2021). https://doi.org/10.1007/s10854-021-05825-5
113. Yadav, A.A., et al.: Efficient NO_2 sensing by Gd-doped SnO_2 nanostructures synthesized via sol-gel technique. Sens. Actuators, B Chem. **318**, 128187 (2020). https://doi.org/10.1016/j.snb.2020.128187
114. Li, J., et al.: High-performance Dy-doped In_2O_3-based H_2S sensors operated at low temperature. Mater. Chem. Phys. **226**, 174–181 (2019). https://doi.org/10.1016/j.matchemphys.2019.01.020
115. Narang, S.B., et al.: Room-temperature ferromagnetism in Tb-doped ZnO nanorods synthesized by hydrothermal route. J. Magn. Magn. Mater. **493**, 165716 (2020). https://doi.org/10.1016/j.jmmm.2019.165716
116. Singh, M.K., et al.: Investigation of RTFM in Er-doped SnO_2 thin films for spintronic applications. Mater. Lett. **308**, 131204 (2022). https://doi.org/10.1016/j.matlet.2021.131204
117. Kumar, R., et al.: Ferromagnetic behavior of Pr-doped TiO_2 nanocrystals: role of oxygen vacancies and electronic structure tuning. J. Appl. Phys. **125**(2), 025107 (2019). https://doi.org/10.1063/1.5063784
118. Patil, V.L., Vanalakar, S.A., Tarwal, N.L., Kim, J.H., Patil, P.S.: Enhanced liquefied petroleum gas sensing performance of spray-deposited Pr-doped ZnO thin films. Sens. Actuators B Chem. **239**, 1185–1193 (2017). https://doi.org/10.1016/j.snb.2016.09.043
119. Dhanalakshmi, R., Ramachandran, K., Natarajan, T.S.: Effect of Nd doping on the structural, optical and gas sensing properties of ZnO nanorods synthesized by hydrothermal method. J. Alloy. Compd. **785**, 256–265 (2019). https://doi.org/10.1016/j.jallcom.2019.01.157
120. Mani, G.K., Rayappan, J.B.B., Krishnan, U.M.: High response and fast recovery ethanol sensor based on La-doped SnO_2 nanoparticles synthesized by hydrothermal route. Appl. Surf. Sci. **321**, 180–186 (2014). https://doi.org/10.1016/j.apsusc.2014.09.148
121. Suryanarayanan, R., Chandramohan, R., Murugesan, K.: Magnetic and optical properties of Gd-doped TiO_2 nanoparticles: evidence for room-temperature ferromagnetism. J. Magn. Magn. Mater. **452**, 264–271 (2018). https://doi.org/10.1016/j.jmmm.2017.12.093
122. Amutha, K., Suriyanarayanan, N., Muralidharan, R.: Structural, optical and magnetic properties of Ho-doped ZnO nanoparticles for spintronic applications. Mater. Chem. Phys. **249**, 123000 (2020). https://doi.org/10.1016/j.matchemphys.2020.123000
123. Thota, S., Singh, S.K.: Magnetic and electronic transport properties of Sm-doped SnO_2 nanoparticles. J. Appl. Phys. **111**(7), 07C313 (2012). https://doi.org/10.1063/1.3677850
124. Ghosh, et al.: Ceram. Int. (2021). https://doi.org/10.1016/j.ceramint.2020.09.186
125. Sharma, M., Singh, D., Bedi, R.K.: Structural, optical and gas sensing properties of Gd-doped NiO thin films. Ceram. Int. **48**(3), 3460–3469 (2022). https://doi.org/10.1016/j.ceramint.2021.10.184
126. Yadav, N., et al.: Mater. Chem. Phys. (2021). https://doi.org/10.1016/j.matchemphys.2021.124931
127. Tripathi, A., et al.: J. Lumin. (2021). https://doi.org/10.1016/j.jlumin.2021.118053
128. Mishra et al.: Mater. Sci. Semicond. Process. (2021). https://doi.org/10.1016/j.mssp.2021.106579

Recent Progress of Nanostructured Rare Earth Metals for Photovoltaic Applications

C. Andal and E. Kavitha

Abstract Nanostructured materials, especially metal oxides and rare earths, have received great interest owing to their capability to increase photovoltaic performance. With their favorable electrical, optical, and chemical properties, the metal oxide nanoparticles (MONPs) have high light absorption capability, enhanced surface area, as well as better charge transport—making them specifically suitable for solar energy applications. Also, rare earth elements with their quantum effects, high surface activity, and stable electronic structures have shown potential in solar cell technology. In this chapter, recent advancements in the production and uses of rare earth elements and metal oxide nanostructures are examined, especially as they relate to solar energy harvesting. It emphasizes how such materials will lead to higher solar cell efficiency with increased light trapping, reduced loss of energy, and increased photon-to-electron conversion. The role of rare earth elements in future solar technologies, including perovskite and dye-sensitized solar cells, is also discussed. Based on the continuing research into synthesis routes, structural manipulation, and photovoltaic performance, the nanomaterials are likely to propel the manufacture of cost-effective and efficient solar technologies. In photovoltaic applications, novel metal oxide combinations with equal molar ratios are used with increase in solar efficiency because of their excellent electrical, physical, and chemical qualities and high light absorption capacity. Similarly nowadays rare earth elements are also examined to create nanomaterials for a range of uses with the same qualities of metal oxides. Rare earth elements were used to synthesis of nanomaterials with various applications particularly in solar cell applications. In this book chapter, the usage of metal oxides compounds and rare earth elements for the synthesis of nanoparticles are discussed especially in solar cell application.

Keywords Metal oxides · Rare earth elements · Solar cell efficiency · Photovoltaic

---

C. Andal (✉) · E. Kavitha
Department of Physics, Dr. M.G.R. Educational and Research Institute, Chennai, Tamil Nadu, India
e-mail: andal.phy@drmgrdu.ac.in

© The Author(s), under exclusive license to Springer Nature Singapore Pte Ltd. 2026
N. Kumar et al. (eds.), *Rare Earth-Doped Metal Oxide Nanostructures*, Engineering Materials, https://doi.org/10.1007/978-981-95-6192-6_3

1 Introduction

Nanomaterials are particles with molecular sizes between 1 and 100 nm. A nanostructure is a microstructure of intermediate size. These particles' distinct mechanical, optical, electrical, and magnetic properties are essential in several disciplines, including material science, physics, and chemistry. Numerous benefits are provided by nanostructured materials with nanometer-sized dimensions, including a large surface area, enhanced mechanical properties, more atoms on the nanoparticle's surface than greater volume particles, and other physical characteristics, as well as tunable electronic properties that are critical for the construction of different sensors [1, 2]. Different dimensions of the nanostructure and its constituent parts are used to categorize them. The nanostructure of nanomaterials is set with different elementary units by varying its dimensions such as zero dimensions (0D) for nanoclusters and nanoparticles, similarly in one dimension (1D) for nanotubes and nanowires, and two dimension (2D) for nanoplates and nanolayers structures. "0D nanostructures" are defined as nanostructures with particles less than 100 nm in all pertinent directions. Since new materials called nanomaterials with nanosize ranging from 1 to 100 nm are used in various fields due to their acoustic, optical, electrical, magnetic, and thermal properties, researchers are very interested in creating novel nanomaterials due to their potential applications in a variety of fields. Figure 1 illustrates how nanotechnology links several application areas, such as chemistry, biology, biotechnology, medicine, pharmaceuticals, food and agriculture, electronics, and the environment [3–5].

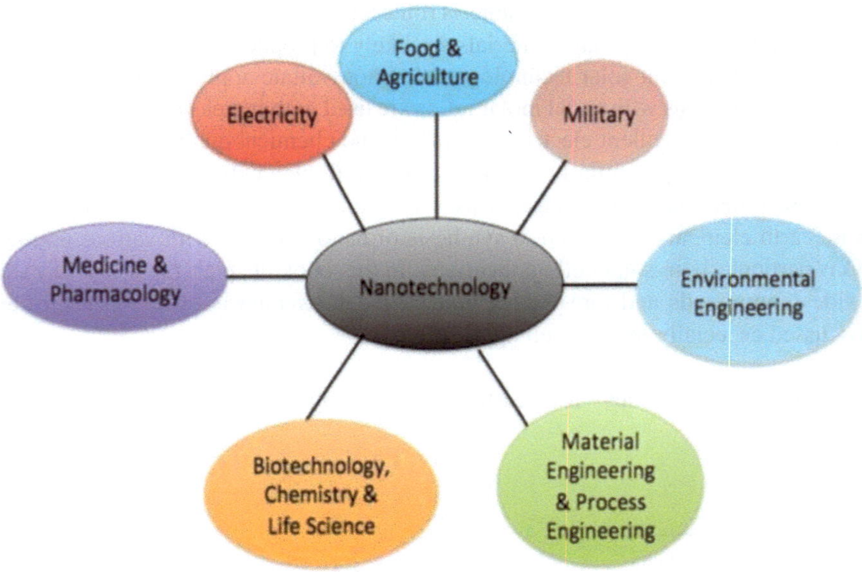

Fig. 1 Applications of nanotechnology in various fields

With their special qualities that can greatly improve photovoltaic devices' performance, nanomaterials have become a game-changer in the realm of solar cell technology. These materials' nanoscale size leads to enhanced surface area-to-volume ratios, tunable optoelectronic characteristics, and quantum confinement effects, all of which enhance solar cell efficiency. For example, size-dependent bandgaps in quantum dots with diameters between 2 and 10 nm enable fine control over the parameters of light absorption and emission [6, 7]. Additionally, nanostructured materials can improve light absorption by increasing light trapping and decreasing reflection. According to studies, nanostructured surfaces can lower reflection across the solar spectrum to less than 1%, while untreated silicon surfaces can reduce reflection to 30–40% [8].

1.1 Metal Oxide Nanoparticles (MONP)

Because of their special physical and chemical characteristics, like heat transfer and thermal conductivity, metal oxide nanoparticles are regarded as one of the most promising nanomaterials in many fields [3]. Metal oxide nanoparticle (MONP) synthesis is a promising area of study in the physical, chemical, and materials sciences. Metal oxide nanoparticle (MONP) production is important since it's utilized to make fuel cells, sensors, micro-electric circuits, catalysts, and more. Under a variety of circumstances, oxygen reacts to create a wide range of metallic oxides with different structural configurations. Nanoscale metal oxides have demonstrated crucial uses in photovoltaic, biomedical, fluorescence, and optical sensors, catalysts, and more [4].

Nanostructured metal oxide's size and shape are the primary factors that alter its surfaces in applications related to optics, mechanics, catalysis, electricity, and magnetism. To explore an increased activity spectrum with uniformly sized pores in the biomedical industry and other disciplines, metal oxide nanomaterial design has garnered a lot of attention. Using metal oxide to create nanomaterials with uniform shapes and small particle sizes is an intriguing career path in scientific study. Nanostructured metal oxides can be created using a wide range of technologies, including lithography, inert gas condensation, co-precipitation, and milling. The goal of all these techniques is to synthesize the desired product with particular chemical and physical features. All these technique methods are used to realize the target product with specific chemical and physical properties should be synthesized in a model-assisted synthesis methodology. A wide range of metal oxide nanoparticles from groups III–IV and VI–XII elements (ZrO_2, MoO_3, Mn_2O_3, CuO, Ag_2O, CoO, PdO, Co_3O_4, NiO, Y_2O_3) and semi-metal oxide groups (In_2O_3, SnO_2, Al_2O_3, and Bi_2O_3) may now be synthesized using a novel solvent-deficient approach. Iron (Fe), aluminum (Al), titanium (Ti), zinc (Zn), nickel (Ni), copper (Cu), tin (Sn), cobalt (Co), magnesium (Mg), cerium (Ce), bismuth (Bi), yttrium (Y), gadolinium (Gd), and most other metallic compounds which are suitable for the synthesis in nanosize also have been selected to create a variety of metal oxide nanoparticles.

Numerous MONPs, including silicon dioxide (SiO_2), nickel oxide (NiO), copper oxide (CuO), iron oxide (Fe_3O_4 and Fe_3O_2), zinc oxide (ZnO), titanium dioxide (TiO_2), and others, have been created using various techniques, and each of these distinct nanomaterials has a unique shape. Nickel oxide (NiO), copper oxide (CuO), iron oxide (Fe_3O_4 andFe_3O_2), zinc oxide (ZnO), etc. using various methods, and all these different nanomaterials show different morphology such as spherical, and triangular, star, nanowire, etc. [5, 6]. So metal oxides are best candidates for the synthesis of nanomaterials due to their several general properties including optical activity, magnetic properties and structural properties by having specific surface area and shape. MONPs have significant improvements in their physical and chemical characteristics because of their high density and limited size. So the formation of nanoparticles has some criteria such as understanding the many ways to create, researching their characteristics and uses are therefore essential. Its superior photoelectric conversion efficiency (PCE) has already been demonstrated even though the short duration of its synthesis [7].

1.2 Rare Earth Metal Elements (REE)

Rare earth elements (REEs) are belonging in the lanthanide series of metals such as lanthanum (La) to lutetium (Lu) along with scandium (Sc), yttrium (Y), promethium (Pm) which include atomic numbers from 57 to 71 in the periodic table. These lanthanides contain radioactive property naturally [8]. Generally, rare earth elements are divided into two categories namely heavy rare earth elements and light rare earth elements. The heavy rare earth elements include the elements such as europium, gadolinium, terbium, dysprosium, holmium, erbium, thulium, ytterbium, lutetium, and yttrium. The light rare earth elements include the elements such as lanthanum, cerium, praseodymium, neodymium, promethium, and samarium. The formation of heavy and light rate earth elements are due to their variation of unpaired and paired electrons in the 4F shell in their electronic configuration. Usually, semiconductors with rare earth metal oxide can be used in a variety of applications. Along with their semiconducting behavior, rare earth elements are used in solar cell applications, in industry to make metal and glass, in medicine for medical imaging, radar or computer screens, and everyday lighting fixtures with their specific characteristics such as high-temperature resistance, toughness, strength, and other mechanical properties [9, 10].

Rare earth elements have the essential properties such as optical, electrical, magnetic, acoustics and photo-electromagnetic properties due to their high surface area effect, quantum effect, small size effect and rich electronic structure which are applicable for these properties. And also, the nature of superconductivity and high chemical activity of the rare earth elements can also greatly progress the development of many new functional materials such as optical materials, luminescent materials, crystal materials, magnetic materials, battery materials, electronic ceramics, engineering ceramics, engineering plastics, packaging materials, precision polishing materials, and also some other functional materials which play an important role in the fields with wide range of many applications [11].

2 Solar Cell—An Introduction

Using the photovoltaic effect, a solar cell or solar panel is a device that directly transforms solar radiation into electrical energy. Due to its growing efficiency, low cost, and the fact that solar cells don't need fuel or chemical reactions to generate electricity, silicon is typically employed in their construction [12, 13]. When sunlight strikes silicon material, electrons are ejected with the applied energy $E = hv$, where v is the threshold frequency of the incident light (sunlight) and h is Planck's constant ($h = 6.626 \times 10^{-34}$ Js). Positive and negative electric plates combine to form photovoltaic cells, often known as solar cells. Photovoltaic cells or solar cells are a combination of positive and negative electric plates that are made up of semiconductor-like silicon materials. When sunlight falls on it, electrons are knocked out of the semiconductor as shown in Fig. 2. An electric current is thus generated through the flow of emitted electrons and generates electricity [14].

In order to transform sunlight into electrical energy that can be distributed to commercial, residential, and industrial customers, solar cells can be grouped together as arrays, which are collections of individual cells. Solar cell panels are the popular term for the smaller form of solar cells. These solar panels transform sunlight energy into electrical energy which are mounted on the rooftops to provide electricity inside the home. In isolated terrestrial locations where electric power sources are either unavailable or more costly to install for electrical power supply, solar cell panels are also utilized. They are particularly utilized in space installations, such as meteorological and communications satellites. Additionally, solar cells are utilized in a few consumer goods like electronics. Solar cells are also used in some consumer

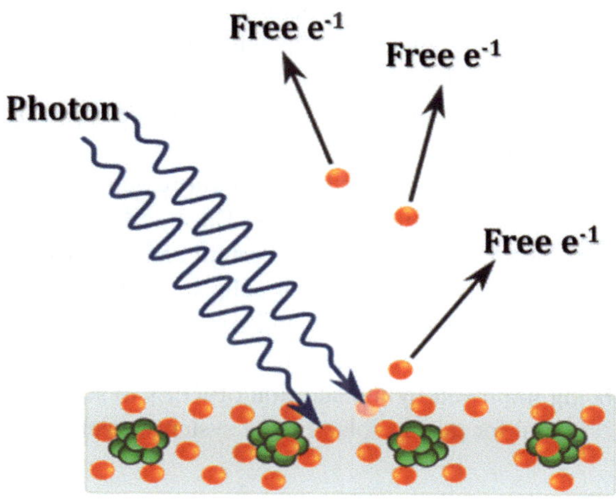

Fig. 2 Demonstration of photoelectric effect

products such as electronic toys, handheld calculators, and portable radios, and also inartificial light devices [15, 16].

2.1 Photovoltaic (PV) Technologies

Solar energy, the primary component of photovoltaic (PV) technology, is the best renewable energy source available on a worldwide scale. The light energy from sunshine is transformed into electrical power in photovoltaic technology. The core of PV technology consists of materials that, when exposed to light, generates electricity through a photovoltaic effect. One of the critical tools for converting solar energy into electrical power is photovoltaic (PV) technology. Solar cells, batteries, controllers, and inverters are the primary components of a photovoltaic system. With various modeling and simulation processes for assessing their efficiency, the development of these components is essential for progressing solar PV power generation. Temperature, irradiance level, cell type, and orientation all have an impact on PV panel efficiency [13, 17].

To boost efficiency, researchers have investigated several PV cell layouts and integration techniques, including floating and building-integrated PVs. PV technology provides a practical means of turning solar into electricity and represents a sustainable future. In photovoltaic cell (PV), the electron moves from valence band to conduction band by light photons for creating pairs of electrons and holes, when they strike the surface of semiconductor material found inside. The PV surface's front side collects these free electrons, which then create an electric current that can be used to create electricity. The photovoltaic effect has been impacted by the properties of semiconductor materials, such as their band gap, absorption coefficient, and charge carrier mobility [18, 19].

Light is absorbed and converted into electric current by the semiconducting materials used in photovoltaic cells. With an accomplishment rate of 26.7%, solar cells made of crystalline semiconducting materials are more efficient [20]. Therefore, it is crucial to produce PV cells with increasing efficiency levels. The researchers are aiming to synthesize rare earth metals that share characteristics with semiconducting materials and have higher efficiency percentages by combining two or more rare earth metals.

Additionally, rare earth elements are essential to the manufacturing of solar panels, particularly thin-film solar cells. These cells are made more durable and efficient by the addition of elements like cerium and dysprosium. Because of their low weight and flexibility, thin-film solar cells are employed in a variety of applications, such as portable solar chargers and rooftop installations. The photovoltaic conversion efficiency, or the rate at which input sunlight is transformed into output power, is being improved by the addition of REEs to solar panels [21]. This enhancement is essential for raising solar energy systems' overall output and lowering their cost.

2.2 *An Overview of Materials Used for Solar Cell*

The different types of solar panel cells are available by using various organic and inorganic metals and their compounds for the manufacturing of solar panels. They are such as

(a) **Monocrystalline and polycrystalline silicon solar cells**—Solar panels primarily use silicon (Si) material because of its band gap energy (1.1 eV), which allows for the highest amount of solar light to be observed and converted into electric current. Single (mono) crystal (crystalline) silicon solar cells are used to create monocrystalline solar panels. This type of solar cell is made of pure silicon that has been shaped into bars and cut into wafers with the necessary cell edges. These wafers need to be rounded and smoothed out to increase the amount of electricity produced. The assembled versions of monocrystalline solar cells are used to create polycrystalline panels. Raw silicon is melted together and then poured into a square mold to create wafers [22].

(b) **III-V semiconductor solar cells**—Semiconductors from alloys of III and V group metals with equal numbers of atoms—that is, Group III elements like boron, aluminum, gallium, and indium—that have three electrons in their outermost orbit are used to make this kind of solar cell. Similarly, the outer shell of Group V elements—which include nitrogen, phosphorus, arsenic, and antimony—has five electrons. Solar cells use a combination of these group elements that exhibit semiconductor activity. Solar photovoltaics are also developed using cadmium and tellurium. Because of their optimum bandgap of 1.45 eV and extended stability, they are regarded as the most effective thin-film material and are combined in a particular ratio to create cadmium telluride solar cells. The additional components utilized in the creation of sophisticated solar photovoltaics. The other materials such as copper, indium, gallium and selenide are used to develop advanced solar photovoltaics for improving its efficiency [23, 24].

(c) **Dye-sensitized solar cells**—Like batteries, these solar cells are made with liquid dyes. An economically based solar cell using liquid dye molecules on a titanium (IV) oxide layer was developed by Gratzel and Regan in 1991. The main plan, which is depicted below, is now referred to as a dye-sensitized solar cell method [25]. One kind of nanomaterial that is utilized in solar photovoltaics to acquire its features is carbon nanotubes (CNT). Because of its hexagonal lattice carbon structure, the conductor material created using CNT allows manufacturers to create solar cells that convert 75% of solar energy into electrical energy [26].

(d) **Perovskites solar cells**—Perovskite solar cell (PSC) is a type of solar cell that includes organic and inorganic materials with nanostructure. Generally the combination of materials is ABX_3, where A and B both are anions with positive ions and X is cation with negative ions. A is a divalent element with 2 electrons in its outer shell and B is tetravalent with 4 electrons in its outer shell elements as metals. And X is a negative ion that is usually either with oxygen or a halogen [27].Perovskite cells have many opto-electrical properties that is useful in solar cells due to their low-cost fabrication, high-efficiency, thin, lightweight and flexible properties. The most commonly perovskite solar cells methylammonium lead trihalide ($CH_3NH_3PbX_3$, where X is a halogen ion such as iodide, bromide, or chloride), formamidinium lead trihalide ($H_2NCHNH_2PbX_3$) and methyl-ammonium lead trihalide and $CH_3NH_3SnI_3$ [28].

(e) **Organic solar cells**—Carbon, hydrogen, nitrogen, and oxygen are some of the organic elements used in these solar cells. Their development is driven by the need for inexpensive production materials. Most organic solar cells use organic molecules with sp2 hybridization or double bonds between carbon atoms. Another substance that could be used in new solar photovoltaics to raise its bandgap is organic dyes. Since inorganic materials are difficult to recycle and reuse, the use of organic dyes makes it easier to produce sustainable solar photovoltaics [29].

3 Development of Solar Cells

The development of solar cell include three types based on the generations and using materials namely first, second and third generation. Monocrystalline and polycrystalline silicon are the primary components of the first generation of silicon crystalline solar cells. Although the technology is highly developed and in widespread usage today, there are drawbacks, including expensive raw material needs, complicated procedures, and high costs [30]. Solar cells based on semiconductor compounds are the second generation. Gallium arsenide, selenium phosphide, and other III-V compounds are among its primary synthetic elements. It is difficult to guarantee the safety and environmental protection of these materials because they are mostly hazardous and damaging [31]. Perovskite solar cells (PSCs), dye-sensitized solar cells (DSSCs), quantum dot solar cells (QDSCs), and nanowire solar cells (NWSCs) are among the novel nanomaterials that comprise the third generation. Third-generation solar cells are distinguished by their high power-conversion efficiency and low manufacturing costs [32]. When combined with thin-film technology, they could have a significant economic impact, making solar energy one of the most affordable and accessible solutions for producing energy in future. Thin-film technologies are a preferable choice for light harvesting because of their adjustable bandgap.

4 Applications of Nanostructured Rare Earth Elements

These days Indium, gallium, and tellurium are examples of rare earth elements that are crucial to solar panels. Due to their special characteristics, rare earth elements maximize solar absorption and conversion into electrical power. For instance, gallium increases solar cells' efficiency in low light, while indium increases their conductivity.

However, obstacles to the broad use of solar energy come from the extraction and availability of rare earth elements.

The rare-earth compound ytterbium oxide (YbO_3) can be produced using several techniques, including hydrothermal, sol-gel, and precipitation. Because of its high refractive index, wide band gap, thermal stability, and chemical inertness, it is utilized in solar cell applications. It is utilized in photovoltaic technologies, particularly in the processes of thermal barrier coating, light-trapping layers, doping agents, and antireflection coatings [33].

Because of their promise of high efficiency and low production costs, perovskite solar cells have emerged as one of the most promising new-generation photovoltaic technologies. Nevertheless, non-radiative charge recombination, mismatched absorption, and light-induced deterioration are some of the challenges facing the present generation of perovskite solar cells that hinder their potential to further enhance power conversion efficiency and operational stability for real-world use. Recently, rare-earth elements have been used to successfully get beyond these limitations because of their special photophysical characteristics [34].

Furthermore, the rare-earth elements' many energetic transition orbits allow them to transform high-energy ultraviolet or low-energy infrared photons into visible light that is receptive to perovskites, extending the spectrum response range and preventing harm from high-energy light. Consequently, the integration of rare-earth metals into perovskite solar cells has shown encouraging potential to increase device stability and efficiency at the same time [35]. Table 1 lists a few rare earth elements that are utilized in the production of solar panels in nearly every nation.

Solar energy is a clean and abundant source of power, and rare earth elements contribute significantly to its harnessing. Neodymium, lanthanum, and dysprosium are used in the production of photovoltaic cells, which convert sunlight into electricity. Neodymium and dysprosium help optimize solar cell efficiency, while lanthanum enhances the stability and durability of the panels.

Table 1 Manufacturing of solar panels with rare earth elements

Rare earth elements	Name of the country	Power conversion efficiency	References
Selenium	Japan, Canada, Belgium, and the US	5.2–7.2%	[36]
Gallium	China, Japan, and Canada	68.9% under a specific condition	[37]
Nickel	Indonesia, followed by the Philippines, Russia, Canada, Australia, and Brazil	20.54%	[38]
Indium	China, South Korea, Japan, and Canada	14–30%	[39]
Tellurium	Japan and China	16–21%	[40]
Neodymium	China	20%	[41]
Lanthanum	China, Australia, US, and Russia	18.4%	[42]

5 Conclusion

The specifics of understanding nanoparticles and their properties are described. It is discussed how metal oxides are used in the synthesis of nanomaterials. Rare earth metals are also well-known candidates for the synthesis of nanomaterials due to their semiconducting properties. These nanoparticles have higher efficiency values and are commonly used in solar cell applications. Developments in thin-film solar cell technology, which lessens the reliance on rare earth materials, are among the recent trends in the dependence of solar energy cells on rare earth materials as well as alternative materials like perovskite and organic polymers. Addressing the environmental and political issues surrounding rare earth metals used in solar cell applications requires an understanding of the historical context, important ideas, primary discussion points, case studies, current trends, difficulties, disputes, and prospects for the future. Addressing the environmental and sustainability implications in future with the fulfillment of global electricity needs requires an understanding of the historical background, key concepts, main discussion points, case studies, current trends, challenges, controversies, and future outlook of rare earth metals with solar cell applications.

References

1. Maciulis, V., Ramanaviciene, A., Plikusiene, I.: Recent advances in synthesis and application of metal oxide nanostructures in chemical sensors and biosensors. Nanomaterials **12**(24), 4413 (2022)
2. Mekuye, B., Abera, B.: Nanomaterials: an overview of synthesis, classification, characterization, and applications. Nano Sel. **4**(8), 486–501 (2023)
3. Findik, F.: Nanomaterials and their applications. Period Eng Nat Sci. **9**(3), 62–75 (2021)
4. Salem, S.S., Hammad, E.N., Mohamed, A.A., El-Dougdoug, W.: A comprehensive review of nanomaterials: types, synthesis, characterization, and applications. Biointerface Res. Appl. Chem. **13**(1), 41 (2022)
5. Mhetre, H.V., Kanse, Y.K., Patil, S.S.: Nanomaterials: applications in electronics. Int. J. Adv. Eng. Nano Technol. **4**(6) (2021)
6. Yu, T., He, W., Zhang, Q., Ma, D.: Advanced nanomaterials and characterization techniques for photovoltaic and photocatalysis applications. Acc. Mater. Res. **4**(6), 507–521 (2023)
7. Wang, L., Teles, M.P., Arabkoohsar, A., Yu, H., Ismail, K.A., Mahian, O., Wongwises, S.: A holistic and state-of-the-art review of nanotechnology in solar cells. Sustain. Energy Technol. **54**, 102864 (2022)
8. Hajra, K., Maity, D., Saha, S.: Recent advancements of metal oxide nanoparticles and their potential applications: a review. Adv. Mater. Lett. **15**(1), 24011740 (2024). https://doi.org/10.5185/amlett.2024.011740
9. Patil, A.S., Patil, A.V., Dighavkar, C.G., Adole, V.A., Tupe, U.J.: Synthesis techniques and applications of rare earth metal oxides semiconductors: a review. Chem. Phys. Lett. **796**, 139555 (2022)
10. Hossain, M.K., Hossain, S., Ahmed, M.H., Khan, M.I., Haque, N., Raihan, G.A.: A review on optical applications, prospects, and challenges of rare-earth oxides. ACS Appl. Electron. Mater. **3**(9), 3715–3746 (2021)

11. Naresh, K., Chandan, K.: Advancements in nanomaterials for solar cell applications: a comprehensive overview of recent developments. Int. J. Res. Sci. Eng. **3**(1), 50–65 (2023). https://doi.org/10.55529/ijrise.31.50.65
12. Ali, N., Hussain, A., Ahmed, R., Wang, M.K., Zhao, C., Haq, B.U., Fu, Y.Q.: Advances in nanostructured thin film materials for solar cell applications. Renew. Sust. Energ. Rev. **59**, 726–737 (2016)
13. Sharma, S., Jain, K.K., Sharma, A.: Solar cells: in research and applications—a review. Mater. Sci. Appl. **6**(12), 1145–1155 (2015)
14. Copeland, A.W., Black, O.D., Garrett, A.B.: The photovoltaic effect. Chem. Rev. **31**(1), 177–226 (1942)
15. Mozumder, M.S., Mourad, A.H.I., Pervez, H., Surkatti, R.: Recent developments in multifunctional coatings for solar panel applications: a review. Sol. Energy Mater. Sol. Cells. **189**, 75–102 (2019)
16. Dallaev, R., Pisarenko, T., Papež, N., Holcman, V.: Overview of the current state of flexible solar panels and photovoltaic materials. Materials **16**(17), 5839 (2023)
17. Andrei, V., Wang, Q., Uekert, T., Bhattacharjee, S., Reisner, E.: Solar panel technologies for light-to-chemical conversion. Acc. Chem. Res. **55**(23), 3376–3386 (2022)
18. Fan, B., Lin, F., Wu, X., Zhu, Z., Jen, A.K.Y.: Selenium-containing organic photovoltaic materials. Acc. Chem. Res. **54**(20), 3906–3916 (2021)
19. Miah, M.H., Khandaker, M.U., Rahman, M.B., Nur-E-Alam, M., Islam, M.A.: Band gap tuning of perovskite solar cells for enhancing the efficiency and stability: issues and prospects. RSC Adv. **14**(23), 15876–15906 (2024)
20. Ru, X., Yang, M., Yin, S., Wang, Y., Hong, C., Peng, F., et al.: Silicon heterojunction solar cells achieving 26.6% efficiency on commercial-size p-type silicon wafer. Joule **8**(4), 1092–1104 (2024)
21. Kharel, P.L., Zamborini, F.P., Alphenaar, B.W.: Enhancing the photovoltaic performance of dye-sensitized solar cells with rare-earth metal oxide nanoparticles. J. Electrochem. Soc. **165**(3), H52 (2018)
22. Enaganti, P.K., Dwivedi, P.K., Srivastava, A.K., Goel, S.: Study of solar irradiance and performance analysis of submerged monocrystalline and polycrystalline solar cells. Prog. Photovolt. Res. Appl. **28**(7), 725–735 (2020)
23. Li, Z., Tan, H.H., Jagadish, C., Fu, L.: III–V semiconductor single nanowire solar cells: a review. Adv. Mater. Technol. **3**(9), 1800005 (2018)
24. Bosi, M., Pelosi, C.: The potential of III-V semiconductors as terrestrial photovoltaic devices. Prog. Photovolt. Res. Appl. **15**(1), 51–68 (2007)
25. O'regan, B., Grätzel, M.: A low-cost, high-efficiency solar cell based on dye-sensitized colloidal TiO2 films. Nature. **353**(6346), 737–740 (1991)
26. Wieland, L., Li, H., Rust, C., Chen, J., Flavel, B.S.: Carbon nanotubes for photovoltaics: from lab to industry. Adv. Energy Mater. **11**(3), 2002880 (2021)
27. Horantner, M.T., Leijtens, T., Ziffer, M.E., Eperon, G.E., Christoforo, M.G., McGehee, M.D., Snaith, H.J.: The potential of multijunction perovskite solar cells. ACS Energy Lett. **2**(10), 2506–2513 (2017)
28. Wu, Y., Li, X., Fu, S., Wan, L., Fang, J.: Efficient methylammonium lead trihalide perovskite solar cells with chloroformamidinium chloride (cl-FACl) as an additive. J. Mater. Chem. A. **7**(13), 8078–8084 (2019)
29. Hoppe, H., Sariciftci, N.S.: Organic solar cells: an overview. J. Mater. Res. **19**(7), 1924–1945 (2004)
30. Baghel, N.S., Chander, N.: Performance comparison of mono and polycrystalline silicon solar photovoltaic modules under tropical wet and dry climatic conditions in east-Central India. Clean Energy. **6**(1), 165–177 (2022)
31. Zhang, Y., Conibeer, G., Liu, S., Zhang, J., Guillemoles, J.F.: Review of the mechanisms for the phonon bottleneck effect in III–V semiconductors and their application for efficient hot carrier solar cells. Prog. Photovolt. Res. Appl. **30**(6), 581–596 (2022)

32. Carvalho Junior, J.A., Nunes, C.L., Machado, W.S., Schiavon, M.A.: Recent progresses in quantum-dot-sensitized solar cells: the role of counter electrodes. Energ. Technol. **12**(8), 2400254 (2024)
33. Liu, D.: Analysis on the luminescence mechanism and application of rare earth luminescent materials. Chem. Manag. **31**, 47–48 (2019)
34. Yun, S., Qin, Y., Uhl, A.R., Vlachopoulos, N., Yin, M., Li, D., et al.: New-generation integrated devices based on dye-sensitized and perovskite solar cells. Energy Environ. Sci. **11**(3), 476–526 (2018)
35. Cang, L., Qian, Z., Wang, J., Chen, L., Wan, Z., Yang, K., Zhang, H., Chen, Y.: Applications and functions of rare-earth ions in perovskite solar cells. Chin. Phys. B. **31**(3), s (2022)
36. Youngman, T.H., Nielsen, R., Crovetto, A., Seger, B., Hansen, O., Chorkendorff, I., Vesborg, P.C.: Semitransparent selenium solar cells as a top cell for tandem photovoltaics. Sol. RRL. **5**(7), 2100111 (2021)
37. Helmers, H., Lopez, E., Höhn, O., Lackner, D., Schön, J., Schauerte, M., et al.: 68.9% efficient GaAs-based photonic power conversion enabled by photon recycling and optical resonance. Phys. Status Solidi (RRL)–Rapid Res. Lett. **15**(7), 2100113 (2021)
38. Kim, J.W., Cho, E., Lee, H.J., Kwon, S.N., Park, J.S., Kim, M., et al.: Enhancing efficiency of inverted perovskite solar cells by sputtered nickel oxide hole-transport layers. Sol. RRL. **8**(4), 2300933 (2024)
39. Gharibshahian, I., Orouji, A.A., Sharbati, S.: Suitable top cell partners for copper indium gallium selenide-based tandem solar cells to achieve> 30% efficiency. Phys. Status Solidi A. **218**(15), 2000796 (2021)
40. Ağca, S., Çankaya, G., Sonmezoglu, S.: Impact of tellurium as an anion dopant on the photovoltaic performance of wide-bandgap cu (in, Ga) Se2 thin-film solar cells with rubidium fluoride post-deposition treatment. Front. Energy Res. **11**, 1215712 (2023)
41. Almakayeel, N., Kaliyannan, G.V., Gunasekaran, R.: Enhancing power conversion efficiency of polycrystalline silicon photovoltaic cells using yttrium oxide anti-reflective coating via electro-spraying method. Ceram. Int. **50**(21), 42392–42403 (2024)
42. Thirupathi, S., Karthikeyan, J., Kesavamoorthy, J., Savarimuthu, J.A.T., Pazhanivel, S., Varathan, E., Murugesan, V.: Enhanced performance of dye-sensitized solar cells by lanthanum doped-TiO_2 Photoelectrodes. Trans. Indian Ceram. Soc. **84**(1), 49–56 (2025)

Enhanced Adsorption and Degradation of Synthetic Dyes Using Rare Earth Metal-Doped Biodegradable Polymers

Amanpreet Kaur, Ayasha Negi, Shubhangee Agarwal, Man Vir Singh, Jitendra Pal Singh, and Neha Bhatt

Abstract The adverse environmental effects and health burdens associated with dye discharge have necessitated the adoption of photocatalytic degradation as an effective method for addressing water contamination. Recent studies in photocatalysis indicate that nanocomposites incorporating rare earth metals are essential for improving the degradation of dyes in sewage water treatment. This paper reviews the significance of degradation of dyes in the presence of light, highlighting its environmental benefits, sustainable characteristics, and versatility in a wide range of fields. The doping of rare earth metals in nanocomposites significantly enhances photocatalytic activity. This is achieved through an extended absorption range of the photocatalysts, which enhances charge segregation and ensures stable frameworks. REMs function as photosensitisers, creating sub-band-energy levels between the conduction band and valence band of nanocrystals (NCs). The REMs-doped NCs demonstrate enhanced efficiency in dye decomposition relative to conventional methods, attributable to alterations in band structures, improved charge separation, and increased surface area. Multiple characterization methods are utilized to examine the structural and morphological properties of REMs-doped NCs, providing significant insights into their photocatalytic activity. Optimization strategies such as morphology control, surface modification, co-doping, and heterojunction formation are crucial for enhancing the photocatalytic performance of REM-doped NCs in photocatalytic reactions. The roles of various rare earth metals, including cerium, praseodymium, neodymium, samarium, dysprosium, europium, gadolinium, erbium, lanthanum, terbium, and ytterbium, have been examined regarding their potential applications in environmental remediation

A. Kaur (✉) · A. Negi · S. Agarwal
Department of Chemistry, School of Sciences, IFTM University, Moradabad, Uttar Pradesh, India
e-mail: amanpreet2225@gmail.com

M. V. Singh
School of Applied and Life Sciences, UIT, Uttaranchal University, Dehradun, Uttarakhand, India

J. P. Singh
Department of Physics, School of Sciences, IFTM University, Uttar Pradesh, Moradabad, India

N. Bhatt
Department of Chemistry, Pt. L.M.S., Rishikesh Campus, SDS University, Tehri Garhwal, Uttarakhand, India

© The Author(s), under exclusive license to Springer Nature Singapore Pte Ltd. 2026
N. Kumar et al. (eds.), *Rare Earth-Doped Metal Oxide Nanostructures*, Engineering Materials, https://doi.org/10.1007/978-981-95-6192-6_4

via photocatalytic degradation of dyes. This review will provide a comprehensive analysis of current progress in the implementation of REMs-doped NCs for the enhancement of dye degradation, contributing to the development of more effective photocatalysts for water purification.

Keywords Photocatalysis · Dyes remediation · Rare earth metals · Environmental concerns

1 Introduction

Water is a limited and crucial resource for humanity, now confronting escalating worldwide issues, notably due to pollution. A significant environmental issue is wastewater tainted with dye as a result of industrial activities [1]. The textile, leather, paper, plastic, and printing industries saw significant expansion in the twentieth century, resulting in substantial quantities of synthetic dyes being released into aquatic ecosystems. Dyes and pigments are used worldwide, with an estimated variety exceeding 10,000 varieties and an annual production of approximately 0.7 million metric tonnes. Textile manufacture consumes significant quantities of water, around 200 L every kilogram of textile produced, with a normal mill using up to 1.6 million gallons daily.

The discharge of dye wastewater poses significant dangers to the environment. These pollutants obstruct sunlight penetration, disrupting the photosynthetic process in aquatic ecosystems, and are often composed of extremely poisonous heavy metals (e.g., lead, chromium) or aromatic compounds. The intricate aromatic ring structure renders the dye molecules very durable, preventing their degradation under standard treatment conditions (e.g., aerobic digestion, photolysis, heat treatment, and oxidation). Moreover, some dyes have mutagenic, carcinogenic, and cytotoxic properties, posing a risk to human health by potentially damaging the liver, kidneys, brain, reproductive system, and central nervous system.

Efforts to tackle these problems have been executed using several physicochemical and biological treatment methods. Among these technologies, adsorption has emerged as one of the most advantageous and cost-effective ways for dye removal, owing to its operational simplicity, efficacy, and economic viability [2]. Activated carbon is the standard adsorbent owing to its remarkable capacity (surface area) and adsorption properties. However, its elevated production costs and regeneration constraints hinder extensive use, particularly for low concentration or stubborn dyes that exhibit inadequate selectivity, such as dispersion and vat dyes [3].

Several attempts have been made in the recent past to formulate better alternative adsorbents that are both more efficient and economical. These include naturally occurring materials (e.g., clays, zeolites), biosorbents, agricultural by-products, industrial wastes, and their corresponding composites [4]. A significant amount of research has concentrated on modifying these materials to improve their surface characteristics, dye adsorption capabilities, and selectivity for different dye types.

Despite the abundance of novel adsorbent materials, there is a scarcity of thorough research examining dye adsorption from a macro viewpoint [5]. This study will categorize dyes according to their health and environmental risks, assess different dye removal methods, and analyze the adsorption isotherms pertinent to dye sorption [6]. It thoroughly encapsulates and synthesizes contemporary advances in alternative adsorbents, emphasizing their physicochemical properties, efficacy, and constraints across diverse environmental conditions [7]. The paper identifies fundamental challenges hindering the transition of these materials from laboratory-based discovery to industrial use. Ultimately, it delineates prospective avenues, including the development of multifunctional adsorbents, scalability constraints, and hybrid integration techniques, which are essential for the future progression of dye remediation technologies [8].

2 Rare Earth Metal-Doped Biodegradable Polymers for Dye Degradation

Nanomaterials have significantly revolutionized wastewater treatment technology, especially in the decolorization of dye effluents. Nanostructures, generally ranging from 1 to 100 nm, exhibit exceptional surface area-to-volume ratios, leading to enhanced physicochemical and catalytic activity relative to their bulk counterparts [9]. Metal oxides (MOs) are extensively researched nanomaterials that have shown efficacy in the degradation of organic contaminants in industrial effluent. These nanometal oxides were extensively used in remediation, optical, and mechanical systems [10]. Certain metal oxides, including ZnO, TiO_2, CeO_2, and La_2O_3, have shown significant efficacy in the photocatalytic destruction of dyes. Lanthanum oxide (La_2O_3), a rare-earth metal oxide, is distinguished by its substantial and broad band gap energy (4.3–5.8 eV) and its distinctive electrical structure. In contrast to other lanthanides, La^{3+} lacks 4f electrons and has an electronic configuration comparable to that of xenon [11]. This enhances its unique characteristics, making it suited for many applications. The many biological actions of rare-earth elements, such as lanthanum, which include anticoagulant, antiemetic, antibacterial, anticancer, and antioxidant properties, enhance their use in biomedical and pharmaceutical domains. Moreover, La_2O_3 has superior economic efficiency relative to other rare-earth oxides, including Y_2O_3 and Gd_2O_3 [12].

La_2O_3 nanoparticles have applications in hydrogen storage, superconductors, optoelectronic materials, laser crystals, LEDs, biosensors, high refractive index optical fibers, and catalysis. Furthermore, owing to its intrinsic characteristics, La_2O_3 rapidly interacts with ambient CO_2 to produce lanthanum carbonate, hence affecting its stability and crystalline qualities [13]. Nanostructured La_2O_3 has been synthesized via hydrothermal, sol–gel, thermal degradation, sonochemical, and co-precipitation methods. The co-precipitation technique is valued for its simplicity, affordability, and lack of need for specialized equipment.

The fabrication of La_2O_3 nanostructures significantly affects the physicochemical characteristics and the activity of the resultant catalysts. Molybdenum (Mo) may be used as a dopant to augment photocatalytic activity [14]. The inclusion of Mo promotes thermal stability, reduces the band gap, increases charge carrier concentration, and broadens the absorbing ability into the visible wavelengths of light. It further stabilizes the electronic band framework of metallic oxides semiconductors. Ikram et al. [15] synthesized undoped and Mo-doped La_2O_3 nanostructures (0.02, 0.04, and 0.06 wt % Mo) using the one-pot co-precipitation technique. The resulting materials were then evaluated for their antibacterial efficacy against *Staphylococcus aureus* and *Escherichia coli*, as well as their ability to degrade methylene blue (MB), a prevalent model dye [15]. The X-ray diffraction (XRD) examination indicated that La_2O_3 included both cubic and hexagonal structures. Doping does not influence the phase structure but markedly enhances the crystallinity. Density functional theory (DFT) and UV–Vis spectroscopy were employed to ascertain the optical properties of the products, confirming a reduction in the energy gap and the emergence of additional strong absorption bands upon Mo embedding [16]. High-resolution transmission electron microscopy (HR-TEM) revealed that La_2O_3 nanostructures were rod-shaped and exhibited increased density with increasing Mo content. Photocatalytic degradation experiments showed that molybdenum doping improved the degradation efficiency of methylene blue [17]. Increased Mo concentrations augmented antibacterial activity, as the inhibition zone measured 5.05 mm against *Staphylococcus aureus* and 5.45 mm against *Escherichia coli* at 6 wt% Mo doping [18]. In addition to these results, Singh et al. [19] assessed the photocatalytic efficacy of many oxides of metals, namely NiO, ZnO, CuO, CeO_2, SnO_2, and Mn_3O_4, concerning the degradation of methylene blue and methyl orange under UV–visible light. Each nanoparticle, at a concentration of 100 mg, has been suspended in 100 mL of dye mixture and subjected to irradiation for 2 h, with samples collected every 20 min [19]. ZnO has been identified as the most efficient catalyst, demonstrating exceptional removal efficacy for the two dyes.

These dynamics demonstrate the efficacy of nanostructured metal oxides, namely Mo-doped La_2O_3, as effective photocatalysts for environmental remediation. The substitution of Mo in La_2O_3 enhances optical and structural characteristics, significantly improves dye degradation and antibacterial activity, and presents promising opportunities for the development of multifunctional nanomaterials in sustainable wastewater treatment applications [20] (Table 4.1).

3 Metal-Incorporated Biopolymeric Nanomaterials

Metal-doped nanomaterials demonstrate improved efficacy in the photocatalytic destruction of azo dyes, principally owing to their customized pore structure, surface crystallography, and superior mechanical strength [31]. The core mechanism of photocatalytic degradation is a photo-induced electron transfer process, closely linked to redox processes. The introduction of metallic nanoparticles alters

Table 4.1 Strength and limitations of different dye degradation techniques

Methods	Strength	Limitations	References
Adsorption	The most advanced technology, effortless operation, and efficient removal of dyes from effluent	The adsorption efficiency is constrained by the environmental conditions and the structure of the adsorbent Secondary pollution is effortless to generate	[21]
Biochemical method	Specificity of biodegradation. Non-toxic impact on microorganisms	Processing speed is slow, and the ground occupies a significant area. Environmental conditions that are advantageous, including pH, temperature, and nutrition	[22]
Chemical oxidation	A prompt and comprehensive response. There is no secondary pollution	Excessive energy consumption Additional chemical reagents are necessary	[23]
Coagulation	Simple process and operation, minimal equipment cost, and high efficacy for hydrophobic dyes	The treatment procedure generates a substantial amount of sediment, which is challenging to dehydrate and inappropriate for the decolorization of hydrophilic pigments	[24]
Membrane separation	Selectivity to dyes, removal of a variety of dyes, and minimal space requirements	High initial, maintenance, and operational expenses The utilization of austere conditions (pressurization and effluent flow restrictions) Short service life and effortless obstruction	[25]
Photocatalysis	Thorough degradation of dyes and effective eradication of obstinate dye molecules High efficiency and energy conservation	It is challenging to distinguish between the active components and intermediate products Bad light transmittance with a poor effect on high concentration dye effluent	[26]
Nanofiltration	Absence of adsorbent depletion	The disposal of suspended effluent is a challenging process	[27]
Ozonation	Effective removal of pigmentation	Elevated electricity expenses necessitate	[28]
Fenton	Effective for both soluble and nonsoluble pigments	Creating aromatic compounds that may be detrimental to both human health and the environment	[29]
UV hydrogen peroxide photocatalysis	Pigments undergo degradation through the utilization of electrons	The cost is elevated by the substantial energy consumption of UV lamps	[30]

the semiconducting characteristics of the solution, facilitating interfacial charge transfer, reducing metal oxidation, and improving electron lifetime, which eventually enhances photocatalytic reactivity [32]. Surface plasmon resonance (SPR), often seen in nanoscale metallic nanoparticles, enhances localized electron density and optimizes charge carrier dynamics. Polymeric nanocomposites containing metallic nanoparticles work as self-stabilizing systems, enhancing dispersion and functional integration [33]. Metal-polymer hybrid materials, such as metal-infused resins, are widely used in industrial catalytic processes. Attaching metallic nanoparticles to polymeric matrices improves structural density and physicochemical stability while also providing unique functional characteristics compared to monomeric metal nanoparticles [34].

The size distribution of metal nanoparticles is typically heterogeneous; however, the introduction of a polymer matrix facilitates particle size homogeneity and enhances overall system stability. This is shown by the integration of bacterial cellulose fibers with gold nanoparticles (AuNPs). Gold nanoparticles (AuNPs), averaging 74.32 nm in size and generated from fresh fruiting bodies of *Flammulinavelutipes* (Enoki mushroom), effectively degrade roughly 75% of methylene blue dye after 4 h. The extensive substrate specificity of AuNPs facilitates the degradation of many organic dyes [35]. The f-orbital electrons of lanthanum (La) interact with different functional groups of biopolymers, resulting in the creation of coordination complexes with increased surface areas. The integration of La into biopolymeric matrices improves photocatalytic efficacy, especially in the destruction of organic contaminants like azo dyes [2]. A lanthanum-chitosan compound achieved 90% degradation of methylene blue in 40 min. This approach led to a substantial decrease in chemical oxygen demand (COD), roughly by a factor of eight, indicating efficient mineralization of the dye [36]. Figure 4.1 represents the Synthesis of Rare Earth Metal-Doped Biodegradable Polymers for photocatalytic dye degradation.

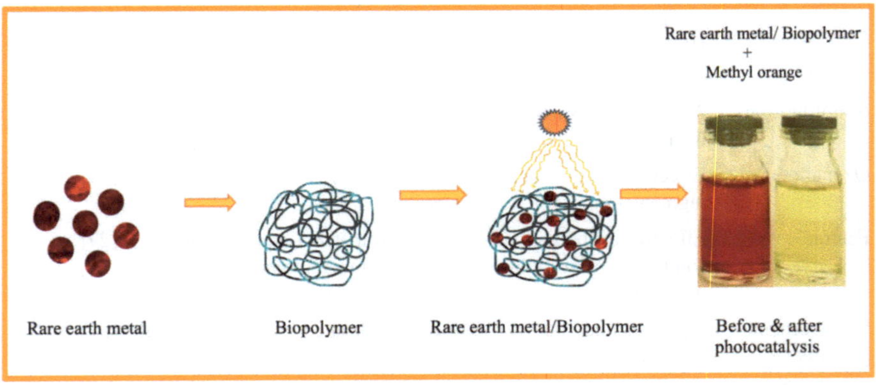

Fig. 4.1 Synthesis of rare earth metal-doped biodegradable polymers for photocatalytic dye degradation

4 Photo-Catalytic Characteristics of Rare Earth-Doped Metallic Oxides

Oxides of different metals are essential constituents in photocatalytic water splitting processes owing to their semiconducting characteristics. Cadmium sulfide (CdS), a metal sulfide, has been thoroughly studied for hydrogen (H_2) production; yet, its photocatalytic effectiveness is inadequate, mainly because of photocorrosion and rapid charge carrier reconstitution [37]. Like metal sulfides, several metal oxides have been investigated for photocatalytic uses. Titanium dioxide (TiO_2) and zinc oxide (ZnO) are prominently researched nanomaterials in this domain. However, their substantial band gaps ($\approx$ 3.2 eV) considerably restrict their capacity to absorb visible light, leading to diminished quantum efficiency in photocatalytic processes [38]. To address these constraints, the integration of foreign dopants, especially rare earth elements, has surfaced as an effective technique to improve quantum yield and overall photo-degradation efficacy. Rare earth doping has shown the ability to enhance charge carrier isolation and inhibit the recombining of photo induced electron–hole pairs, thereby enhancing photocatalytic efficiency [39]. The mechanical justification for using rare earth elements as dopants is thoroughly detailed in the following section.

5 Rare Earth Metal Doped Metal Oxides

The distinct physicochemical characteristics of rare earth metal-doped metal oxides, such as their intense redox activity, excellent thermal stability, and customized electronic structures, have made them become cutting-edge functional materials. Rare earth elements like La, Ce, Eu, or Nd may be doped into metal oxides like TiO_2, ZnO, or CeO_2 to greatly increase their surface area, optical absorption, and catalytic activity. Improved charge carrier mobility and separation are made possible by these changes, which are essential for uses in environmental cleanup, gas sensing, and photocatalysis. By introducing localized energy states inside the bandgap, rare earth ions improve reaction kinetics in surface-driven processes and allow for more efficient use of visible light.

5.1 Gd-Doped Metal Oxides Nanoparticles

Multiple investigations have shown the efficacy of rare earth metal doping, namely gadolinium (Gd), in augmenting the photocatalytic efficiency of semiconductor nanomaterials used for splitting of water and breakdown of dyes. Mahalakshmi et al. indicated that the introduction of Gd ions into TiO_2 enhances its band gap energy and promotes superior charge carrier separation, thereby diminishing electron–hole pair recombination [40]. Moreover, Gd doping leads to a decrease in TiO_2 nanoparticle dimensions, hence augmenting surface area and catalytic efficacy. The introduction of 0.03 wt% Gd doping in TiO_2 markedly enhanced its degrading efficiency, owing to the half-filled 4f orbital of Gd^{3+} ions, which efficiently capture electrons and facilitate interfacial charge transfer at the solid–liquid interface [41]. Although ZnO is acknowledged as a promising semiconductor for photocatalysis, its actual effectiveness is often limited by the fast recombining of photogenerated electron–hole pairs. Kalam et al., investigated Gd-doped ZnO nanoparticles, discovering improved charge separation, diminished particle size, and augmented surface area. Their research also emphasized the impact of calcination duration on photocatalytic efficacy. ZnO doped with Gd and calcined at 500 °C for 7 h demonstrated excellent photocatalytic activity; however, extended calcination (9 h) resulted in particle aggregation, increased crystallite size, and decreased surface area, thereby reducing activity [42]. Photoluminescence (PL) spectroscopy further validated that Gd doping causes a shift in emission spectra, signifying reduced electron–hole recombination. Selvaraj et al. produced Gd-doped ZnO nanorods by a sonochemical process and assessed their photocatalytic effectiveness for the breakdown of methylene blue (MB) under ultraviolet illumination. No deterioration was found after 300 min in the absence of a catalyst. Nonetheless, the use of Gd-doped ZnO markedly enhanced photocatalytic activity. A doping concentration of 3 wt% Gd resulted in a 92.76% degradation of MB after 300 min [43]. The research revealed that alkaline circumstances (pH 10) augmented photocatalytic activity, presumably owing to heightened production of hydroxyl radicals that promote oxidation. At pH 10, total (100%) degradation of MB dye was accomplished within 270 min using 3% Gd-doped ZnO. These data indicate that photocatalytic effectiveness is significantly influenced by the pH of the solution. Zhang et al. investigated the size-dependent photocatalytic properties of Gd-doped bismuth ferrite ($BiFeO_3$, BFO), a substance with an appropriate band gap of around 2.2 eV [44]. While pure BFO exhibits photocatalytic activity, it is hindered by elevated recombination rates of charge carriers. Gd doping led to a progressive diminution in particle size and an enhancement in outer area. Studies on the deterioration of Rhodamine B (RhB) indicated that pure BFO destroyed just 22.3% of RhB under visible light after 270 min, whereas Gd-doped BFO (at 1%, 3%, and 5%) had degradation rates of 34.2%, 56.8%, and 42.1%, respectively [45]. The 3% Gd-doped BFO had the greatest photocatalytic efficacy. SEM investigations revealed that increased Gd concentration resulted in reduced particle size while simultaneously enhancing agglomeration, which may impede photocatalytic activity by restricting access to active sites. Therefore, determining an ideal dopant concentration is crucial for enhancing photocatalytic performance [46].

In a separate investigation, Al-Hamdi et al. produced Gd-doped SnO_2 nanoparticles and assessed their efficacy in degrading phenol under sunshine and exposure to visible light. The photo catalysis efficacy of Gd-doped SnO_2 was evaluated against other rare earth-doped TiO_2 systems, namely those doped with Ce, Nd, and La. Among them, Gd-doped TiO_2 demonstrated exceptional performance, underscoring Gd's efficacy as a powerful dopant for augmenting photocatalytic processes [47].

5.2 Ce-Doped Metal Oxides Nanoparticles

Titanium dioxide (TiO_2) is acknowledged as a proficient photocatalyst; however, its extensive band gap constrains its photocatalytic efficacy during visible light exposure. Shi et al. found that the incorporation of cerium (Ce) into TiO_2 markedly improves its photocatalytic efficacy [48]. Cerium, owing to its partly occupied 4f and 4d orbitals, serves as an effective catalytic center and promotes photocatalytic interactions with organic molecules. The existence of empty orbitals in Ce facilitates efficient interaction with diverse functional groups. The researchers generated Ce-doped TiO_2 nanoparticles by integrating varying amounts of Ce, thereafter subjecting them to calcination at 500 °C [49]. X-ray diffraction (XRD) research illustrated that an increase in Ce levels induces to larger and less strong diffraction peaks of TiO_2, indicating a decrease in crystallite size. The diminishment of nanoparticle size results in an increased specific surface area, consequently augmenting redox activity and raising charge transfer capacity of photo-generated electron–hole pairs, eventually elevating photocatalytic performance [50].

The UV–Visible wavelengths of absorption of Ce-doped TiO_2 displayed a redshift, broadening the absorption edge into the visible light spectrum (400–600 nm), indicating enhanced photocatalytic activity under the visible light [51]. This differs from clean TiO_2, which mostly absorbs in the ultraviolet spectrum. Mora et al. examined the photocatalytic characteristics of cerium-doped ZnO, which intrinsically has a broad band gap of 3.37 eV [52]. Although ZnO is a promising photocatalyst, its efficacy is limited by the rapid charge carrier recombination. Rare-earth elements doping with elements like Ce are recognized for generating photo-induced charge transport mechanisms and inhibiting recombination of electron–hole. Moreover, rare-earth doping may decrease the gap between the bands, so promoting the excitation of electrons from the valence band to the conduction band [53]. Mora et al. produced Ce-doped ZnO nanoparticles with varying Ce amount in solution (0.5, 1.39, 2.55, 3.28, 3.71, and 4.14%). Studies on photocatalytic degradation indicated that an increase in Ce content improved photocatalytic activity, reaching an optimum concentration of 3.28%. Photocatalytic effectiveness decreased above this concentration under UV–Visible irradiation [52]. At a Ce doping level of 3.28%, the degradation of Direct Red 23 (DR-23) was completed in 70 min, while other concentrations required 110 min, demonstrating that this doping level provides the most effective photocatalytic performance. Meshram et al. produced cerium-doped ZnO nanoparticles for the breakdown of MB. Their results demonstrated that Ce doping lowers the gap between the bands and reduces particle size, both of which enhance photocatalytic efficiency [54]. The degradation efficiency improved from 49% with undoped ZnO to 69% with 3% Ce-doped ZnO in 30 min, underscoring the boost in degradation kinetics [55]. Liu et al. produced cerium-doped titanium dioxide nanosheets by a one-pot hydrothermal technique and evaluated their photocatalytic efficacy via the deterioration of RhB under UV–Visible irradiation. Fluorescence spectroscopy revealed that augmenting Ce concentration to 0.5% decreased fluorescence intensity, signifying inhibited electron–hole recombination and improved charge separation [29]. Nonetheless, a further rise in Ce concentration resulted in elevated fluorescence intensity, indicating heightened recombination rates and diminished photocatalytic efficiency [56]. The sample with least doping, i.e., 2% Ce doping had the lowest photocatalytic degradation efficiency under UV–Visible light, highlighting the need of adjusting dopant amount [57].

5.3 Sm-Doped Metal Oxides Nanoparticles

Milanova et al. indicated that doping titanium dioxide (TiO_2) with samarium (Sm), produced by an ultrasonic-assisted sol–gel technique utilizing acetic acid as both solvent and catalyst, results in improved photocatalytic efficacy [58]. Their research demonstrated that the photocatalytic efficacy of TiO_2 fluctuates markedly with the concentration of the Sm dopant. A rise in Sm percentage corresponded with an enlarged surface area of the resultant nanocrystals. The photocatalytic activity was assessed using Alizarin Red S (ARS) dye as a model contaminant. TiO_2 doped with 1% Sm had the maximum activity, with around 93% dye degradation [59]. This improvement is due to reduced particle size, augmented BET surface area, and efficient inhibition of electron–hole recombination. Arasi et al. produced Sm-doped TiO_2 by a hydrothermal technique and examined its photocatalytic efficacy in degrading Rhodamine B (RhB). The studies, using 60 mg of catalyst in 60 mL of RhB solution, demonstrated a significant increase in degradation efficiency from 88% (pure TiO_2) to 95.76% after Sm doping. These results further validate that the introduction of Sm into TiO_2 elevates photocatalytic activity by improving quantum efficiency [60].

Benammaret al. investigated Sm-doped into TiO_2 nanoparticles for the photocatalytic breakdown of methanol. The presence of Sm was seen to augment photocatalytic efficiency, possibly attributable to the capability of Sm^{3+} ions to capture photogenerated charge carriers, hence facilitating electron/hole transfer to the retain oxygen molecules on the outer layer [61]. The study employing Diffuse Reflectance Spectroscopy (DRS) revealed a shift in the wavelength in the absorption spectra subsequent to Sm inclusion, indicating that electrons were effectively trapped in the conduction band by adsorbed oxygen, while photogenerated holes were efficiently captured by hydroxyl groups or water molecules. This mechanism facilitates the synthesis of hydroxyl radicals, which enhances the level of photocatalytic activity [62].

Faraz et al. sought to examine the impact of Sm doping levels on the photocatalytic capabilities of ZnO nanoparticles by using a gel-combustion process to produce Sm-doped ZnO nanoparticles. This work demonstrated that the inclusion of Sm generates surface imperfections advantageous for photocatalysis and causes a redshift in the absorption edge, enabling Sm to absorb visible light. Samarium-doped ZnO nanoparticles with 1, 3, and 5% Sm concentrations showed band gap energies of 3.25, 3.10, and 3.05 eV each, therefore demonstrating that the gap between the bands decreases with an increase in Sm concentration [63]. The photocatalytic activity was assessed by degrading Malachite Green (MG) dye under visible light exposure. The 3% Sm-doped ZnO had the greatest photocatalytic efficiency. Exceeding this ideal concentration of Sm led to a reduction in photocatalytic activity, presumably owing to enhanced electron–hole recombination and a drop in defect density [64]. The results indicate that low levels of Sm^{3+} effectively capture photogenerated electrons and inhibit recombination, thus prolonging charge carrier lifetimes. UV–Vis absorption spectra reveal that 3% Sm doping leads to an absorption onset at approximately 400 nm, confirming enhanced visible light responsiveness and a diminished energy of band gaps relative to undoped ZnO [65]. In summary, samarium doping in TiO_2 and ZnO significantly improves photocatalytic activity by augmenting surface area, inhibiting charge recombination, and broadening light absorption into the visible spectrum. Nevertheless, the concentration of Sm must be carefully controlled, since excessive doping may result in detrimental effects on photocatalytic effectiveness [66].

5.4 Dy-Doped Metal Oxides

Farooq et al. examined the augmentation of photocatalytic activity in ZnO nanoparticles by doping with dysprosium ions (Dy^{3+}). The Dy-doped ZnO nanoparticles were produced by a sonochemical technique, and their photocatalytic efficacy was assessed utilizing an azo dye as a model contaminant. Scanning Electron Microscopy (SEM) study demonstrated that undoped ZnO nanoparticles had irregular shape and considerable aggregation, resulting in enlarged grain size and decreased surface area both adverse to photocatalytic effectiveness [67]. Doping ZnO with Dy^{3+} ions markedly reduced nanoparticle aggregation and modified the grain size, leading to a considerable increase in surface area. This decrease in aggregation not only enhanced the dispersion of nanoparticles but also reduced photo-generated charge recombination and photo-corrosion, consequently increasing the availability of photoactive sites and improving photocatalytic activity [68]. Phuruangrat et al. generated gadolinium-doped molybdenum trioxide (Gd-doped MoO_3) nanobelts and dysprosium-doped zinc tungstate (Dy-doped $ZnWO_4$) nanorods by a hydrothermal technique. The degradation of methylene blue (MB) dye when exposed to visible light was used to measure the photocatalytic effectiveness of the produced materials. Pure MoO_3 nanobelts demonstrated restricted photocatalytic efficacy, achieving just 77.34% degradation of MB in 30 min [69]. Upon doping with 3% Dy^{3+}, the degradation efficiency significantly enhanced to 97.76%, underscoring the essential function of Dy^{3+} in improving light absorption and charge carrier separation. Dy-doped $ZnWO_4$ nanorods, at a 3% doping concentration, achieved 97.46% destruction of MB after 30 min under visible light, underscoring their exceptional efficiency.

These studies collectively illustrate that rare-earth doping, specifically Dy^{3+} and Gd^{3+}, into diverse semiconductor matrices (ZnO, MoO_3, $ZnWO_4$) significantly enhances photocatalytic performance by augmenting surface area, minimizing agglomeration, inhibiting electron–hole recombination, and broadening light absorption into the visible spectrum [70].

5.5 *Er-Doped Metal Oxides*

Ahmad et al. produced ZnO nanoparticles doped by erbium (Er^3) by utilizing homogeneous precipitation approaches to increase the photocatalytic efficiency of undoped ZnO. Scanning Electron Microscopy (SEM) demonstrated a consistent morphology characterized by a smaller particle size distribution and the absence of apparent aggregation in the doped samples [71]. The crystal structure remained invariant after Er^{3+} doping. UV–Vis absorption analyses indicated a little wavelength shift in the absorption edge, implying enhanced visible light absorption. The photocatalytic process was assessed by the deterioration of methylene blue (MB) dye. Pure ZnO attained a degradation efficiency of 69.28% in 1 h, but Er-doped ZnO demonstrated improved efficiencies of 86.3%, 97.7%, and 91.3% for doping concentrations of 2.1%, 4.8%, and 6.3%, respectively [72]. A doping concentration of 4.8% yielded the maximum photocatalytic activity; any further increase resulted in a decrease, perhaps owing to the formation of recombination centers generated by excess dopants.

Bhatia et al. used a straightforward combustion technique to produce Er-doped ZnO nanoparticles at different concentrations 2.0, 2.5, 3.0, and 3.5 atomic weight percent, referred to as EZ1 through EZ4. X-ray diffraction (XRD) verified the presence of the hexagonal wurtzite phase in all samples, but FESEM examination indicated that Er doping resulted in a reduction of dimensions of particles. The assessment of photocatalytic degradation was conducted using Direct Red 31 (DR-31) dye [73]. Among all samples, EZ2 (2.5 at. wt% Er) had the best degrading efficiency at 99.1%. Elevating Er^{3+} concentration above the optimal level led to decreased activity, maybe due to dopant aggregation or excessive defect generation serving as recombination sites.

Jain et al. investigated the photocatalytic properties of Er-doped ZnO nanoparticles produced by a solid-state reaction technique. Doping concentrations between 1 and 5% by weight were examined for the deterioration of Remazol Red Brilliant F3B (RR180) dye. XRD examination revealed a little leftward shift in peak locations with elevated Er content, indicating lattice deformation and suppression of particle development [74]. A continuous reduction in band gap energy was seen with increased Er^{3+} concentrations, enhancing electron excitation under visible light. Optimal photocatalytic performance was observed at 3 wt % Er, with further increases leading to reduced efficiency, possibly owing to light penetration blockage and enhanced charge carrier recombination at elevated dopant concentrations [75].

These findings together demonstrate that Er^{3+} doping in ZnO nanoparticles may markedly improve photocatalytic performance by decreasing particle size, augmenting surface area, moving absorption into the visible spectrum, and improving charge carrier dynamics. Nonetheless, a crucial threshold of dopant concentration exists, beyond which photocatalytic activity diminishes, underscoring the need for optimizing dopant levels [76].

5.6 Pr-Doped Metal Oxides

Chiou et al. produced praseodymium (Pr^{3+}) doped TiO_2 nanoparticles by the sol–gel process and examined their photocatalytic capabilities. The produced nanoparticles underwent calcination at different temperatures, resulting in a phase change from anatase to rutile. Anatase TiO_2, with a band gap of around 3.2 eV, is recognized for its enhanced photocatalytic efficacy relative to the rutile phase [77]. Pr^{3+} doping in the TiO_2 matrix markedly improved photocatalytic activity by promoting increased photo-generated charge transfer between the valence and conduction bands. UV–Visible absorption spectroscopy showed a wavelength shift in the absorption edge from 350 to 520 nm with increasing Pr content, signifying enhanced visible-light absorption. The largest shift occurred at a doping dose of 0.072 mol % Pr. Beyond this concentration, the absorption edge reverted to shorter wavelengths, indicating a decline in light-harvesting efficiency [78].

The band gap of undoped TiO_2 was determined to be 3.20 eV, but Pr doping demonstrated a progressive decrease in the gap between the bands, ultimately reaching 2.80 eV. This constriction is advantageous for visible light photocatalysis since it facilitates the excitation of electrons. Nevertheless, at Pr^{3+} concentrations beyond the optimal threshold, the duration of charge separation decreased owing to heightened recombination, adversely impacting photocatalytic effectiveness [79].

The photocatalytic activity was assessed using phenol as the target contaminant. Pure TiO_2 attained a degradation efficiency of just 42%. The addition of Pr^{3+} enhanced the degradation rate, reaching a maximum of 94.4% for 0.072 mol % Pr-doped TiO_2. Increasing the dopant percentage to 0.22 mol% led to a little reduction in performance to 91.3%, either owing to the emergence of recombination sites or excess dopant obstructing charge mobility [80].

In summary, Pr^{3+} doping augments the photocatalytic efficacy of TiO_2 by enhancing light absorption into the visible spectrum, reducing the gap between bands, and enhancing charge transport dynamics. Maintaining an adequate dopant concentration is essential to prevent adverse consequences, including diminished charge separation efficiency and spectral blue shifts [81].

5.7 Nd-Doped Metal Oxides

Neodymium (Nd^{3+}) doping has garnered significant interest in enhancing the photocatalytic efficacy of semiconductors of metal oxide owing to its capacity to influence particle size, band gap energy, and charge transport recombination dynamics.

Mzoughi et al. produced neodymium-doped zirconium dioxide (ZrO_2) nanoparticles using the sol–gel technique [82]. X-ray diffraction (XRD) examination verified that the material maintained its tetragonal crystalline form after doping. Scanning electron microscopy (SEM) demonstrated a significant reduction in particle size from 43 nm in pure ZrO_2 to 20 nm in Nd-doped ZrO_2. The photocatalytic efficacy was assessed using methyl orange (MO) dye. Pure ZrO_2 exhibited deterioration mostly in the UV spectrum with an efficiency of 68%, but 9% Nd-doped ZrO_2 showed enhanced activity in the visible spectrum, achieving 99% degradation efficiency after 120 min [43]. This improvement is due to the drop in band gap energy from 2.8 eV (pure) to 2.6 eV (doped), which enhances light absorption and charge carrier dynamics.

Zhang et al. similarly generated Nd-doped ZnO nanoparticles using the sol–gel method. The XRD results validated the wurtzite structure of ZnO. The photocatalytic efficacy was evaluated against Congo Red dye under visual illumination [45]. The findings demonstrated a significant association between Nd doping concentration and photocatalytic efficacy, with 4% Nd-doped ZnO attaining the greatest degradation rate of 93.7% after 30 min of visible light exposure. The performance enhancement is mostly attributable to improved light absorption and effective charge carrier separation resulting from Nd inclusion [83].

Alam et al. synthesized Nd-doped ZnO nanoparticles by the process of precipitation and assessed their photocatalytic efficacy against methylene blue (MB) dye. Neodymium concentrations varying from 3 to 15 wt% were evaluated. Fluorescence spectroscopy revealed a change in the emission band from ultraviolet (367 nm for pure ZnO) to the faint blue range (380–474 nm) after Nd doping. The spectrum alterations signify defect creation and shallow donor levels linked to the dopant, facilitating charge separation and reducing recombination rates [84]. Thus, the photocatalytic activity escalated in direct correlation with the concentration of Nd doping, validating Nd's function in augmenting photo-efficiency via the modulation of recombination and the enhancement of surface-active sites. In conclusion, Nd^{3+} doping in ZrO_2 and ZnO enhances photocatalytic degradation under visible light by cutting the band gap, decreasing particle size, and facilitating improved electron–hole separation. The suitable dopant concentration is essential for enhancing the photocatalytic efficacy of these metallic oxide systems [85].

5.8 Eu-Doped Metal Oxides

The incorporation of rare earth elements like europium (Eu^{3+}) has been investigated to enhance the photocatalytic efficacy of metal oxide semiconductors by altering their optical, morphological and structural characteristics [86]. Using the technique of precipitation, Kumar et al. produced europium-doped zinc oxide nanoparticles. X-ray diffraction (XRD) investigation verified that both doped and pure ZnO maintained the hexagonal wurtzite structure, indicating no substantial structural change due to doping. Methyl orange (MO) dye photocatalytic degradation experiments conducted under UV light have demonstrated that photocatalytic efficacy rises with increasing Eu content. The improvement was ascribed to the enhanced production of hydroxyl ($OH^{\cdot}$) and hydroperoxyl (HO_2^-) radicals, potent oxidants generated by UV irradiation as a result of effective charge carrier formation and separation. Pure ZnO decomposes 75.7% of MO in 180 min, but 1% Eu-doped ZnO achieved 95.3% degradation under the same circumstances [87].

In a similar manner, Korake et al. used a low-temperature microwave-assisted method to create Eu-doped ZnO. The XRD pattern validated the hexagonal wurtzite phase of ZnO. The band gap decreased with the augmentation of Eu concentration, decreasing from 3.34 eV in pure ZnO to 3.00 eV in 0.5% Eu-doped ZnO. Brunauer–Emmett–Teller (BET) surface area studies indicated a rise from 42 m^2/g for pure ZnO to 49m^2/g for 0.2% Eu-doped ZnO, indicating a decrease in nanoparticle size [88]. The photocatalytic degradation of MO under UV light enhanced with increasing Eu concentration, with 0.2 mol % Eu-doped ZnO exhibiting optimum activity 91% degradation of MO in 180 min due to enhanced surface area and charge carrier separation [89]. Bouras et al. [90] examined Eu-doped SnO_2 nanostructures produced by a hydrothermal technique. The research demonstrated that doping with Eu modified the shape of the nanomaterials, shifting from nanorods (pure SnO_2) to nanosheets and then reverting to nanorods at higher doping concentrations. XRD measurements indicated a little rightward shift in diffraction peaks due to doping, implying lattice strain or replacement [90]. Photoluminescence (PL) spectra revealed two discrete emission bands one in the UV area at 350 nm (weak) and another in the visible region at 666 nm (strong), the latter ascribed to oxygen vacancies and electron transfer from SnO_2 to Eu^{3+} ions. Despite the elevated Eu content resulting in enhanced PL intensity, normally signifying an increased recombination rate, the 10 wt% Eu-doped SnO_2 yet demonstrated substantial photocatalytic degradation efficiency of (RhB). This was mostly attributable to the augmented surface area and UV absorbance, which offset the heightened recombination rate [91]. In conclusion, Eu^{3+} doping in ZnO and SnO_2 markedly improves photocatalytic performance under visible or UV light. The enhancements are mostly attributed to decreased band gaps, augmented surface area, defect-induced charge separation, and morphological alterations, all of which facilitate improved formation and usage of reactive oxygen species for dye degradation [92] (Table 4.2; Figs. 4.2 and 4.3).

Table 4.2 Synthesis and degradation methods of various rare earth doped metal oxide nanoparticles

Material	Method	Degradation efficiency%	References
2.5% Er:ZnO	Combustion	99.1	[86]
1% Er:ZnO	Solid state reaction	82	[87]
2% Er:ZnO	Solid state reaction	90	[87]
3% Er:ZnO	Solid state reaction	99	[87]
4% Er:ZnO	Solid state reaction	70	[87]
0.4% Er: CeO_2	Hydrothermal	94	[88]
Pr:TiO_2	Acid-peptized sol–gel	94.4	[77]
Pure ZrO_2	Polymer-assisted	68	[37]
Nd:ZrO_2	Sol–gel	99	[43]
4% Nd:ZnO	Sol–gel	93.7	[43]
1% Eu:ZnO	Precipitation	75.7	[51]
0.2% Eu:ZnO	Microwave-assisted	91	[51]
1% Eu:SnO_2	Hydrothermal	74	[51]
3% Eu:SnO_2	Hydrothermal	70	[51]
5% Eu:SnO_2	Hydrothermal	60	[51]
7% Eu:SnO_2	Hydrothermal	85	[51]
10% Eu:SnO_2	Hydrothermal	90	[51]
2% Eu:Bi2WO_6	Hydrothermal	73.7	[53]
5% Eu:Bi2WO_6	Hydrothermal	96.2	[53]
Eu:$BiVO_4$	Hydrothermal	93.6	[54]
Nd:TiO_2	Hydrothermal	100	[84]
Er:TiO_2	Solvothermal	98.78	[54]
4% Pr:Bi_2O_3	Acrylamide polymerization	70.4	[74]
0.05% Sm:TiO_2	Microwave-assisted sol–gel method	98	[59]
0.1% Sm:TiO_2	Microwave-assisted sol–gel method	96	[59]
0.6%Gd:SnO_2	Sol–gel	100	[60]
3%Gd:ZnO	Sonochemical	92.76%	[73]
1%Gd:$BiFeO_3$	Sol–gel	34.2%	[74]
3%Gd:$BiFeO_3$	Sol–gel	56.8%	[74]
5%Gd:$BiFeO_3$	Sol–gel	42.8%	[74]
0.5% Ce:ZnO	Combustion method	85	[48]
2.5% Ce:ZnO	Combustion method	95	[48]
3.28% Ce:ZnO	Combustion method	100	[48]
3% Ce:ZnO	Hydrothermal	69	[48]
1% Ce:ZnO	Combustion method	51	[52]

(continued)

Table 4.2 (continued)

Material	Method	Degradation efficiency%	References
CeO2-TiO_2	Polymer-assisted method	24	[52]
3% Ce:ZnO	Combustion method	69	[52]
5% Ce:ZnO	Combustion method	64	[52]
1% Sm:TiO_2	Modified-ultrasonic assisted sol–gel	69	[60]
4.8% Er:ZnO	Homogenous precipitation	97.3	[47]
3% Dy:$ZnWO_4$	Hydrothermal	97.46	[73]

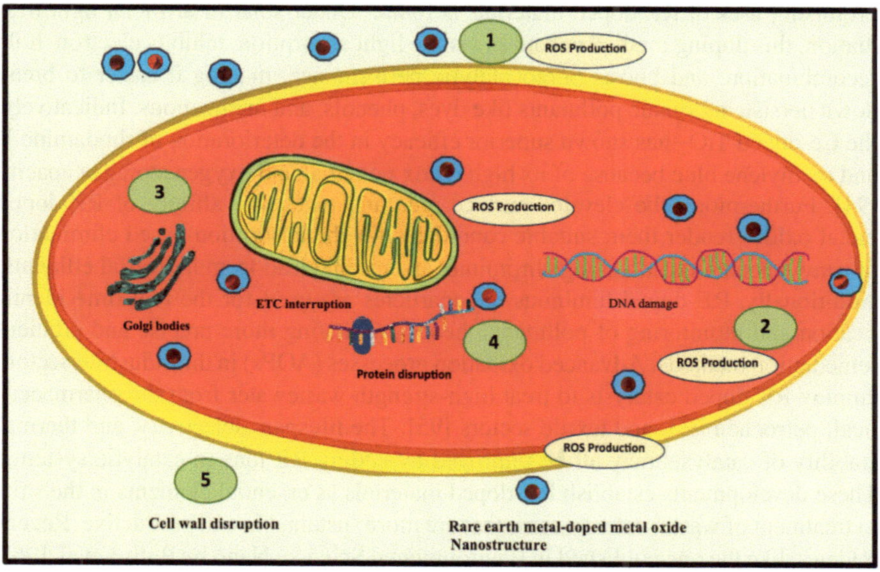

Fig. 4.2 Antibacterial mechanism of rare earth metal-doped nanostructure

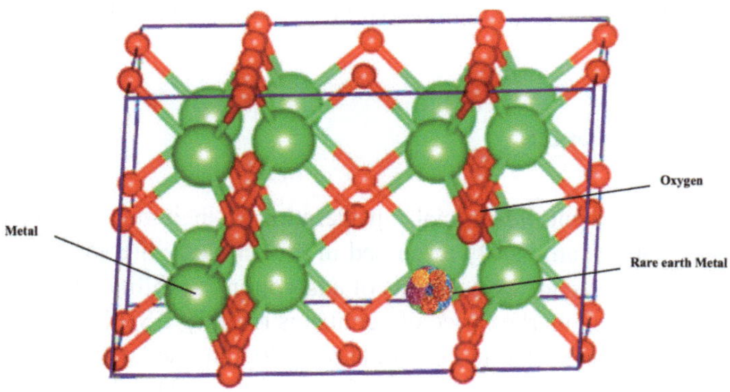

Fig. 4.3 Structure of rare earth metal-doped nanoparticle metal oxide

6 Application of RE-Doped Materials in Wastewater Treatment and Industries

Rare earth (RE)-doped materials are attracting a lot of interest in industrial and wastewater treatment applications because of their distinct optical characteristics, electronic structures, and catalytic capabilities. These substances are excellent choices for eliminating a range of organic and inorganic contaminants from wastewater because of their improved photocatalytic and adsorption capabilities [93]. In photocatalysis, when elements like cerium (Ce), lanthanum (La), europium (Eu), and terbium (Tb) are doped into semiconductors like TiO_2 or ZnO, one of the most promising uses of RE-doped materials is found. Under solar or artificial light irradiation, this doping method enhances visible light absorption, inhibits electron–hole recombination, and boosts photocatalytic performance, making it easier to break down persistent organic pollutants like dyes, phenols, and medications. Indicatively, the Ce-doped TiO_2 has shown superior efficacy in the deterioration of rhodamine B and methylene blue because of its high redox potential and oxygen storage capacity [94]. Furthermore, the elevated surface area and metal ion affinity of RE-doped metal oxides render them suitable candidates for the adsorption-based elimination of heavy metal ions, including chromium, arsenic and lead from industrial effluents. Additionally, RE-doped luminous nanoparticles are used for the real-time identification and monitoring of pollutants, hence facilitating more precise and efficient remediation methods. Advanced oxidation processes (AOPs) in the industrial sectors employ RE-doped catalysts to treat high-strength wastewater from the pharmaceutical, petrochemical, and textile sectors [95]. The lifespan, selectivity, and thermal stability of catalysts may all be improved by adding RE ions to catalytic systems. These developments establish RE-doped materials as essential elements in the shift to treatment of waste water systems that are more sustainable and productive. Recent reviews, like the one published in Environmental Science: Nano by Raliya et al. [96], indicate that the use of RE-doped nanomaterials in wastewater remediation is not only technically possible but also consistent with green chemistry principles, indicating their potential importance in future industrial and environmental applications.

7 Economic Aspects

The expense of adsorbents is a crucial consideration for their practical use. Nevertheless, cost calculations are hardly discussed in the scholarly literature. Most studies emphasize laboratory-scale assessments of adsorption capability, often overlooking economic factors. The market price of zeolites is roughly US$0.03 to US$0.12 per kilogram, Fuller's earth is around US$0.04 per kilogram, and commercial activated carbon is marketed at approximately US$2.00 to US$2.20 per kilogram. However, straight comparisons of adsorbent pricing might be deceptive without assessing their individual features and adsorption efficiency [97].

The term "low-cost adsorbents" often refers only to the raw market price of certain materials. In real situations, the economic feasibility of these materials is often undermined by costly regeneration and treatment procedures. A thorough assessment must take into account several considerations, including local availability, transportation expenses, treatment and modification needs, recyclability, and the overall longevity of the adsorbent [98].

To improve adsorption capacity and selectivity, pre-treatment techniques such as drying, autoclaving, or chemical modification (e.g., cross-linking with organic or inorganic agents) are often used. These therapies may markedly elevate operating expenses. The regeneration process may need considerable energy or solvent consumption, and the proper disposal or handling of depleted adsorbents is a significant and sometimes neglected challenge [99].

Although some adsorbents may seem expensive initially, their long-term reusability and ease of regeneration might make them more cost-effective across their whole existence. Consequently, a life-cycle cost analysis encompassing synthesis, use, regeneration, and disposal is crucial for assessing the actual cost-effectiveness of an adsorbent.

Chitosan, cyclodextrin, and their composites have superior adsorption capabilities. Nonetheless, their extensive use is impeded by substantial starting expenses and intricate synthesis processes. Notwithstanding their promise, there is an absence of comprehensive evaluations of these materials from a whole life-cycle view point [100].

8 Future Prospects

The use of rare earth element (REE)-doped biodegradable polymers in the adsorption and degradation of synthetic dyes offers a viable approach for sustainable wastewater treatment. With the escalation of industrial pollution, especially from textile and dye sectors, the establishment of productive and environment-friendly remediation approaches has become vital. In this context, integrating the unique physicochemical characteristics of rare earth elements with biodegradable polymer matrices is an appealing approach. These hybrid materials have significant adsorption capabilities, remarkable photocatalytic activity, and environmental friendliness, making them suitable for multifunctional water treatment applications [101]. Future research is anticipated to concentrate on the design and synthesis of materials that concurrently provide adsorption, photocatalysis, and antibacterial properties. The integration of rare earth elements, like lanthanum or cerium, into biodegradable polymers has shown an improvement in photocatalytic degradation of colors while maintaining biocompatibility. These multifunctional materials may substantially advance the development of next-generation wastewater treatment systems with improved performance and adaptability [102]. A primary focus of current research is the improvement of photocatalytic efficiency. Doping biodegradable polymers with rare earth elements may provide intermediate band gap energy levels, enhancing the absorption of light and facilitating efficient charge transport separation, therefore expediting dye degradation rates. Future research may investigate sophisticated doping techniques and the use of innovative polymer matrices to enhance these capabilities further [103]. The implementation of green synthesis techniques for the production of REE-doped biodegradable polymers is expected to increase. Employing plant-derived extracts or bio-waste for nanoparticle synthesis can reduce environmental impact and is consistent with the global focus on sustainable chemical practices. These environmentally sustainable methods are poised to significantly contribute to the advancement of this domain [104]. Research must focus on the scalability, durability, and versatility of these materials for practical applications in water treatment systems. Examinations of their efficacy in intricate and fluctuating wastewater matrices, with their enduring environmental consequences, will be crucial. Furthermore, the issues associated with the restricted supply and elevated costs of rare earth elements must be recognized. To address these issues, further research might concentrate on identifying alternative, earth-abundant dopants with similar effectiveness, as well as on devising efficient rare earth element recovery and recycling methods from used materials [105]. In conclusion, REE-doped biodegradable polymers provide significant promise for enhancing sustainable wastewater treatment. Achieving this potential requires multidisciplinary cooperation, advancements in materials science, and the use of green chemistry concepts to address current limits and enhance practical applications in environmental restoration [106].

9 Conclusion

Rare earth doping has become a flexible and efficient method for band gap technology in transition metal oxides, facilitating improved photocatalytic performance. This paper examines critical elements of materials development and doping stability related to rare earth-doped oxides of metals. The integration of rare earth elements into metal oxide matrices causes a red shift in the absorption spectra, broadening light absorption into the visible range, an essential aspect for solar-driven applications. Moreover, methodologies for synthesizing materials characterized by high crystallinity and extensive specific surface areas, as assessed by BET analysis, are comprehensively examined. The photocatalytic efficiency of metal oxides is markedly enhanced by rare earth doping, either by band gap reduction or diminishing charge carrier reshuffling. Doping with suitable rare earth ions significantly improves the separation of photo-generated electron–hole pairs, successfully preventing their recombination and enhancing photocatalytic effectiveness. These doped systems frequently exhibit an apparent reduction in the optical band gap energy, improving light consumption. A detailed description of the photocatalytic performance of several rare earth-doped metallic oxides is presented in table format for comparative analysis. Although the majority of investigations so far have focused on dye degradation as a case study reaction, the potential use of these materials for photocatalytic hydrogen (H_2) production remains underexplored. Not with-standing their advantageous band alignments and beneficial charge transfer properties essential for effective photocatalytic water splitting, these materials have not been extensively used in this application. In conclusion, even though the development of rare earth-doped metal oxide photocatalysts has advanced significantly, there is still a great deal of untapped potential in this field. A significant difficulty that persists is the standardization of photocatalytic efficiency measurements to provide valid comparisons and assess economic viability for practical implementation. This analysis provides a framework for further investigations focused on maximizing the potential of rare earth-doped nanomaterials for photochemical hydrogen production.

References

1. Lonngren, K.E., Bai, E.W.: On the global warming problem due to carbon dioxide. Energy Policy **36**(4), 1567–1568 (2008)
2. Mehtab, A., Ahmed, J., Alshehri, S.M., Mao, Y., Ahmad, T.: Rare earth doped metal oxide nanoparticles for photocatalysis: a perspective. Nanotechnology **33**(14), 142001 (2022)
3. Heard, B.P., Brook, B.W., Wigley, T.M., Bradshaw, C.J.: Burden of proof: a comprehensive review of the feasibility of 100% renewable-electricity systems. Renew. Sustain. Energy Rev. **76**, 1122–1133 (2017)
4. Brown, V.M., Jordan, D.H.M., Tiller, B.A.: The effect of temperature on the acute toxicity of phenol to rainbow trout in hard water. Water Res. **1**(8–9), 587–594 (1967)
5. Bak, T., Nowotny, J., Rekas, M., Sorrell, C.C.: Photo-electrochemical hydrogen generation from water using solar energy. Materials-related aspects. Int. J. Hydrogen Energy **27**(10), 991–1022 (2002)
6. Deng, Y., Yeo, B.S.: Characterization of electrocatalytic water splitting and CO_2 reduction reactions using in situ/operando Raman spectroscopy. ACS Catal. **7**(11), 7873–7889 (2017)
7. Esposito, D.V., Chen, J.G., Birkmire, R.W., Chang, Y., Gaillard, N.: Hydrogen production from photo-driven electrolysis of biomass-derived oxygenates: a case study on methanol using Pt-modified WO3 thin film electrodes. Int. J. Hydrogen Energy **36**(16), 9632–9644 (2011)
8. Jose, V.: Photobiological hydrogen as a renewable fuel. Holistic Appr. Environ. **11**(2), 67–71 (2021)
9. Ghirardi, M.L., Dubini, A., Yu, J., Maness, P.C.: Photobiological hydrogen-producing systems. Chem. Soc. Rev. **38**(1), 52–61 (2009)
10. Baykara, S.Z.: Experimental solar water thermolysis. Int. J. Hydrogen Energy **29**(14), 1459–1469 (2004)
11. Helseth, L.E.: Electrical energy harvesting from water droplets passing a hydrophobic polymer with a metal film on its back side. J. Electrostat. **81**, 64–70 (2016)
12. Yılmaz, S., Selim, H.: A review on the methods for biomass to energy conversion systems design. Renew. Sustain. Energy Rev. **25**, 420–430 (2013)
13. Saidur, R., Abdelaziz, E.A., Demirbas, A., Hossain, M.S., Mekhilef, S.: A review on biomass as a fuel for boilers. Renew. Sustain. Energy Rev. **15**(5), 2262–2289 (2011)
14. Ahmad, T., Farooq, U., Phul, R.: Fabrication and photocatalytic applications of perovskite materials with special emphasis on alkali-metal-based niobates and tantalates. Ind. Eng. Chem. Res. **57**(1), 18–41 (2018)
15. Ikram, S., Ashraf, F., Alzaid, M., Mahmood, K., Amin, N., Haider, S.A.: Role of nature of rare earth ion dopants on structural, spectral, and magnetic properties in spinel ferrites. J. Supercond. Novel Magn. **34**, 1745–1751 (2021)
16. Tentu, R.D., Basu, S.: Photocatalytic water splitting for hydrogen production. Curr. Opin. Electrochem. **5**(1), 56–62 (2017)

17. Luan, J., Li, Y.: Photocatalytic water splitting for hydrogen production with Gd2MSbO7 (M= Fe, In, Y) photocatalysts under visible light irradiation. Materials **8**(1), 16–30 (2014)
18. Miseki, Y., Sayama, K.: Photocatalytic water splitting for Solar hydrogen production using the carbonate effect and the Z-scheme reaction. Adv. Energy Mater. **9**(23), 1801294 (2019)
19. Singh, A.K., Mathew, K., Zhuang, H.L., Hennig, R.G.: Computational screening of 2D materials for photocatalysis. J. Phys. Chem. Lett. **6**(6), 1087–1098 (2015)
20. Raizada, P., Soni, V., Kumar, A., Singh, P., Khan, A.A.P., Asiri, A.M., Thakur, V.K., Nguyen, V.H.: Surface defect engineering of metal oxides photocatalyst for energy application and water treatment. J. Materiomics **7**(2), 388–418 (2021)
21. Naldoni, A., Allieta, M., Santangelo, S., Marelli, M., Fabbri, F., Cappelli, S., Bianchi, C.L., Psaro, R., Dal Santo, V.: Effect of nature and location of defects on bandgap narrowing in black TiO_2 nanoparticles. J. Am. Chem. Soc. **134**(18), 7600–7603 (2012)
22. Prévot, M.S., Sivula, K.: Photoelectrochemical tandem cells for solar water splitting. J. Phys. Chem. C **117**(35), 17879–17893 (2013)
23. Navarro, R.M., Sanchez-Sanchez, M.C., Alvarez-Galvan, M.C., Del Valle, F., Fierro, J.L.G.: Hydrogen production from renewable sources: biomass and photocatalytic opportunities. Energy Environ. Sci. **2**(1), 35–54 (2009)
24. Li, P., Tian, Y., Du, M., Xie, Q., Chen, Y., Ma, L., Huang, Y., Yin, Z., Xu, H., Wu, X.: Mechanism of rotenone toxicity against *Plutella xylostella*: new perspective from a spatial metabolomics and lipidomics study. J. Agric. Food Chem. **71**(1), 211–222 (2022)
25. Ali, M., Sreekrishnan, T.R.: Aquatic toxicity from pulp and paper mill effluents: a review. Adv. Environ. Res. **5**(2), 175–196 (2001)
26. Klavarioti, M., Mantzavinos, D., Kassinos, D.: Removal of residual pharmaceuticals from aqueous systems by advanced oxidation processes. Environ. Int. **35**(2), 402–417 (2009)
27. Robinson, T., McMullan, G., Marchant, R., Nigam, P.: Remediation of dyes in textile effluent: a critical review on current treatment technologies with a proposed alternative. Biores. Technol. **77**(3), 247–255 (2001)
28. Ogugbue, C.J., Sawidis, T.: Bioremediation and detoxification of synthetic wastewater containing triarylmethane dyes by *Aeromonas hydrophila* isolated from industrial effluent. Biotechnol. Res. Int. **2011**(1), 967925 (2011)
29. Liu, D., Huang, J., Tao, X., Wang, D.: One-step synthesis of C-Bi_2WO_6 crystallites with improved photo-catalytic activities under visible light irradiation. RSC Adv. **5**(81), 66464–66470 (2015)
30. Barrera, A., Tzompantzi, F., Campa-Molina, J., Casillas, J.E., Pérez-Hernández, R., Ulloa-Godinez, S., Velásquez, C., Arenas-Alatorre, J.: Photocatalytic activity of Ag/Al_2O_3–Gd_2O_3 photocatalysts prepared by the sol–gel method in the degradation of 4-chlorophenol. RSC Adv. **8**(6), 3108–3119 (2018)
31. Brites, F.F., Santana, V.S., Fernandes-Machado, N.R.C.: Effect of support on the photocatalytic degradation of textile effluents using Nb_2O_5 and ZnO: photocatalytic degradation of textile dye. Top. Catal. **54**, 264–269 (2011)
32. Li, H., Wang, G., Zhang, F., Cai, Y., Wang, Y., Djerdj, I.: Surfactant-assisted synthesis of CeO_2 nanoparticles and their application in wastewater treatment. RSC Adv. **2**(32), 12413–12423 (2012)
33. Farooq, U., Chaudhary, P., Ingole, P.P., Kalam, A., Ahmad, T.: Development of cuboidal $KNbO_3$@ α-Fe_2O_3 hybrid nanostructures for improved photocatalytic and photoelectrocatalytic applications. ACS Omega **5**(32), 20491–20505 (2020)
34. Akhoondi, A., Ghaebi, H., Karuppasamy, L., Rahman, M.M., Sathishkumar, P.: Recent advances in hydrogen production using MXenes-based metal sulfide photocatalysts. Synthesis and Sintering **2**(1), 37–54 (2022)
35. Ran, J., Zhang, J., Yu, J., Jaroniec, M., Qiao, S.Z.: Earth-abundant cocatalysts for semiconductor-based photocatalytic water splitting. Chem. Soc. Rev. **43**(22), 7787–7812 (2014)
36. Khan, M.M., Adil, S.F., Al-Mayouf, A.: Metal oxides as photocatalysts. J. Saudi Chem. Soc. **19**(5), 462–464 (2015)

37. Yan, Y., Xia, B. Y., Ge, X., Liu, Z., Fisher, A., Wang, X.: A flexible electrode based on iron phosphide nanotubes for overall water splitting. Chem. A Eur. J. **21**(50), 18062–18067 (2015)
38. Ju, L., Dai, Y., Wei, W., Li, M., Liang, Y., Huang, B.: One-dimensional cadmium sulphide nanotubes for photocatalytic water splitting. Phys. Chem. Chem. Phys. **20**(3), 1904–1913 (2018)
39. Iwashina, K., Iwase, A., Ng, Y.H., Amal, R., Kudo, A.: Z-schematic water splitting into H_2 and O_2 using metal sulfide as a hydrogen-evolving photocatalyst and reduced graphene oxide as a solid-state electron mediator. J. Am. Chem. Soc. **137**(2), 604–607 (2015)
40. Paramasivam, I., Jha, H., Liu, N., Schmuki, P.: A review of photocatalysis using self-organized TiO_2 nanotubes and other ordered oxide nanostructures. Small **8**(20), 3073–3103 (2012)
41. Mangrulkar, P.A., Joshi, M.M., Tijare, S.N., Polshettiwar, V., Labhsetwar, N.K., Rayalu, S.S.: Nano cobalt oxides for photocatalytic hydrogen production. Int. J. Hydrogen Energy **37**(13), 10462–10466 (2012)
42. Kalam, A., Al-Sehemi, A.G., Ashrafuzzaman, M., Allami, S.A.S., Sharif, A.M., Yadav, P., Assiri, M.A., Du, G.: Synthesis of gadolinium doped ZnO nanomaterials using the modified-solvothermal method and studied the effect of gadolinium on the structural, morphological, and optical properties. J. Inorgan. Organometallic Polymers Mater. **33**(10), 3076–3086 (2023)
43. Selvaraj, S., Mohan, M.K., Navaneethan, M., Ponnusamy, S., Muthamizhchelvan, C.: Synthesis and photocatalytic activity of Gd doped ZnO nanoparticles for enhanced degradation of methylene blue under visible light. Mater. Sci. Semicond. Process. **103**, 104622 (2019)
44. Danish, M.S.S., Estrella, L.L., Alemaida, I.M.A., Lisin, A., Moiseev, N., Ahmadi, M., Nazari, M., Wali, M., Zaheb, H., Senjyu, T.: Photocatalytic applications of metal oxides for sustainable environmental remediation. Metals **11**(1), 80 (2021)
45. Zhang, X., Zhou, G., Zhang, H., Wu, C., Song, H.: Characterization and activity of visible light–driven TiO_2 photocatalysts co-doped with nitrogen and lanthanum. Transition Met. Chem. **36**, 217–222 (2011)
46. Medhi, R., Marquez, M.D., Lee, T.R.: Visible-light-active doped metal oxide nanoparticles: review of their synthesis, properties, and applications. ACS Appl. Nano Mater. **3**(7), 6156–6185 (2020)
47. Chen, H., Wang, L.: Nanostructure sensitization of transition metal oxides for visible-light photocatalysis. Beilstein J. Nanotechnol. **5**(1), 696–710 (2014)
48. Yu, J., Yu, J.C., Cheng, B., Zhao, X.: Photocatalytic activity and characterization of the sol-gel derived Pb-doped TiO_2 thin films. J. Sol-Gel Sci. Technol. **24**, 39–48 (2002)
49. Natarajan, T.S., Natarajan, K., Bajaj, H.C., Tayade, R.J.: Enhanced photocatalytic activity of bismuth-doped TiO_2 nanotubes under direct sunlight irradiation for degradation of Rhodamine B dye. J. Nanopart. Res. **15**(5), 1–18 (2013)
50. Kumaresan, L., Mahalakshmi, M., Palanichamy, M., Murugesan, V.: Synthesis, characterization, and photocatalytic activity of Sr^{2+} doped TiO_2 nanoplates. Ind. Eng. Chem. Res. **49**(4), 1480–1485 (2010)
51. Kim, S.P., Choi, M.Y., Choi, H.C.: Photocatalytic activity of SnO_2 nanoparticles in methylene blue degradation. Mater. Res. Bull. **74**, 85–89 (2016)
52. Mora, J.R., Flores-Carrasco, G., Juárez, H., Pacio, M., Olvera, M.D.L.L., Rabanal, M.E.: Ce-doped ZnO nanonails synthesized by a simple thermal evaporation method for photocatalytic degradation. Opt. Mater. **157**, 116156 (2024)
53. Meshram, S.P., Adhyapak, P.V., Pardeshi, S.K., Mulla, I.S., Amalnerkar, D.P.: Sonochemically generated cerium doped ZnO nanorods for highly efficient photocatalytic dye degradation. Powder Technol. **318**, 120–127 (2017)
54. Lone, I.H., Kalam, A., Ahmed, J., Alhokbany, N., Alshehri, S.M., Ahmad, T.: Quenching assisted reverse micellar synthesis and electrical properties of high surface area BiFeO3 nanoparticles. J. Nanosci. Nanotechnol. **20**(6), 3823–3831 (2020)
55. Li, F.B., Li, X.Z., Hou, M.F., Cheah, K.W., Choy, W.C.H.: Enhanced photocatalytic activity of $Ce3^{+}$-TiO_2 for 2-mercaptobenzothiazole degradation in aqueous suspension for odour control. Appl. Catal. A **285**(1–2), 181–189 (2005)

56. Saqib, N.U., Adnan, R., Shah, I.: A mini-review on rare earth metal-doped TiO_2 for photocatalytic remediation of wastewater. Environ. Sci. Pollut. Res. **23**, 15941–15951 (2016)
57. Kuzhalosai, V., Subash, B., Shanthi, M.: A novel sunshine active cerium loaded zinc oxide photocatalyst for the effective degradation of AR 27 dye. Mater. Sci. Semicond. Process. **27**, 924–933 (2014)
58. Milanova, M., Tsvetkov, M.: Rare earths doped materials. Crystals **11**(3), 231 (2021)
59. Wang, C., Zeng, T., Zhu, S., Gu, C.: Synergistic mechanism of rare-earth modification TiO_2 and photodegradation on benzohydroxamic acid. Appl. Sci. **9**(2), 339 (2019)
60. Arasi, S.E., Madhavan, J., Antony Raj, M.V.: Effect of samarium (Sm3+) doping on structural, optical properties and photocatalytic activity of titanium dioxide nanoparticles. J. Taibah Univ. Sci. **12**(2), 186–190 (2018)
61. Benammar, I., Salhi, R., Deschanvres, J.L., Maalej, R.: The effect of rare earth element (Er, Yb) doping and heat treatment on suspension stability of Y_2O_3 nanoparticles elaborated by sol-gel method. J. Market. Res. **9**(6), 12634–12642 (2020)
62. Xia, W., Pei, Z., Leng, K., Zhu, X.: Research progress in rare earth-doped perovskite manganite oxide nanostructures. Nanoscale Res. Lett. **15**(9), 1–55 (2020)
63. Faraz, M., Naqvi, F.K., Shakir, M., Khare, N.: Synthesis of samarium-doped zinc oxide nanoparticles with improved photocatalytic performance and recyclability under visible light irradiation. New J. Chem. **42**(3), 2295–2305 (2018)
64. Depner, S.W., Kort, K.R., Banerjee, S.: Precursor control of crystal structure and stoichiometry in twin metal oxide nanocrystals. Cryst. Eng. Comm. **11**(5), 841–846 (2009)
65. Vaidya, S., Ahmad, T., Agarwal, S., Ganguli, A.K.: Nanocrystalline oxalate/carbonate precursors of Ce and Zr and their decompositions to CeO_2 and ZrO_2 nanoparticles. J. Am. Ceram. Soc. **90**(3), 863–869 (2007)
66. Ahmad, T., Ganguli, A.K.: Structural and dielectric characterization of nanocrystalline (Ba, Pb) ZrO_3 developed by reverse micellar synthesis. J. Am. Ceram. Soc. **89**(10), 3140–3146 (2006)
67. Farooq, U., Ahmed, J., Alshehri, S.M., Ahmad, T.: High-surface-area sodium tantalate nanoparticles with enhanced photocatalytic and electrical properties prepared through polymeric citrate precursor route. ACS Omega **4**(21), 19408–19419 (2019)
68. Farooq, U., Phul, R., Alshehri, S.M., Ahmed, J., Ahmad, T.: Electrocatalytic and enhanced photocatalytic applications of sodium niobate nanoparticles developed by citrate precursor route. Sci. Rep. **9**(1), 4488 (2019)
69. Phuruangrat, A., Thipkonglas, S., Thongtem, T., Thongtem, S.: Hydrothermal synthesis and characterization of Dy-doped MoO_3 nanobelts for using as a visible-light-driven photocatalyst. Mater. Lett. **195**, 37–40 (2017)
70. Phul, R., Khan, M.M., Sardar, M., Ahmed, J., Ahmad, T.: Multifunctional electrochemical properties of synthesized non-precious iron oxide nanostructures. Crystals **10**(9), 1–14 (2020)
71. Ahmad, T., Khatoon, S., Coolahan, K., Lofland, S.E.: Solvothermal synthesis, optical and magnetic properties of nanocrystalline Cd1− xMnxO (0.04< x= 0.10) solid solutions. J. Alloy. Compd. **558**, 117–124 (2013)
72. Phul, R., Perwez, M., Ahmed, J., Sardar, M., Alshehri, S., Alhokbany, N., Majeed Khan. M.A., Ahmad, T.: Efficient multifunctional catalytic and sensing properties of synthesized ruthenium oxide nanoparticles. Catalysts **10**(7), 1–12 (2020)
73. Senol, S.D.: Hydrothermal derived nanostructure rare earth (Er, Yb)-doped ZnO: structural, optical and electrical properties. J. Mater. Sci. Mater. Electron. **27**, 7767–7775 (2016)
74. Ahmad, T., Phul, R.: Magnetic iron oxide nanoparticles as contrast agents: hydrothermal synthesis, characterization and properties. Solid State Phenom. **232**, 111–145 (2015)
75. Jain, S.K., Farooq, U., Jamal, F., Kalam, A., Ahmad, T.: Hydrothermal assisted synthesis and structural characterization of Zn doped SnO_2 nanoparticles for catalytic reduction of 4-nitrophenol. Mater. Today Proc. **36**, 717–723 (2021)
76. Zhou, Y.: Controllable design, synthesis and characterization of nanostructured rare earth metal oxides. Phys. Sci. Rev. **5**(3), 20180084 (2020)

77. Chiou, C.H., Juang, R.S.: Photocatalytic degradation of phenol in aqueous solutions by Pr-doped TiO_2 nanoparticles. J. Hazard. Mater. **149**(1), 1–7 (2007)
78. Nasi, R., Esposito, S., Freyria, F.S., Armandi, M., Gadhi, T.A., Hernandez, S., Rivolo, P., Ditaranto, N., Bonelli, B.: Application of reverse micelle sol–gel synthesis for bulk doping and heteroatoms surface enrichment in mo-doped TiO_2 nanoparticles. Materials **12**(6), 937 (2019)
79. Lang, J., Han, Q., Yang, J., Li, C., Li, X., Yang, L., Zhang, Y., Gao, M., Wang, D., Cao, J.: Fabrication and optical properties of Ce-doped ZnO nanorods. J. Appl. Phys. **107**(7) (2010)
80. Ahmad, T., Shahazad, M., Phul, R.: Hydrothermal synthesis, characterization and dielectric properties of zirconia nanoparticles. Mater. Sci. Eng. Int. J **1**(3), 100–104 (2017)
81. Mu, W., Xie, X., Li, X., Zhang, R., Yu, Q., Lv, K., Wei, H., Jian, Y.: Characterizations of Nb-doped WO_3 nanomaterials and their enhanced photocatalytic performance. RSC Adv. **4**(68), 36064–36070 (2014)
82. Mzoughi, M., Anku, W.W., Oppong, S.O., Shukla, S.K., Agorku, E.S., Govender, P.P.: Neodymium doped ZrO_2-graphene oxide nanocomposites: a promising photocatalyst for photodegradation of Eosin Y dye. Adv. Mater. Lett. **7**(12), 946–950 (2016)
83. Murugadoss, G., Jayavel, R., Kumar, M.R.: Structural and optical properties of highly crystalline Ce, Eu and co-doped ZnO nanorods. Superlattices Microstruct. **82**, 538–550 (2015)
84. Alam, U., Khan, A., Ali, D., Bahnemann, D., Muneer, M.: Comparative photocatalytic activity of sol–gel derived rare earth metal (La, Nd, Sm and Dy)-doped ZnO photocatalysts for degradation of dyes. RSC Adv. **8**(31), 17582–17594 (2018)
85. Khade, G.V., Suwarnkar, M.B., Gavade, N.L., Garadkar, K.M.: Sol–gel microwave assisted synthesis of Sm-doped TiO_2 nanoparticles and their photocatalytic activity for the degradation of methyl orange under sunlight. J. Mater. Sci. Mater. Electron. **27**, 6425–6432 (2016)
86. Choudhury, B., Borah, B., Choudhury, A.: Extending photocatalytic activity of TiO_2 nanoparticles to visible region of illumination by doping of cerium. Photochem. Photobiol. **88**(2), 257–264 (2012)
87. Kumar, M., Singh, G., Chauhan, M.S.: Europium (Eu3+)-doped ZnO nanostructures: Synthesis, characterization, and photocatalytic, chemical sensing and preliminary assessment of magnetic properties. Ceram. Int. **47**(12), 17023–17033 (2021)
88. Korake, P.V., Kadam, A.N., Garadkar, K.M.: Photocatalytic activity of Eu3+-doped ZnO nanorods synthesized via microwave assisted technique. J. Rare Earths **32**(4), 306–313 (2014)
89. Bhatia, S., Verma, N.: Photocatalytic activity of ZnO nanoparticles with optimization of defects. Mater. Res. Bull. **95**, 468–476 (2017)
90. Bouras, K.: Re-doped SnO_2 oxides for efficient UV-Vis to infrared photon conversion: application to solar cells. Doctoral dissertation, Université de Strasbourg) (2016)
91. Ahmed, J., Mao, Y.: Ultrafine iridium oxide nanorods synthesized by molten salt method toward electrocatalytic oxygen and hydrogen evolution reactions. Electrochim. Acta **212**, 686–693 (2016)
92. Yuan, J., Choo, E.S.G., Tang, X., Sheng, Y., Ding, J., Xue, J.: Synthesis of ZnO–Pt nanoflowers and their photocatalytic applications. Nanotechnology **21**(18), 185606 (2010)
93. Basseem, M., Emam, A.A., Kamal, F.H., Gamal, A.M., Abo Faraha, S.A.: Novel functionalized of ZnO with Sm^{3+}, La^{3+}, and Sr^{2+}/ZnO single and tri-doped nanomaterials for photocatalytic degradation: synthesis, DFT, kinetics. J. Mater. Sci. **58**(33), 13346–13372 (2023)
94. Shahid, T., Mirza, M., Kosar, S., Ramzan, A., Ismail, A., Fareed, R., Husain, S., Mustafa, S., Yasmeen, N., Ameer, A., Nawaz, A., Safdar, M.: A facile synthesis of metal vanadate incorporated with metal oxide/rare earth metal oxide composite for photocatalytic application. Next Mater. **4**, 100191 (2024)
95. Dhanda, E., Nain, A., Dahiya, S.: A highly effective PANI@(Ce-Er) dual doped ZnO (PCEZ) nanocomposite as UV-light driven photocatalyst for degradation of MB dye: synthesis and characterization. Phys. Scr. **99**(8), 085043 (2024)
96. Raliya, R., Saharan, V., Dimkpa, C., Biswas, P.: Nanofertilizer for precision and sustainable agriculture: current state and future perspectives. J. Agric. Food Chem. **66**(26), 6487–6503 (2017)

97. Akhter, F., Ahmed, J., Pinjaro, M.A., Shaikh, F.A., Arain, H.J., Ahsan, M.J.: Metal-doped carbon dots (MCDs) as efficient nano-adsorbents for detection, monitoring, and degradation of wastewater pollutants: recent progress, challenges, and future prospects. Water Air Soil Pollut. **234**(11), 672 (2023)
98. Ayman, I., Aadil, M., Shahid, M., Khalid, A., Samama, H., Somaily, H.H., Warsi, M.F., Mohamed, R.M.: Wet synthesis of magnetically retrievable Mn/Nd co-doped cobalt ferrites for visible light-driven photocatalytic annihilation of azo dye. Mater. Sci. Semiconductor Process. **178**, 108462 (2024)
99. Hossain, M.K., Hossain, S., Ahmed, M.H., Khan, M.I., Haque, N., Raihan, G.A.: A review on optical applications, prospects, and challenges of rare-earth oxides. ACS Appl. Electron. Mater. **3**(9), 3715–3746 (2021)
100. Kumari, S., Sharma, R., Thakur, N., Kumari, A.: Removal of organic and inorganic effluents from wastewater by using degradation and adsorption properties of transition metal-doped nickel ferrite. Environ. Sci. Pollut. Res. **31**(34), 46526–46545 (2024)
101. Patil, A.S., Patil, A.V., Dighavkar, C.G., Adole, V.A., Tupe, U.J.: Synthesis techniques and applications of rare earth metal oxides semiconductors: a review. Chem. Phys. Lett. **796**, 139555 (2022)
102. Chen, W., Ball, A.S., Cole, I., Yin, H.: Metal-doped carbon dots as fenton-like catalysts and their applications in pollutant degradation and sensing. Sustainability **17**(8), 3642 (2025)
103. Uysal, E., Dursun, H.N., Uysal, E.C., Gürmen, S.: Synthesis of low pH stable, highly photocatalytic alginate hydrogels for methylene blue removal from wastewater. Turk. J. Chem. **49**(2), 154–175 (2025)
104. Hossain, M.K., Raihan, G.A., Akbar, M.A., Kabir Rubel, M.H., Ahmed, M.H., Khan, M.I., Hossain, S., Sen, S.K., Ebne Jalal, M.I., El-Denglawey, A.: Current applications and future potential of rare earth oxides in sustainable nuclear, radiation, and energy devices: a review. ACS Appl. Electron. Mater. **4**(7), 3327–3353 (2022)
105. Subramanian, Y., Le Minh, T., Bahruji, H., Sarifuddin, W.S., Mahadi, A.H., Gubendiran, R.K.,.Muhammed Ali, S.A., Raj, V., Yassin, H., Irvine, J.T.S., Azad, A.K.: Heteroanionic synthesis of lanthanum/neodymium-based titanium oxycarbide: a novel approach with multiple objectives for clean energy and pollutant-free environment. Clean Energy **9**(1), 89–107 (2025)
106. Ahmad, T., Phul, R., Khatoon, N., Sardar, M.: Antibacterial efficacy of Ocimum sanctum leaf extract-treated iron oxide nanoparticles. New J. Chem. **41**(5), 2055–2061 (2017)

Role of Artificial Intelligence and Machine Learning in Rare Earth-Doped Metal Oxide Nanoparticles

Priyanka Rawat, Shweta Chaudhary, Naveen Kumar, and Kirti Babber

Abstract Rare earth-doped metal oxide nanoparticles (RE-MO NPs) are notable for their enhanced electronic, optical, magnetic, and catalytic properties, influenced by the incorporation of elements like Ce, Eu, La, and Gd into oxides such as TiO_2, ZnO, and Fe_2O_3. These dopants help create defect states and optimize band structures, which boosts performance in various fields including photocatalysis and nanomedicine. However, the rational design of RE-MO NPs is challenged by complex interactions between composition, structure, and synthesis. This review discusses the transformative role of Artificial Intelligence (AI) and Machine Learning (ML) in addressing these challenges through predictive modeling and optimization techniques. It explores various AI/ML methods, including supervised and unsupervised learning, deep learning, and generative models, for tasks like property prediction and materials discovery. The paper also highlights future advancements such as autonomous labs and explainable AI to enhance the understanding of structure–property relationships. Overall, it emphasizes the significant impact of AI/ML in accelerating the discovery and intelligent design of rare earth nanomaterials for advanced applications.

Keywords Artificial intelligence · Machine learning · Deep learning · XAI · Nanoparticles · Rare earth-doped metal oxides

P. Rawat (✉)
Department of Computer Science, UIE, Chandigarh University, Mohali 140413, Punjab, India
e-mail: priyankaktwr@gmail.com

S. Chaudhary
Department of English, UILAH, Chandigarh University, Mohali 140413, Punjab, India

N. Kumar
Department of Physics, Galgotias University, Greater Noida 203201, Uttar Pradesh, India

K. Babber
Department of Electronics and Communication Engineering, UIE, Chandigarh University, Mohali 140413, Punjab, India

© The Author(s), under exclusive license to Springer Nature Singapore Pte Ltd. 2026

N. Kumar et al. (eds.), *Rare Earth-Doped Metal Oxide Nanostructures*, Engineering Materials, https://doi.org/10.1007/978-981-95-6192-6_5

1 Introduction

Rare earth (RE)-doped metal oxide nanoparticles (MO-NPs) are a crucial functional nanomaterial class because of their special electronic, optical, and catalytic properties. Doping with rare earth metals like europium (Eu), cerium (Ce), yttrium (Y), gadolinium (Gd), and lanthanum (La) alters the host metal oxide framework—most often ZnO, TiO_2, CeO_2, or Fe_2O_3—by creating lattice defects, promoting charge separation, red-shifting band gaps, and increasing stability [1, 2]. These alterations allow wide-ranging application in photocatalysis, gas sensing, luminescence, biomedical diagnostics, and environmental detoxification. In-spite of the author's great promise, optimizing the performance of RE-doped MO-NPs is a formidable challenge. They are governed by a high-dimensional interaction of variables such as dopant type and content, synthesis temperature, pH, ratio of precursors, calcination time, and post-processing conditions. Traditional trial-and-error experiment-based approaches are costly and time-consuming, with slow discovery cycles and non-repeatable outcomes. In this context, artificial intelligence (AI) and machine learning (ML) offer appropriate solutions to facilitate research and development at a fast pace. Such data-driven approaches have the capability to identify concealed trends in vast and multi-dimensional data reservoirs, model structure–property relationships, optimize synthesis conditions, and even identify novel compositions with a pre-specified functionality [3, 4]. RE-MO NP optimization and discovery are extremely interactive combinations of processing, compositional, and structural parameters. Traditional trial-and-error experimental approaches are usually time-consuming, labor-intensive, and limited in capability, especially when exploring large compositional and processing design spaces. Machine learning (ML) and artificial intelligence (AI) offer powerful tools to overcome the above limitations by enabling data-driven modeling, prediction, and inverse design of materials with pre-selected qualities. These approaches have the capability to significantly reduce the number of necessary physical experiments by optimizing the optimum synthesis conditions and dopant compositions with the help of predictive algorithms trained on existing experimental or simulated data. In addition, deep learning techniques like convolutional neural networks (CNNs) and generative models have the ability to reveal latent correlations between synthesis inputs and functional outputs, e.g., band gap tuning or photocatalytic activity, that are otherwise hard to interpret with standard statistical approaches [5]. As illustrated by the latest developments in autonomous materials discovery platforms, AI/ML integration can change the game for RE-MO NP research by speeding up innovation cycles and enhancing reproducibility in synthesis and property tuning. Machine learning models like random forests, support vector machines, deep neural networks, and generative adversarial networks (GANs) are increasingly being used throughout materials science, for instance, in catalysis, energy storage, and nanomedicine [5, 6]. Yet, the use of AI and ML for RE-doped MO-NPs is still comparatively not as investigated as for drug discovery or the design of bulk materials. Small datasets, non-standardized descriptors, and the absence of integrated databases pose

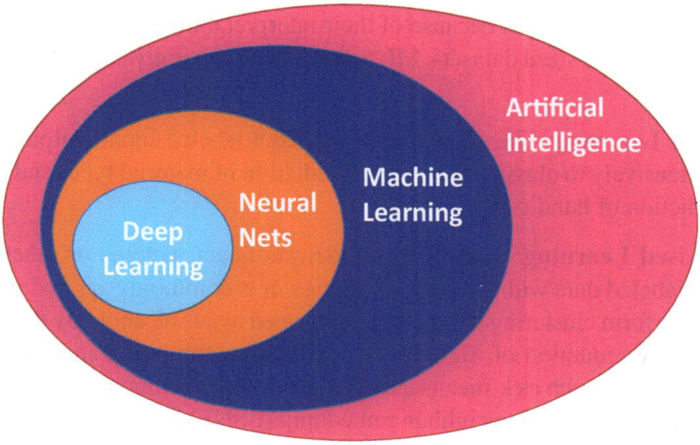

Fig. 1 Hierarchical relationship between artificial intelligence (AI), machine learning (ML), and deep learning (DL)

challenges. Still, with increasing interest in FAIR (Findable, Accessible, Interoperable, Reusable) data principles and automatic experimentation, the area is on the verge of making fast strides.

2 Overview of Artificial Intelligence and Machine Learning

2.1 Definition and Scope

Artificial intelligence (AI) is a wide area of computer science that deals with creating systems that can execute tasks that usually demand human intelligence, including reasoning, learning, problem-solving, and decision-making [7]. Machine learning (ML) is a subfield of methods in AI that enable computers to learn patterns from data and make decisions or predictions without being directly programmed for a specific task. Deep Learning (DL) is another subarea of ML that employs artificial neural networks with extensive layers (i.e., deep architectures) to learn hierarchical representations and features from large data sets [8]. Figure 1 effectively showcases the nested relationship among AI, ML, and DL, helping readers grasp the conceptual distinctions and overlaps.

2.2 Traditional ML Algorithms

A number of classical machine learning (ML) algorithms have been extensively used in materials science, especially to forecast physicochemical properties of metal oxides and doped materials because of their interpretability, efficiency, and efficacy on small to medium-sized datasets. ML tasks may be generally categorized into three types:

Supervised Learning: The algorithm trains from labeled input–output pairs. It is applied extensively to classification (e.g., prediction of material type) and regression (e.g., prediction of band gap).

Unsupervised Learning: The algorithm detects latent patterns or inherent structures in unlabeled data without prior categories. It is commonly applied in materials science to perform clustering similar types of doped material, anomaly detection, and dimensionality reduction of high-dimensional spectroscopic or imaging data. For example, methods such as k-means clustering, principal component analysis (PCA), and t-distributed stochastic neighbor embedding (t-SNE) have been effectively used to cluster nanoparticle compositions, remove noise in spectroscopic datasets, and visualize material property spaces [9].

Reinforcement Learning (RL): An agent acquires skills to make a series of choices by interacting with an environment in order to maximize the total reward. While this is not as common in materials science, RL is surfacing in such fields as autonomous synthesis planning [10]. Figure 2 categorizes machine learning into three primary paradigms: supervised learning, unsupervised learning, and reinforcement learning. Supervised learning includes classification algorithms (e.g., SVM, Random Forest) and regression models (e.g., Linear and Lasso Regression) trained on labeled data. Unsupervised learning techniques, such as K-Means and PCA, uncover hidden patterns in unlabeled data through clustering and association rules. Reinforcement learning, divided into model-based and model-free approaches, enables agents to learn optimal actions through interaction with an environment using algorithms like Q-learning and SARSA.

Simple relationships between input features like dopant concentration or particle size and output variables like electrical conductivity or band gap energy are modeled using linear regression. These models are relatively interpretable and suitable for small to moderate datasets, which are common in experimental nanomaterials research [11].

Decision trees divide data into branches along feature thresholds, providing interpretable, rule-based predictions that are helpful in classifying material response. Random forests, which is an ensemble technique with multiple decision trees, enhances prediction accuracy and minimizes overfitting by methods of aggregation and randomization. Support vector machines (SVMs) determine the best hyperplanes in high-dimensional feature spaces to undertake binary or multi-class classification and have been successfully employed for material phase identification and toxicity classification [12]. k-Nearest Neighbors (k-NN) is a non-parametric, simple method that predicts based on the closest points in the training set and can be best utilized for property estimation when data are scarce [13].

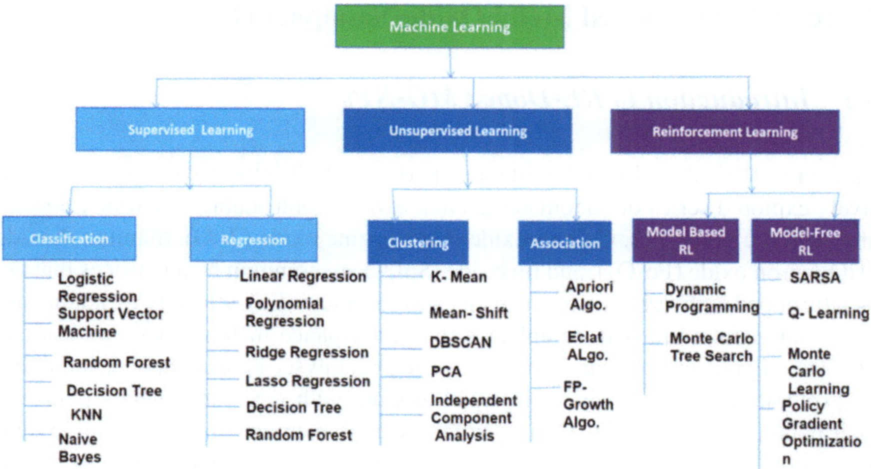

Fig. 2 Classification of machine learning

2.3 *Deep Learning*

Deep learning models employ several layers of connected neurons to identify intricate, high-level abstractions within uncooked input data. Deep learning models are particularly potent when utilized with big data or unstructured data such as microscope images, spectroscopy, or text synthesis logs. Convolutional neural networks (CNNs) are best suited for image-based data analysis (e.g., TEM/XRD), where spatial structure matters. They have been successfully employed to predict morphology, particle size distribution, and zeta potential of nanoparticles [14]. Autoencoders are unsupervised deep learning architectures that learn effective, compact data representations by mapping the input onto a lower-dimensional latent space and recovering it. They are used extensively for dimensionality reduction, denoising, and feature learning, especially when dealing with high-dimensional data, e.g., in spectroscopic, imaging, or structural data. Autoencoders have also been used in materials science to analyze Raman and UV–Vis spectra, remove redundancy from high-throughput screening data, and extract intrinsic patterns from structural datasets [15, 16]. Their capacity to learn nonlinear relationships makes them a suitable alternative to conventional methods such as PCA in dealing with nanomaterials of complex nature. Generative Adversarial Networks (GANs) are composed of two networks—a discriminator and a generator—and are used for generating synthetic material structures, spectra, or microscopy images. They are also used in reverse materials design to search for desired properties achieved by compositions or processing routes [6].

These approaches have the benefit of automatic learning of features, which minimizes dependence on hand-designed descriptors and allows for improved performance in difficult tasks.

3 Rare Earth-Doped Metal Oxide Nanoparticles

3.1 Introduction to RE-Doped MO-NPs

Rare earth (RE)-doped metal oxide nanoparticles (MO-NPs) are one of the extensively explored series of nanomaterials that exhibit controllable electronic, optical, and catalytic properties. Cerium oxide (CeO_2), zinc oxide (ZnO), titanium dioxide (TiO_2), iron oxide (Fe_2O_3), and tin oxide (SnO_2) are common host matrices that are usually doped with rare earth elements such as europium (Eu), yttrium (Y), lanthanum (La), and gadolinium (Gd) to enhance the performance in functional applications [2]. The significance of metal oxides as photocatalysts is accentuated, including an exploration into the manner in which doping with rare earth metals improves their photocatalytic activity. Doping with rare earth elements increases visible light absorption and reduces charge carrier recombination due to their unique 4f orbital characteristics. The paper discusses several synthesis methodologies and the role of rare earth doping on key features such as band gap and charge dynamics. It concludes with possible future directions for optimizing the photocatalytic activity of metal oxides, with the intention to spur further investigation in this field [1]. Additionally, the strategic importance of rare earth metals—such as lanthanides, scandium, and yttrium—is elaborated as these have widespread applications in industries such as catalysis, electronics, and optics. Their novel 4f electron structure and other physicochemical properties make them very promising for electrocatalysis. However, there is a lack of in-depth reviews on their modification methods and the latest developments. To bridge this gap, the review explores the latest applications of rare earth-modified metal oxides in processes like wastewater treatment, hydrogen peroxide production, hydrogen evolution reaction (HER), carbon dioxide reduction reaction (CO_2RR), nitrogen reduction reaction (NRR), and machine learning-assisted investigations. It also explores electrocatalytic processes and performance-determining factors, aiming to guide the development of advanced multifunctional electrocatalysts for energy and environmental applications. The inclusion of RE ions into the crystal lattice of metal oxides alters their structural and physicochemical characteristics. RE ions may cause lattice strain, create oxygen vacancies, increase charge carrier separation, and modify band gap energy levels—altering both surface and bulk material properties [17].

3.2 Applications of RE-Doped MO-NPs

Rare earth (RE) doped metal oxide nanoparticles (MONPs) display excellent physicochemical properties due to which they are in high demand in various scientific and industrial applications. Their distinctive 4f electron arrangements, superior coordination ability, and good luminescence make them suitable for new functionalities when used in metal oxide matrices. Below are some of the major application areas of RE-doped MO-NPs, supported by recent studies:

Photocatalysis: RE doping enhances photocatalytic performance by narrowing band gaps, minimizing charge carrier recombination, and broadening light absorption into the visible spectrum. For example, introducing dopants such as Ce^{3+}, Eu^{3+}, and La^{3+} into TiO_2 or ZnO promotes the creation of oxygen vacancies, which serve as active sites for photocatalytic reactions, thereby increasing the degradation rates of pollutants like methylene blue and phenol when exposed to visible light [18, 19]. Cerium- and erbium-doped zirconium dioxide showed improved visible-light-driven photocatalytic activity for methylene blue degradation, demonstrating potential for environmental cleanup applications [20]. Saqib et al. highlights that doping titanium dioxide (TiO_2) with rare earth metals enhances its photocatalytic efficiency for wastewater treatment. This enhancement is achieved by reducing the band gap, enabling visible light absorption, and inhibiting the anatase-to-rutile phase transformation, thereby improving the degradation of organic pollutants under visible light irradiation [21].

Gas Sensing: RE-doped SnO_2, ZnO, and Fe_2O_3 exhibit enhanced sensitivity, selectivity, and reduced detection thresholds for gases such as NO_2, NH_3, H_2S, and volatile organic compounds (VOCs). Doping results in a higher number of oxygen vacancies and improved surface electron mobility, facilitating better gas adsorption and quicker response/recovery times [22, 23]. Doping SnO_2 with transition metals like iron (Fe) has emerged as a powerful technique to enhance its gas-sensing performance. When Fe is introduced into the SnO_2 structure, it significantly boosts the material's ability to detect gases—particularly hydrogen sulfide (H_2S) [24]. The ethanol molecule can be differentiated from other potentially dangerous gases such as ammonia and aldehyde using sensors based on SnO_2-AuNPs-GO nanocomposites. Furthermore, this sensory nanocomposite can lower power consumption and enhance the practical application of gas detection since the detection mechanism may operate at low temperatures [25].

Luminescence and Display Devices: Rare earth (RE) elements make up an extensive list of the most effective and flexible stimulants. In addition, RE ions are particularly appropriate for use in visible-display technologies owing to their unique tendency to emit red, green, and blue light [26]. RE elements such as Eu^{3+} and Tb^{3+} are widely utilized in optical and display technologies due to their sharp, vivid emission lines. Eu^{3+} is known for its strong red emission (~ 612 nm), while Tb^{3+} emits a green light (~ 545 nm). These doped materials find extensive applications in light-emitting diodes (LEDs), display screens, and X-ray phosphors [27, 28].

Solar Energy Conversion: RE-doped oxides function as upconversion and down-conversion layers within photovoltaic devices to optimize solar energy efficiency. Ions like Yb^{3+}, Er^{3+}, and Tm^{3+} absorb low-energy photons and re-emit higher-energy light that silicon-based solar cells can capture, improving overall efficiency [29].

Biomedicine and Drug Delivery: Additionally, biocompatible RE-doped nanoparticles, including La_2O_3, Gd_2O_3, and Fe_2O_3, have demonstrated potential in drug delivery systems and as MRI contrast agents. These materials show remarkable stability, low cytotoxicity, and improved bioimaging capabilities. For instance, Gd^{3+}-doped Fe_2O_3 nanoparticles serve dual purposes for MRI and hyperthermia therapies [30]. RE nanoparticles have emerged as a painless diagnostic method for monitoring and detecting different biological processes, luminescent imaging may be crucial for understanding the origin and course of several illnesses. This technology has been accepted as an excellent way to discover the morphological, anatomical, and physiological traits in tissues as well as a broad spectrum of bio-samples from cells to beings, in addition to providing an excellent sensitivity and capacity of real-time monitoring [31, 32]. Another significant development in the medical area that offers more precise physiological and clinical information with excellent specificity and sensitivity for ailments diagnosis and treatment is the integration of luminous or MRI techniques into a single system. Moreover, Lanthanide Nano particles have been extensively used in dual-mode [33].

Environmental and Catalytic Applications: RE-doped catalysts have exhibited exceptional activity in environmental applications such as wastewater treatment and CO_2/N_2 reduction processes. Their strong adsorption capabilities and electronic interactions with contaminants render them ideal for heterogeneous catalysis and advanced oxidation techniques [34]. According to a study the Re-doped NiFeOx catalyst offers higher charge transfer and long-term operational robustness in alkaline environments along with improved electronic configuration and increased structural stability. Thus, by providing an affordable, high-performance catalyst that can maintain high current output and can tackle significant obstacles in large-scale water electrolysis therefore opening door for feasible hydrogen synthesis through water splitting [35]. In addition, the study done by Geetha et al. suggests that doping $ZnWO_4$ nanoparticles with rare earth elements (La and Ce) significantly increases their photocatalytic efficacy of degrading methylene blue dye. This advancement offers an effective, eco-friendly wastewater treatment technique that supports sustainable water management and healthier aquatic habitats by improving pollutant breakdown under light irradiation [36].

The versatile properties of RE-doped metal oxide nanoparticles (MO-NPs) position them as promising contenders in various cutting-edge industries. Their capacity to adjust band gaps, enhance surface activity, and improve optical characteristics allows them to tackle complex challenges in clean energy, sensing, medical diagnostics, and environmental sustainability. With the incorporation of machine learning for material discovery and refinement, the future applications of RE-doped nanomaterials are poised to be even more effective and impactful.

3.3 Why AI/ML Is Needed for RE-MO NPs

Though RE-doped MO-NPs display promising properties, their synthesis is hampered by the huge and intricate design space consisting of synthesis parameters, dopant amounts, and host-dopant pairs. Conventional trial-and-error laboratory practices not only are time-consuming but also not efficient in identifying optimal composition or processing conditions [10]. AI and ML methods bring a paradigm shift with data-driven discovery and optimization. Such approaches can simulate nonlinear input parameters (e.g., temperature, pH, dopant concentration) versus output property (e.g., band gap, degradation efficacy) relationships by training from historical experimental or computational data [5]. Additionally, ML methods can quickly screen thousands of potential dopant-host pairs, determine feature relevance, and forecast material behavior prior to experimental confirmation—saving enormous time and expense experimentally [37]. ML is also important in processing high-dimensional and unstructured data, for example, spectroscopic data or electron micrograph images, whereby deep neural models such as CNNs are able to learn significant features automatically [17]. In addition, generative models (for instance, GANs and autoencoders) enable inverse materials design where researchers are able to specify desirable properties and produce possible synthesis routes or compositions to exhibit it [6].

Hence, the incorporation of AI/ML in RE-MO NP research holds great promise to speed up material discovery, streamline synthesis, enhance reproducibility, and reveal concealed structure–property relationships that are otherwise hard to decipher.

4 Various Descriptors for ML Modeling

Correct and representative descriptors are of utmost importance for machine learning (ML) model success in predicting material characteristics and facilitating the discovery of new doped metal oxide nanoparticles. These descriptors—quantitative or qualitative values that extract key material characteristics—should accurately represent the physicochemical, structural, and electronic characteristics critical to the application of interest.

4.1 Structural Descriptors

Structural descriptors are numerical formulations of the geometric and chemical properties of a material. They play a vital role in ML models to predict and optimize the properties of RE-doped MO-NPs. The important structural descriptors are:

Ionic Radius

The ion size plays a significant role in lattice distortion when doping, impacting phase stability and carrier mobility. Precise ionic radii are of importance in modeling such effects. There has been a recent extension of Shannon's database of ionic radii using machine learning to predict radii for ions with rare oxidation states and coordination numbers [38].

Electronegativity

This characteristic has an influence on bond polarity and charge distribution in the material. Differences in electronegativity between dopant and host atoms can introduce large modifications to electronic structure and defect formation energies. Research has established that electronegativity differences play a central role in controlling the oxygen vacancy concentration in perovskite oxides.

Oxygen Vacancies

These defects are major drivers of electrical conductivity tuning, photocatalytic activity, and magnetic properties. Machine learning has been recently used to predict the concentration of oxygen vacancies in perovskite oxides, showing that parameters such as electronegativity and cation environment are key determinants [17].

4.2 Optical Descriptors

Electronic and optical behavior descriptors are essential in modeling the functional performance of rare earth (RE) doped metal oxide nanoparticles. Descriptors capture how dopants affect material properties like light absorption, charge dynamics, and electronic structure—properties paramount to application in catalysis, photovoltaics, and sensors. Band gap energy is a basic property that dictates optical absorption, photocatalysis efficiency, and electronic transport. Precise prediction of the band gap in RE-doped oxides allows for the property tailoring of materials for particular opto-electronic devices. Density functional theory (DFT) has recently been coupled with machine learning to improve doped semiconductor band gap predictions [39].PL intensity informs us about defect states, charge carrier recombination rates, and the optical material quality. RE doping modulates defect concentrations and energy transfer mechanisms, which affect the PL behavior. A recent study of ZnO nanostructures with rare earth ion doping showed remarkable changes in PL emissions, highlighting the significance of this descriptor in assessing optical features [40]. XPS gives surface chemistry, oxidation state, and chemical environment information that is essential for exploring doping effects. It is particularly useful in deciding on the successful integration of RE ions into lattices of hosts and in assessing surface concentrations of defects [41]. Charge carrier mobility, recombination lifetime, and diffusion length are important to device performance in photocatalytic and optoelectronic devices. They are measurable using time resolved spectroscopy or calculations. Recent review papers underscore the importance of understanding these dynamics to enhance photoelectrode materials [42].

4.3 Data Sources and Curation

Precise data gathering and preprocessing are fundamental to the development of robust ML models. Key data sources are:

DFT-based databases such as Materials Project, Open Quantum Materials Database (OQMD), and AFLOW, that provide computed electronic structures, formation energies, and band gaps [43, 44]. Experimental literature is a valuable source of data for materials science research, holding rich information on doping levels, synthesis conditions, and performance metrics. Exclusion of such information can be done through curation or by employing natural language processing (NLP) tools. Recent advances in NLP have rendered exemption of data from peer-reviewed journals much easier and more accurate. For instance, Jiang et al. [45] outline the ground breaking contribution of NLP and large language models (LLMs) to automating data extraction processes in materials science. Similarly, Lee et al. [46] describe an autonomous extraction pipeline that applies convolutional neural networks and LLMs to extract tabular scientific text data, indicating the capability of AI-based approaches to facilitate data acquisition for machine learning. Image- and spectrum-based inputs for deep learning models like CNNs [47] are given by data from XRD, TEM, UV-Vis, PL, and FTIR methods.

5 Applications of AIML in RE-Doped MO-NPs

Machine learning (ML) techniques have been employed to forecast significant properties of RE-MO NPs, such as band gap energies, magnetic responses, and catalytic efficiencies. For example, Santos et al. [48] utilized automated ML techniques to improve the accuracy of luminescence thermometry via Nd^{3+}-doped YAG nanoparticles by a factor of 5.5 compared to conventional methods. Equally, Singh et al. [49] created a descriptor-based ML model that is used to predict the enthalpies of formation of rare earth intermetallic compounds to enable the design of new materials with specified thermodynamic properties. The design of new materials with tailored properties depends on optimizing synthesis parameters. Liu et al. [50] illustrated the application of ML algorithms, random forest, and support vector machines, to forecast the phase and particle size of iron oxides as a function of synthesis conditions like pH, temperature, and precursor concentration. Their models were up to 96% accurate in phase prediction, highlighting the promise of ML in informing experimental synthesis. Evaluation of the RE-MO NPs' toxicity is crucial for their safe use, particularly in biomedical applications. Sayed et al. [51] suggested an ML model that employed a binary version of the sine tree-seed algorithm as a feature selection model and support vector machines as a classification model to predict the probable toxicity of metal oxide nanoparticles. This strategy benefits the elucidation of how nanoparticle properties correspond to their biological impact, enabling safer material development. RE-doped ZnO nanoparticles have been found to demonstrate potential in environmental remediation, especially for photocatalytic degradation of contaminants. Synthesis and characterization of such nanoparticles have been studied recently as a potential tool for degrading contaminants such as tetracycline, malachite green, and crystal violet. Although these researches are mainly experimental in nature, AI/ML methods could further enhance the photocatalytic activity by predicting synthesis conditions and doping levels [52]. In the biomedical field, RE-doped iron oxide nanoparticles have been researched for magnetic hyperthermia and magnetic resonance imaging (MRI) applications. Laha et al. [53] discussed the application of RE ions in improving magnetic properties of iron oxide nanostructures with emphasis on cancer theranostics applications. The inclusion of AI/ML methods can also contribute to designing nanoparticles with optimized characteristics for desired biomedical applications.

Jhaveri et al. [54] identifies several key challenges in applying machine learning (ML) in engineering. These include integrating diverse sensors and ensuring data quality, as real-world systems often face issues like varying sampling rates and noise. Scalability and real-time performance are also critical, as engineering applications require efficient processing that traditional models may not provide. Furthermore, the interpretability of ML models is essential for trust and timely interventions, while limited labeled data poses a significant obstacle in industrial settings. Lastly, integrating ML with legacy systems presents practical difficulties, emphasizing the need for adaptable and resource-efficient ML solutions in engineering contexts. Jiao et al. [55] suggested models and framework that consists of four main branches: Classical Mechanics-inspired, Electromagnetism-inspired, Statistical Physics-inspired, and Quantum Mechanics-inspired approaches. The Classical Mechanics-inspired category features techniques like Geometric Deep Learning and Dynamical Systems Neural Networks, which enhance learning through geometric structures and capture the dynamics of physical systems. The Electromagnetism-inspired paradigm introduces Optical Neural Networks that utilize optical implementations for efficient computation. The Statistical Physics-inspired category is the most comprehensive, including models that mimic entropy, chaos, and energy-based learning, with various subtypes such as Generative Adversarial Networks (GANs) and Boltzmann Machines. Finally, the Quantum Mechanics-inspired approaches encompass Quantum Machine Learning and Quantum Deep Learning, which leverage quantum computation to enhance learning algorithms and include innovative architectures like Quantum Recurrent Neural Networks.

Table 1 consolidates recent progress in using machine learning (ML) and deep learning (DL) methods for the characterization, design, and optimization of rare earth (RE)-doped metal oxides. Different methods like regression models, convolutional neural networks (CNNs), support vector regression (SVR), and generative models have been used in different domains including property prediction, optimization of synthesis, photocatalysis, and toxicity prediction. The applications cover a wide range from band gap prediction to automatic image analysis and inverse material design, showcasing the diverse applicability of data-driven techniques in expediting RE-based material discovery and development.

Table 1 Overview of machine learning and deep learning applications in rare earth (RE)-doped metal oxide research

Application area	ML/DL technique used	Material focus	Outcome	References
Property prediction	CGCNN, SVR, random forest	La-, Ce-, Nd-doped oxides	Accurate prediction of band gap, dielectric constants, and defect levels	[56]
Synthesis optimization	NLP, decision trees, Bayesian opt	Doped ZnO, TiO_2	Identification of optimal synthesis parameters (e.g., temperature, dopant concentration)	[10]
Photocatalysis	Regression models, RF	RE-doped TiO_2, CeO_2, ZnO	Prediction of photocatalytic degradation efficiency; improved dopant selection strategies	[9]
Image-based characterization	CNNs	TEM, SEM, XRD data of RE-MO NPs	Automated feature extraction and phase recognition from imaging/spectral data	[47]
Multi-objective discovery	Bayesian optimization, ensemble models	Various RE-doped metal oxides	Accelerated discovery balancing multiple target metrics (e.g., band gap, stability, optical properties)	[17]
Bandgap prediction	Random forest, XGBoost	TiO_2, ZnO doped with RE ions	Predict band gap of doped metal oxides using descriptors	[57]
Photocatalytic activity	ANN, SVM, kNN	CeO_2, ZnO, TiO_2 with RE dopants	Model photocatalytic degradation under visible light	[5]
Material discovery	GANs, autoencoders, Bayesian optimization	Metal oxides with La, Y, Gd, Eu	Generate new dopant–host combinations for improved properties	[6]

(continued)

Table 1 (continued)

Application area	ML/DL technique used	Material focus	Outcome	References
Inverse design	Variational autoencoder + DNN	RE-doped Y_2O_3, ZnO, TiO_2	Predict synthesis conditions for target band gap or PL	[58]
Synthesis optimization	Active learning + random forest	Sm/Eu doped perovskites and TiO_2	Reduce number of experiments by guiding synthesis	[59]
Property prediction	SVR, DNN	Binary/ternary oxides including REs	Predict dielectric constants and formation energy	[11]
Image segmentation	CNN, U-net	RE-doped CeO_2, ZnO nanoparticle imaging	Analyze morphology and particle size from HRTEM images	[14]
Toxicity prediction	Random forest, gradient boosting	CeO_2, Gd_2O_3, La_2O_3, Y_2O_3 nanoparticles	Predict nanotoxicity of doped oxides	[3]

6 Challenges

Within materials science, especially in the field of RE-MO NPs, large, high-quality datasets are not readily available. Limited availability causes hindrance to training robust ML models and may result in overfitting and poor generalization. Additionally, variability in experimental data collection protocols adds noise, making model development more challenging [60]. The lack of standardized and widely accepted descriptors in materials science presents a major challenge for the effective application of machine learning. Descriptors—quantitative features that represent material properties—are essential for developing predictive models. However, inconsistency in how descriptors are defined and used across studies makes it difficult to compare results, undermines reproducibility, and limits the generalizability of models. Several studies have highlighted the need for descriptors that are interpretable, causally linked to target properties, and reproducible across datasets and research groups [61]. Most ML models, particularly deep learning models, are "black boxes" that give predictions without clear reasoning. Such lack of transparency hampers the adoption and trust in AI in materials science, where revealing underlying mechanisms is essential [62]. Dynamic phenomena like doping diffusion and phase transformations are still difficult to capture. Most classical ML models have difficulty coping with the temporal nature of material behavior, with the need for the incorporation of time-dependent data and sophisticated modeling approaches [63].

7 Future Perspectives

Through these advancements, researchers can speed up the discovery and optimization of materials with designed properties, in response to key challenges in nanoparticles. The combination of AI, robotics, and automation is opening the door to automated laboratories that are able to conduct self-driving experiments. They can design, carry out, and analyze experiments by themselves, speeding up materials discovery and optimization vastly. For RE-MO NPs, such systems can search through enormous compositional spaces and synthesis parameters with ease to determine optimal conditions for targeted properties. The integration of Density Functional Theory (DFT) simulations with machine learning (ML) and experiments offers a multiscale strategy for materials modeling. The hybrid approach forecasts material properties at multiple scales, enhancing the accuracy and applicability of models. For RE-MO NPs, this technique can provide data for electronic structures, defect states, and surface interactions, which can facilitate the design of materials with desired functionalities [64]. Digital twins—virtual models of real systems—are becoming important materials science tools. By integrating real-time information with AI, digital twins can forecast and simulate the behavior of doped materials under a variety of conditions, enabling proactive decision-making. In RE-MO NPs, digital twins are able to simulate the effect of doping on structural, electronic, and optical behavior so that rapid prototyping and optimization can be achieved [65].

Development of XAI models is needed for deducing mechanistic insight into material behavior. XAI methods seek to render ML models more transparent, enabling researchers to comprehend and trust the predictions of ML models. For RE-MO NPs research, XAI can clarify the interconnection of dopant properties, synthesis conditions, and properties generated, which will lead to rational material design. Capture of dynamic phenomena like doping diffusion and phase transitions is still not easy. Classical ML models tend to have difficulty with the temporal nature of material behavior, requiring incorporating time-dependent information and sophisticated modeling methods. Embedding temporal dynamics into ML models can make prediction more effective for RE-MO NPs, especially for purposes with variable operation conditions. Symbolic Regression (SR) serves as a crucial link between data, machine learning models, and scientific theories, offering analytical equations for scenarios where only empirical methods exist. Challenges such as high computational costs and complex equations necessitate the development of new computational tools to enhance SR efficiency. Incorporating advanced neural network architectures, like Generative Adversarial Networks, could aid in uncovering hidden physics within small datasets. Additionally, addressing data pre-processing issues and utilizing parallel, GPU-based implementations could significantly reduce computational time. Overall, SR holds significant promise for transforming science and technology, demonstrating potential in generating models that rival established empirical relations [66].

8 Conclusion

Rare earth (RE) doped metal oxide nanoparticles are a frontier category of nanomaterials possessing remarkable structural, electronic, optical, and catalytic properties. Their tunability, by means of well-controlled doping with dopants such as Eu, Ce, La, and Gd, facilitates enhanced functionalities in photocatalysis, luminescence, gas sensing, and biomedical applications. Discovery, design, and optimization of these materials continue to be hampered by the sheer multidimensional search space of dopant–host pairs, synthesis conditions, and performance metrics. This article discussed the transformative power of artificial intelligence (AI) and machine learning (ML) for addressing these issues. AI/ML software can precisely predict properties like band gap energy, catalytic activity, and magnetism; refine synthesis conditions like calcination temperature and pH; and even facilitate inverse design for finding new RE-doped materials. The incorporation of deep learning methods such as autoencoders, generative models, and explainable AI only improves our capability to decipher intricate structure–property relations and suggest novel material compositions. In our research, we particularly aim to apply AI and ML strategies to predict and optimize the functional behavior of RE-doped metal oxide nanoparticles. Through the curation of accurate datasets from experimental literature and DFT databases, the extraction of valid structural and electronic descriptors, and the construction of predictive and generative ML models, we can expedite the discovery of high-performance nanomaterials. Our strategy uses supervised learning for property prediction, active learning for experiment guidance, and autoencoders for feature reduction and inverse mapping.

The findings of this study illustrate the great potential of integrating AI and ML methods with materials-domain expertise in materials science. Interdisciplinary collaboration between materials scientists, data scientists, and engineers will be essential for the development of open, standardized datasets, interpretable models, and autonomous research infrastructure in future. In adopting these innovations, not only will materials discovery be accelerated but also the way to sustainable, intelligent design of next-generation functional nanomaterials will be opened.

References

1. Mehtab, A., Ahmed, J., Alshehri, S.M., Mao, Y., Ahmad, T.: Rare earth doped metal oxide nanoparticles for photocatalysis: a perspective. Nanotechnology **33**(14), 142001 (2022)
2. Zhu, X., Zhang, J., Liu, J., Zhang, Y.: Recent progress of rare-earth doped upconversion nanoparticles: synthesis, optimization, and applications. Adv. Sci. **6**(22), 1901358 (2019)
3. Winkler, D.A.: Role of artificial intelligence and machine learning in nanosafety. Small **16**(36), 2001883 (2020)
4. Le, T., Epa, V.C., Burden, F.R., Winkler, D.A.: Quantitative structure–property relationship modeling of diverse materials properties. Chem. Rev. **112**(5), 2889–2919 (2012)
5. Chen, C., Zuo, Y., Ye, W., Li, X., Ong, S.P.: Learning properties of ordered and disordered materials from multi-fidelity data. Nat. Comput. Sci. **1**(1), 46–53 (2021)

6. Kim, B., Lee, S., Kim, J.: Inverse design of porous materials using artificial neural networks. Sci. Adv. **6**(1), eaax9324 (2020)
7. Vasudevan, R., Pilania, G., Balachandran, P.V.: Machine learning for materials design and discovery. J. Appl. Phys. **129**(7) (2021)
8. LeCun, Y., Bengio, Y., Hinton, G.: Deep learning. Nature **521**(7553), 436–444 (2015)
9. Ward, L., Dunn, A., Faghaninia, A., Zimmermann, N.E.R., Bajaj, S., Wang, Q., Montoya, J., Chen, J., Bystrom, K., Dylla, M., Chard, K., Asta, M., Persson, K.A., Snyder, G.J., Foster, I., Jain, A.: Matminer: an open source toolkit for materials data mining. Comput. Mater. Sci. **152**, 60–69 (2018)
10. Butler, K.T., Davies, D.W., Cartwright, H., Isayev, O., Walsh, A.: Machine learning for molecular and materials science. Nature **559**(7715), 547–555 (2018). https://doi.org/10.1038/s41586-018-0337-2
11. Pilania, G., Wang, C., Jiang, X., Rajasekaran, S., Ramprasad, R.: Accelerating materials property predictions using machine learning. Sci. Rep. **3**(1), 2810 (2013)
12. Furxhi, I., Murphy, F., Mullins, M., Arvanitis, A., Poland, C.A.: Practices and trends of machine learning application in Nanotoxicology. Nano **10**(1), 116 (2020)
13. Rupp, M., Tkatchenko, A., Müller, K.-R., von Lilienfeld, O.A.: Fast and accurate modeling of molecular atomization energies with machine learning. Phys. Rev. Lett. **108**(5), 058301 (2012)
14. Horwath, J.P., Zakharov, D.N., Mégret, R., Stach, E.A.: Understanding important features of deep learning models for segmentation of high-resolution transmission electron microscopy images. Npj Comput. Mater. **6**, 108 (2020)
15. Ghiringhelli, L.M., Vybiral, J., Levchenko, S.V., Draxl, C., Scheffler, M.: Big data of materials science: critical role of the descriptor. Phys. Rev. Lett. **114**(10), 105503 (2015)
16. Ma, J., Sheridan, R.P., Liaw, A., Dahl, G.E., Svetnik, V.: Deep neural nets as a method for quantitative structure–activity relationships. J. Chem. Inf. Model. **55**(2), 263–274 (2015)
17. Li, Y., Liu, Q., Li, T., Bi, H., Shen, Z.: Recent achievements in rare earth modified metal oxides for environmental and energy applications: a review. Chin. Chem. Lett. **110698** (2024)
18. Nasr, M., Balme, S., Eid, C., Habchi, R., Miele, P., Bechelany, M.: Enhanced visible-light photocatalytic performance of electrospun rGO/TiO_2 composite nanofibers. J. Phys. Chem. C. **121**(1), 261–269 (2017)
19. Pathak, T.K., Coetsee-Hugo, E., Swart, H.C., Swart, C.W., Kroon, R.E.: Preparation and characterization of Ce doped ZnO nanomaterial for photocatalytic and biological applications. Mater. Sci. Eng. B. **261**, 114780 (2020)
20. Gionco, C., Paganini, M.C., Giamello, E., Sacco, O., Vaiano, V., Sannino, D.: Rare earth oxides in zirconium dioxide: how to turn a wide band gap metal oxide into a visible light active photocatalyst. J. Energy Chem. **26**(2), 270–276 (2017)
21. Saqib, N.U., Adnan, R., Shah, I.: A mini-review on rare earth metal-doped TiO_2 for photocatalytic remediation of wastewater. Environ. Sci. Pollut. Res. **23**, 15941–15951 (2016)
22. Wang, C., Shang, M., Wei, H., Zhang, M., Zou, W., Meng, X., Chen, W., Shao, H., Lai, Y.: Specific and sensitive on-site detection of Cr(VI) by surface-enhanced Raman spectroscopy. Sensors Actuators B Chem. **346**, 130594 (2021)
23. Bhujel, K., Ningthoujam, S.S., Singh, L.R., Rai, S.: Effect of solution aging on properties of spin coated zinc oxide thin films. Mater. Today Proc. **46**(Part 14), 6419–6422 (2021)
24. Khan, L.A., Ali, S., Ali, N., Zhu, L., Zulfiqar, S., Hussain, S.A., et al.: Investigation of the Fe-doped SnO_2 NPs with enhanced H2S gas sensing performance. Ceram. Int. **51**(5), 6783–6792 (2025)
25. Hakami, M., Khairi, N.A.S., Yusof, N.A., Abdullah, J., Isha, A.: Development of SnO_2-AuNPs-GO modified interdigitated electrode gas sensors for ethanol detection. IEEE Sensors J. **25**(13), 24360–24368 (2025)
26. Ballato, J., Lewis, J.S., Holloway, P.: Display applications of rare-earth-doped materials. MRS Bull. **24**(9), 51–56 (1999)
27. Binnemans, K.: Interpretation of europium(III) spectra. Coord. Chem. Rev. **295**, 1–45 (2015)
28. Najihah, M.Z., Noor, I.M., Winie, T.: Long-run performance of dye-sensitized solar cell using natural dye extracted from *Costus woodsonii* leaves. Opt. Mater. **123**, 111915 (2022)

29. Derouiche, M., Salhi, R., Baklouti, S.: Efficient up-conversion ZnO co-doped (Er, Yb) nanopowders synthesized via the sol-gel process for photovoltaic applications. Materials **15**(21), 7828 (2022)
30. Hatami, E., Nagesh, P.K.B., Chauhan, N., Jaggi, M., Chauhan, S.C., Yallapu, M.M.: In situ nanoparticle self-assembly for combination delivery of therapeutics to non-small cell lung cancer. ACS Appl. Bio Mater. **5**(3), 1104–1119 (2022)
31. Gu, Z., Yan, L., Tian, G., Li, S., Chai, Z., Zhao, Y.: Recent advances in design and fabrication of upconversion nanoparticles and their safe theranostic applications. Adv. Mater. **25**, 3758–3779 (2013)
32. Wang, F., Banerjee, D., Liu, Y., Chen, X., Liu, X.: Upconversion nanoparticles in biological labeling, imaging, and therapy. Analyst **135**, 1839–1854 (2010)
33. Mnasri, W., Parvizian, M., Ammar-Merah, S.: Design and synthesis of luminescent lanthanide-based bimodal nanoprobes for dual magnetic resonance (MR) and optical imaging. Nano **11**(2), 354 (2021)
34. Saviano, L., Brouziotis, A.A., Padilla Suarez, E.G., Siciliano, A., Spampinato, M., Guida, M., et al.: Catalytic activity of rare earth elements (REEs) in advanced oxidation processes of wastewater pollutants: a review. Molecules **28**(17), 6185 (2023)
35. Lyu, X., Dileep, N.P., Pushkar, Y., Pupucevski, M., Lattimer, J., Colon-Mercado, H., et al.: Enhancing durability and activity toward oxygen evolution reaction using single-site re-doped NiFeOx catalysts at ampere-level. Chem. Eng. J. **507**, 160518 (2025)
36. Geetha, G.V., Sivakumar, R., Slimani, Y., Sanjeeviraja, C., Kannapiran, E.: Rare earth (RE: La and Ce) elements doped $ZnWO_4$ nanoparticles for enhanced photocatalytic removal of methylene blue dye from aquatic environment. Phys. B Condens. Matter. **639**, 414028 (2022)
37. Jha, D., Ward, L., Paul, A., Liao, W.-K., Choudhary, A., Wolverton, C., Agrawal, A.: ElemNet: deep learning the chemistry of materials from only elemental composition. Sci. Rep. **8**(1), 17593 (2018)
38. Baloch, A.A.B., Alqahtani, S.M., Mumtaz, F., Muqaibel, A.H., Rashkeev, S.N., Alharbi, F.H.: Extending Shannon's ionic radii database using machine learning. Phys. Rev. Mater. **5**(4), 043804 (2021)
39. Gou, F., Ma, Z., Yang, Q., Du, H., Li, Y., Zhang, Q., You, W., Chen, Y., Du, Z., Yang, J., He, N., Luo, J., Liu, Z., Tian, Z., Mao, M., Liu, K., Yu, J., Zhang, A., Min, F., Sun, K., Xuan, N.: Machine learning-assisted prediction and control of bandgap for organic–inorganic metal halide perovskites. ACS Appl. Mater. Interfaces. **17**(12), 18383–18393 (2025)
40. Zeidan, M.M., Abedrabbo, S.: Enhancing photoluminescence spectra for doped ZnO using neutron irradiation. ACS Omega. **8**(19), 16722–16728 (2023)
41. Hasanah, A.U., Ikbal, M.S., Tahir, D.: Advances in rare earth-doped ZnO photocatalysts: enhancing photogenerated electron-hole pairs for radical atom generation. Chem. Biol. Eng. Rev. **11**(3) (2024). https://doi.org/10.1002/cben.202300058
42. Kafizas, A., Godin, R., Durrant, J.R.: Charge carrier dynamics in metal oxide photoelectrodes for water oxidation. In: Mi, Z., Wang, L., Jagadish, C. (eds.) Semiconductors and Semimetals, vol. 97, pp. 3–46 (2017)
43. Jain, A., Ong, S.P., Hautier, G., Chen, W., Richards, W.D., Dacek, S., Cholia, S., Gunter, D., Skinner, D., Ceder, G., Persson, K.A.: Commentary: the materials project: a materials genome approach to accelerating materials innovation. APL Mater. **1**(1), 011002 (2013)
44. Saal, J.E., Kirklin, S., Aykol, M., Meredig, B., Wolverton, C.: Materials design and discovery with high-throughput density functional theory: the open quantum materials database (OQMD). JOM **65**, 1501–1509 (2013)
45. Jiang, X., Wang, W., Tian, S., Wang, H., Lookman, T., Su, Y.: Applications of natural language processing and large language models in materials discovery. Npj Comput. Mater. **11**(1), 79 (2025)
46. Lee, S., Heinen, S., Khan, D., von Lilienfeld, O.A.: Autonomous data extraction from peer reviewed literature for training machine learning models of oxidation potentials. Mach. Learn.: Sci. Technol. **5**(1), 015052 (2024)

47. Himanen, L., Geurts, A., Foster, A.S., Rinke, P.: Data-driven materials science: status, challenges, and perspectives. Adv. Sci. **6**(21), 1900808 (2019)
48. Santos, E.P., Pugina, R.S., Hilário, E.G., Carvalho, A.J.A., Jacinto, C., Rego-Filho, F.A.M.G., Canabarro, A., Gomes, A.S.L., Caiut, J.M.A., Moura, A.L.: Towards accurate real-time luminescence thermometry: an automated machine learning approach. Sensors Actuators A Phys. **362**, 114666 (2023)
49. Singh, P., Del Rose, T., Vazquez, G., Arroyave, R., Mudryk, Y.: Machine-learning enabled thermodynamic model for the design of new rare-earth compounds. Acta Mater. **229**, 117759 (2022)
50. Liu, J., Zhang, Z., Li, X., Zong, M., Wang, Y., Wang, S., Chen, P., Wan, Z., Liu, L., Liang, Y., Wang, W., Wang, S., Guo, X., Saldanha, E.G., Rosso, K.M., Zhang, X.: Machine learning assisted phase and size-controlled synthesis of iron oxide particles. Chem. Eng. J. **473**, 145216 (2023)
51. Sayed, G.I., Alshater, H., Hassanien, A.E.: Predicting the potential toxicity of the metal oxide nanoparticles using machine learning algorithms. Soft. Comput. **28**(17), 10235–10261 (2024)
52. Dhiman, P., Rana, G., Kumar, A., Dawi, E.A., Sharma, G.: Rare earth doped ZnO nanoparticles as Spintronics and photo catalyst for degradation of pollutants. Molecules **28**(6), 2838 (2023)
53. Laha, S.S., Thorat, N.D., Singh, G., Sathish, C.I., Yi, J., Dixit, A., Vinu, A.: Rare-earth doped iron oxide nanostructures for cancer theranostics: magnetic hyperthermia and magnetic resonance imaging. Small **18**(11), 2104855 (2022)
54. Jhaveri, R.H., Revathi, A., Ramana, K., Raut, R., Dhanaraj, R.K.: A review on machine learning strategies for real-world engineering applications. Mob. Inf. Syst. **2022**(1), 1833507 (2022)
55. Jiao, L., Song, X., You, C., Liu, X., Li, L., Chen, P., et al.: AI meets physics: a comprehensive survey. Artif. Intell. Rev. **57**(9), 256 (2024)
56. Xie, T., Grossman, J.C.: Crystal graph convolutional neural networks for an accurate and interpretable prediction of material properties. Phys. Rev. Lett. **120**(14), 145301 (2018)
57. Das, P., Ganguly, S., Agarwal, T., Maity, P., Ghosh, S., Choudhary, S., Gangopadhyay, S., Maiti, T.K., Dhara, S., Banerjee, S., Das, N.C.: Heteroatom doped blue luminescent carbon dots as a nano-probe for targeted cell labeling and anticancer drug delivery vehicle. Mater. Chem. Phys. **237**, 121860 (2019)
58. Bragança, H., Zeng, H., Dias, A.C., Wang, J., Wang, Z.: Magnetic-gateable valley exciton emission. Npj Comput. Mater. **6**, 90 (2020)
59. Lookman, T., Balachandran, P.V., Xue, D., Yuan, R.: Active learning in materials science with emphasis on adaptive sampling using uncertainties for targeted design. Npj Comput. Mater. **5**, 21 (2019)
60. Oviedo, F., Ferres, J.L., Buonassisi, T., Butler, K.: Interpretable and explainable machine learning for materials science and chemistry. Acc. Mater. Res. **2**(8), 597–608 (2021)
61. Ward, L., Agrawal, A., Choudhary, A., Wolverton, C.: A general-purpose machine learning framework for predicting properties of inorganic materials. Npj Comput. Mater. **2**(1), 1–7 (2016)
62. Arrieta, A.B., Díaz-Rodríguez, N., Del Ser, J., Bennetot, A., Tabik, S., Barbado, A., Garcia, S., Gil-Lopez, S., Molina, D., Benjamins, R., Chatila, R., Herrera, F.: Explainable artificial intelligence (XAI): concepts, taxonomies, opportunities and challenges toward responsible AI. Inf. Fusion. **58**, 82–115 (2020)
63. Gao, C., Lan, X., Li, N., Yuan, Y., Ding, J., Zhou, Z., Xu, F., Li, Y.: Large language models empowered agent-based modeling and simulation: a survey and perspectives. Hum.Ities Soc. Sci. Commun. **11**(1), 1259 (2024)
64. Goswami, L., Deka, M.K., Roy, M.: Artificial intelligence in material engineering: a review on applications of artificial intelligence in material engineering. Adv. Eng. Mater. (2023)
65. Jablonka, K.M., Ongari, D., Moosavi, S.M., Smit, B.: Big-data science in porous materials: materials genomics and machine learning. Chem. Rev. **120**(16), 8066–8129 (2020)
66. Angelis, D., Sofos, F., Karakasidis, T.E.: Artificial intelligence in physical sciences: symbolic regression trends and perspectives. Arch. Comput. Methods Eng. **30**(6), 3845–3865 (2023)

Gas Sensing Technologies for Rare Earth-Doped Metal Oxide Nanostructures: Advances in Energy Harvesting and Environmental Sensing

Payal Patial, Manish Deshwal, Shonak Bansal, Krishna Prakash, and Sujit Kumar Chaudhary

Abstract Gas sensing technologies have become integral to diverse fields such as environmental monitoring, industrial safety, and healthcare diagnostics. Among various sensor materials, metal oxide semiconductors (MOS) are widely used due to their cost-effectiveness, sensitivity, and ease of fabrication. Recent advancements have shown that doping these MOS materials with rare earth elements (REEs) significantly enhances their gas sensing performance. REE doping improves key attributes like surface reactivity, electron mobility, and oxygen ion exchange, resulting in greater sensitivity and selectivity for target gases. Enhanced catalytic activity, increased surface basicity, and the ability to inhibit electron–hole recombination further contribute to improved sensor response. Rare earth-doped MOS sensors have demonstrated efficient detection of gases such as ethanol, acetone, ammonia, and formaldehyde at lower operating temperatures, with faster response and recovery times. Moreover, the integration of REEs supports the development of self-powered and miniaturized sensing devices. This chapter explores the mechanisms, material modifications, and performance benefits associated with rare earth-doped MOS nanostructures, offering insights into their growing importance in next-generation gas sensing technologies.

Keywords Gas sensing · Rare earth metal · Selectivity · Sensitivity

P. Patial (✉) · M. Deshwal · S. Bansal
Department of Electronics and Communication Engineering, University Institute of Engineering, Chandigarh University, Mohali, Punjab 140413, India
e-mail: payal.patial@gmail.com

K. Prakash
Department of ECE, NRI Institute of Technology, Vijayawada, India

S. K. Chaudhary
IMEC, Leuven, Flemish Region, Belgium

© The Author(s), under exclusive license to Springer Nature Singapore Pte Ltd. 2026
N. Kumar et al. (eds.), *Rare Earth-Doped Metal Oxide Nanostructures*, Engineering Materials, https://doi.org/10.1007/978-981-95-6192-6_6

1 Introduction

Gas sensing technologies have emerged as critical components in enhancing industrial processes, ensuring health and safety standards, and advancing environmental protection initiatives. These sensor-based intelligent systems have become increasingly integrated into everyday applications related to safety concerns, including medical diagnostic procedures, air quality assessment, food industry processes, detection of hazardous gases (toxic, flammable, and explosive), and energy efficiency optimization. Fundamentally, gas sensors function as electronic devices designed to detect, identify, and quantify gas concentrations in various environments [1]. The field encompasses numerous sensing technologies classified according to their sensing elements, including acoustic-based gas sensor, MOS-based gas sensor, capacitance-based gas sensor, calorimetric-based gas sensor, electrochemical gas sensor, and optical gas sensors, each serving specific applications across industries [2]. The genesis of chemo-resistive gas sensors can be traced to 1952 when Brattain and Bardeen introduced the first MOS sensors, demonstrating that semiconductor materials exhibit resistance changes in response to atmospheric conditions [3]. Contemporary classification systems, such as that proposed by Liu and colleagues, made two main categories out of gas sensors that depend on the electrical characteristic variations and those utilizing other property changes [4]. The former includes technologies employing CNTs, polymers, and MOS, while the latter encompasses acoustic, calorimetry, gas chromatography, and optical techniques. In contemporary applications, MOS-based gas sensors have gained important research focus because of their versatility, cost-effectiveness, and compact dimensions. In 1938, Wagner and associates may have initially observed conductivity changes in metal oxides on exposure to reducing gas [5], Seiyama and colleagues formally proposed using metal oxide semiconductors for gas detection in 1962 [6]. In_2O_3, NiO, SnO_2, WO_3, and ZnO are various metal oxides that reflects exceptional capabilities in the field of gas sensing [7]. The fundamental mechanism involves molecules of gas and its adsorbing and desorbing, altering conductivity through electronic transmission. These changes are significantly influenced by nanostructure size and geometry. Optimizing properties of metal like band gap, dopant intrusion, and fermi level and network connectivity is essential for enhancing sensor sensitivity [8].

Zinc oxide (ZnO) was the first material where researcher found that the electrical conductivity increases when exposed to reducing gases, which is a fundamental principle behind modern gas sensing technology [5]. Research has yielded various approaches to enhance gas sensing performance and address limitations of pristine MOS sensing materials. In order to increase the gas sensing performance, different types of enhancement techniques can be implemented, like formation of porous structure which increases the surface area, modification in the surface area with different metal oxides, using rare metals, and utilizing polymers to enhance the gas sensing [9]. Doping is the process of intrusion of dopants in the base metals that represent another frequently employed and successful methodology, with noble metals (Ru, Rh, Pd, Au, Ag, Os, Ir, Pt) commonly utilized as dopants [10].

Recent literature suggests that rare earth elements may significantly enhance the sensing effectiveness of MOS gas sensors, though research in this domain remains limited [11]. Consequently, this chapter focuses on rare earth-doped MOS and applications in gas sensing technologies, exploring their potential to advance this critical field.

Rare Earth Element Doping in Metal Oxide Semiconductors for Enhanced Gas Sensing. Rare earth metal doping significantly enhances metal oxide semiconductor (MOS) performance in gas sensing applications by reducing electron–hole pair recombination rates [12]. Research executed by Longamani et al., represented that the rare earth-doped ZnO synthesized by wet chemical techniques shows improved photocatalytic activity because it captures electrons which prevents recombination of charge carriers [13]. Additionally, rare earth oxides (REOs) function as efficient catalysts for dehydrogenation and hydrocarbon ring opening reactions, thereby improving the performance of ZnO-based sensors in the field of gas sensing [14]. Ge and collaborators established ZnO with specific concentrations of CeO_2 (1 and 5 at.%) demonstrated enhanced sensing performance for acetone and alcohol at 100 ppm concentrations [15].

The exceptional gas sensing sensitivity of rare earth-doped materials derives from their expanded surface area and accelerated oxygen ion mobility, factors that enhance free electron conductivity [16]. Moreover, these materials exhibit increased surface basicity, further amplifying their gas sensing capabilities. For instance, ZnO nanowires incorporating rare earth elements reflect improved ethanol gas detecting capability attributed to their elevated surface basicity [11]. In_2O_3-based sensors have been developed for detecting diverse gases including ammonia (NH_3), nitrogen dioxide (NO_2), hydrogen sulfide (H_2S), ethanol, and formaldehyde (HCHO) [17]. As an n-type semiconductor with a substantial band gap ranging from 3.5 to 3.8 eV, In_2O_3 readily accommodates rare earth element doping due to the comparable ionic radii between RE^{X+} ions and In^{3+}, though this substitution introduces increased lattice strain [18]. Han and colleagues reported that cerium-doped In_2O_3 porous nanospheres demonstrated enhanced methanol gas sensing performance, while Anand et al. observed that terbium-doped In_2O_3 nanoparticles improved ethanol gas sensing capabilities [11]. Comprehensive studies have involved doping In_2O_3 with various rare earth elements including thulium, erbium, lanthanum, ytterbium, and cerium. Han and colleagues demonstrated that rare earth element-doped three-dimensionally ordered macroporous (3DOM) In_2O_3 structures, prepared via crystal templating methods, exhibited superior ethanol gas sensing performance [19]. In_2O_3 exists in three structural configurations: cubic (c- In_2O_3), hexagonal corundum (h- In_2O_3), and orthorhombic Rh_2O_3-type structures, with the latter stable only under high temperature and pressure conditions [16]. Zhang and colleagues reported that rare earth elements (La, Pr, Sm, Ce, and Yb) doping into porous In_2O_3 nanosheets lead to a good sensing response toward detection of acetone [20]. Similarly, Yang et al. determined doping of rare earth elements (Pr, Yb, Dy) to mesostructured In_2O_3 shows better sensing for ammonia [21]. Molybdenum trioxide (MoO_3), has a wide band gap exhibiting electrochromic, thermochromic, and photochromic properties, and has also benefited from rare earth doping. Cerium modified MoO_3 nanobelts exhibit exceptional sensitivity for trimethylamine (TMA), a significant industrial compound [22]. Different types of MOS like nanofibers of SnO_2, NiO, Co nano ferrite, and rare earth-doped perovskite materials, show better gas sensing response at low operating temperatures [23]. There are a lot of positive aspects of using resistance-based gas sensors; their elevated temperature demands restrict usage in autonomous sensing applications. Recent advancements have incorporated PENGs, TENGs, and MEMS based gas sensors for room-temperature detection of pollutant gases. TENG-based sensors utilizing pure ZnO generate piezoelectric outputs that function simultaneously as power sources and gas sensing signals [24]. Chu and colleagues fabricated self-powered gas sensors based on PENGs using yttrium-doped ZnO nanorod arrays, achieving 58% sensitivity for carbon monoxide detection [25]. These developments represent significant progress in the application of rare earth element doping to enhance metal oxide semiconductor performance in advanced gas sensing technologies.

2 Rare Earth Elements: Properties and Applications in Gas Sensing

Rare earth elements (REEs) comprise a distinctive group of 17 elements, consisting of the lanthanide series (cerium, praseodymium, neodymium, promethium, samarium, europium, gadolinium, terbium, dysprosium, holmium, erbium, thulium, ytterbium, and lutetium) plus three additional elements: lanthanum, scandium, and yttrium. Their nomenclature originated in the eighteenth century when these elements were initially discovered as components of complex oxides, commonly referred to as "earth" at that time. The perception of scarcity during their discovery period contributed to their designation as "rare earth" elements. Within this group, light lanthanides (cerium through europium) and lanthanum demonstrate greater natural abundance compared to heavy lanthanides (gadolinium through lutetium), resulting in significantly higher market prices for the latter group. The distinctive electron configuration of rare earth elements governs their chemical, metallurgical, and physical characteristics. While these elements can exhibit variable valence states, they predominantly exist in the trivalent form (RE^{3+}). A defining feature of REEs is the presence of multiple unpaired electrons, which enhances numerous properties of these elements. Compared to other periodic table elements such as copper, silver, and gold which have melting points varying around 1000 °C, REEs demonstrate substantially higher melting points (approximately 1000 °C or higher). The unpaired electrons in REEs contribute to their remarkable magnetic, electrical, luminescent, and catalytic properties, making them valuable for various technological applications. As global demand for advanced gas sensing technologies continues to grow across sectors such as environmental protection, industrial safety, medical diagnostics, and smart building applications, research into rare earth element-doped metal oxide semiconductors represents a promising frontier for technological innovation and scientific discovery in this critical field.

3 Factors Influencing Gas Sensing Mechanisms

3.1 Chemical Composition

Semiconducting metal oxides have been extensively researched for simple gas detection at elevated temperatures. Materials for gas sensor applications require diverse properties. While numerous MOS exhibit advantageous characteristics, limited selection satisfies all requirements. Consequently, researchers employ various composite materials including metal oxide-metal oxide combinations, MOS enhanced with additional elements, and ternary compounds [26].

3.2 Temperature and Humidity

MOS-based gas sensors typically function at elevated temperatures, with their response exhibiting strong temperature dependence. A sensor achieves optimal response for a specific gas at temperatures where equilibrium between adsorption and desorption rates is established. Research by Cao et al. demonstrated response curves for pure and 2% Nd-doped ZnO nanorods exposed to 100 ppm ethanol, revealing that response values at 348 °C operating temperature reached 20.3 and 37.8, respectively, for the undoped and Nd-doped specimens [27]. At lower temperatures, reduced kinetic reaction rates result in extended response and recovery periods. The interaction between water vapor and semiconductor surfaces can significantly impact sensor performance. Humidity levels may affect baseline measurements and sensitivity. Water vapor influence can be analyzed from two perspectives. Initially, water vapor adsorption reduces available adsorption sites during humid conditions, thereby decreasing the quantity of adsorbed oxygen species on semiconductor surfaces. Additionally, water vapor can function as electron donors to semiconductors, potentially altering the semiconductor's baseline resistance. Generally, water adsorption substantially reduces gas sensitivity. For instance, Qi et al. developed a Sm_2O_3-doped SnO_2 sensor and evaluated its response to C_2H_2 across 11–95% humidity levels, demonstrating decreased sensor response with increasing humidity [28].

3.3 Nanostructures

Gas sensor characteristics based on MOSs (including stability, sensitivity, selectivity, and response time) can be significantly enhanced by reducing particle size to nanoscale dimensions. Decreasing grain size to nanoscale proportions represents one of the most effective approaches for improving gas-sensing properties. Grain boundary barriers provide the primary gas effect on sensor conductivity, enabling comprehensive control of sensing mechanisms. When grain size becomes the determining factor rather than grain boundaries, it governs the sensing material's sensitivity. For grains with diameter (D) greater than or equal to twice the space-charge layer thickness (L), the space-charge layer creates conduction channels around each neck. Conductivity is influenced by barriers at grain boundaries and cross-sectional regions, enhancing sensitivity. When grains experience complete depletion of mobile charge carriers, conduction pathways between grains diminish, resulting in dramatically reduced conductivity. In cases where $D < 2L$, grain size regulates the sensing material's sensitivity.

Li et al. depicted that cerium doped MoO_3 nanobelts reduced grain size from 28 to 23 nm. This change improved performance of trimethylamine gas sensing [23]. Similarly, Qin et al. showed that neodymium doping in SnO_2 suppressed crystal growth, as evidenced by scanning electron microscopy (SEM) imaging of Ce-doped SnO_2 [29]. Consequently, Nd functions as an effective grain growth inhibitor by reducing grain size, thereby increasing the available surface area for surface reactions, which improves alcohol detection sensitivity. These studies also highlight that, besides grain size, factors like grain shape and the aspect ration (D/L) significantly influence gas sensor attributes of MOS material. Here D refers to particle size and L represents layer thickness [30].

3.4 Morphology

The morphology and structure of rare earth (RE)-doped metal oxide semiconductors (MOSs) are predominantly determined by the synthesis technique employed, which significantly impacts their gas sensing capabilities. Careful selection of the synthesis process is essential to achieve the desired level of nanostructuring in terms of morphology, grain/particle size, and the crystalline and electronic structures of sensing materials. Various methodologies for nanostructuring REE-doped MOSs include hydrothermal processing, electrospinning, microwave-assisted synthesis, solvothermal methods, and colloidal crystal templating [31]. The structure of sensing materials is vital in determining their gas-sensing behavior, as different forms offer various benefits. Zero-dimensional (0D) and one-dimensional (1D) nanostructures are effective because of their high surface-to-volume ratios. In contrast, two-dimensional (2D) and three-dimensional (3D) materials usually have lower ratios, which limits their sensing efficiency. Among these, nanowire and nanotube-based designs have shown significant improvements in gas detection applications [32]. Bai et al. found that both pure and rare-earth-element (REE)-doped In_2O_3, with dopants like Ce, Tm, Eu, Er, and Tb, easily formed 1D nanotube structures. Notably, Tb-doped In_2O_3 nanotubes had rougher surfaces than other variants, a feature that improved their gas-sensing abilities [33]. Similarly, Han et al. discovered that cerium doping in In_2O_3 nanospheres created many interparticle voids, leading to a porous structure with a large active surface area. This structure enhanced performance in methanol sensing [18]. Bharathi et al. reported morphological transformation from nanorods to nanoflowers upon incorporating gadolinium at varying concentrations (2, 4, 6, and 8%) into the ZnO matrix. This incorporation reduced nanorod dimensions while increasing the surface-to-volume ratio. These surfaces facilitate chemical reactions between target gases and zinc oxide, improving gas sensing capability. Among the tested concentrations, 4% Gd-doped ZnO nanoflowers demonstrated superior gas sensing response toward xylene at ambient temperature [34]. Zhu et al. documented that pure NiO structures comprise numerous nanosheets, whereas cerium doping transforms the structure to incorporate nanowires embedded within nanosheets. These nanosheets contain multiple open regions facilitating rapid gas molecule diffusion, resembling cobweb-like formations [35]. Kasirajan et al. observed that pure ZnO exhibits irregularly shaped nanoflower structures that undergo modification upon rare earth element doping. Specifically, neodymium-doped ZnO transformed into nanorod-like structures, enhancing ammonia gas detection capabilities [36].

4 Gas Sensing Mechanism

The sensor works on the principle of reaction of gas molecules with surface species of metal oxide, which further changes the energy levels [37]. The surface of the SMO absorbs the anions even in the presence of air, removing electrons and creating a depletion layer surrounding the grains. The Debye layer or space charge layer is another name for the depletion layer. Back-to-back space charge layers occur at grain borders, forming a potential barrier for electrons crossing the grains. The resistance of n-type metal oxides is increased as a result of this. Equations (6.1) and (6.2) depicts the temperature-dependent oxygen reactions with metal oxide [38].

Reaction with Oxygen

for 150–300 °C

$$MO + O + e \rightarrow MO - O^- \tag{6.1}$$

for 30–150 °C

$$MO + 2O + e \rightarrow MO - O_2^- \tag{6.2}$$

As a response of the foregoing reactions, electrons are extracted, and the resistance of the sensor material increases. The electrons get trapped by the grains of SMO, when anions are removed entirely in the case of a reducing gas purging. It minimizes the resistance of the sensing element by reducing the depletion layer.

Reaction with reducing gas as in Eq. (6.3)

$$R + MO - O \rightarrow MO + RO + e^- \tag{6.3}$$

Reaction with oxidizing gas as in Eq. (6.4)

$$MO - O^- + O + e^- \rightarrow MO - O^- \tag{6.4}$$

In case of oxidizing gas purging over SMO, the electrons are ejected out of the SMO thereby resulting in enhancement of the depletion band, and as a result, the resistance of metal oxides increases. Comparable adsorption of oxygen onto the surfaces of semiconducting metal oxides has been seen for both n- and p-types [39]. The conduction process, on the other hand, is entirely different. Ionosorbed oxygen depletes the surface of carriers in n-type metal oxides by removing electrons, making the surface area more resistant. The ionosorbed oxygen produces an accumulating layer (holes) in case of p-type SMO, making the surface more conducting. Therefore, for the identical target gas, conductivity behaves differently in n-type and p-type metal oxides. Metal oxide gas sensing method, which has been briefly described above, doesn't really tell the entire picture. In fact, the modulating depletion layer impacts both the quantity and mobility of charge carriers [40]. The question arises whether the sensing is being caused by anion vacancies or by incorporation of oxygen atoms. This is also a hot topic of debate at present [41].

4.1 Aspects of Gas Sensor

The fundamental problem for gas sensors is to selectively identify hazardous and flammable molecules at low concentrations. As a result, there is an increasing demand of sensing devices for gases with desirable characteristics including selectivity, stability, robustness, and sensitivity. As a result, understanding the foundations of gas sensor operation is necessary to increase gas sensing performance. Many aspects influence gas sensing qualities, including reaction kinetics, analyte diffusivity on the sensor element's surfaces, sensor morphology, and so on. A gas sensor's basic operation is depicted by three components in the gas sensing process which are receptor function, transducer function, and utility. As a result, customizing each of these factors alone or simultaneously can change or modify the properties of gas sensors.

4.1.1 Receptor Function

The receptor function is defined by the interactions of gas molecules with the sensor surface. The receptor function transforms chemical information into a form of energy that transducers can measure [42]. Adsorption, ion exchange, and electrochemical reactions are all examples of chemical interactions between a sensor's surfaces and an analyte molecule. When oxygen adsorbs on the sensor's surface in the presence of air, it induces anion vacancies. Due to this a depletion region grows at the surface of the grain which results in increment in the work function [43]. With the appearance of electrons in the conduction band, oxygen species reacts with the reducing gas and results in decrement of the depletion region around the sample surface and lowers down the work function. Due to the oxidizing gas's extraction of electrons from the sensor's conduction band, the width of the depletion region increases. The conductance response of the sensor is thought to be caused by variation in the height of the potential barrier [44]. The intrinsic electrical characteristics of metal oxides, as well as deviations from chemical stoichiometry, have a significant impact on the corresponding receptor function. Metal oxides have an inherent flaw called oxygen vacancy. Oxygen anion vacancies induce a space charge layer. Mobility of the primary charge carrier can be determined by the variation in conduction behavior. Metal oxides with a stoichiometric composition have a lower conductivity. Doping or creating defects can improve the conductivity of metal oxides. As a result, doping or the addition of chemicals to a foreign receptor can alter its function. Pd on ZnO, for example, causes more change in the work function because Pd is more reactive. When compared to pure ZnO, Pd doped ZnO has a better sensitivity.

4.1.2 Transducer Function

The transduction function is concerned with the efficient transfer of molecular information into a signal that can be read for example resistance. It is generally referred to the motion of electrons among grain boundaries. The transducer function addresses the variation in resistance with change in work function. The phenomenon is well understood by exploring the effect of changes in resistance and in other words change in work function of the ZnO, which is well and extensively studies gas sensors. By extracting electrons from the crystal or when the ZnO crystal loses electrons, it attains negatively charged anion species at the surface which further disrupt the lattice of ZnO. Because of dangle bonds or unsaturated sites, new energy levels emerge. The following Eq. (6.5) represents induction of negatively charged anion species:

$$\frac{\beta}{2}\mathrm{O}_2(g) + \alpha \mathrm{e}^- \rightarrow O_\beta^{-\alpha}(s) \tag{6.5}$$

where, the value of $\beta = 1$ for mono or single atom and $\beta = 2$ for polyatomicity, whereas the value α is equal to 1or 2, depending upon the oxidation state of oxygen atom. On the surface of ZnO, s stands for site. The creation of an electric field at the surface is caused by the formation of an electron depletion layer on the surface caused by adsorbed charged oxygen species. The energy bands shift upward for anions. In other terms, the entrapment of electrons is caused by induction of depletion regions causing the sensor's resistance to rise. When a depletion surface is exposed to reducing gases (such as CO), the gas is oxidized, as shown by the reaction in Eq. (6.6) below.

$$\beta CO_{(g)} + O_{\beta}^{-\alpha} \rightarrow CO_{2(g)} + \alpha e^{-} \tag{6.6}$$

As a result of the aforesaid reaction, trapped electrons are released into the sensor's conduction band, lowering the sensor's resistance. An electrical signal is created by interpreting a change in chemical data.

4.2 *Electron Transfer Mechanisms in N-Type Metal Oxide Semiconductor Gas Sensors*

The fundamental principles of gas detection in semiconductor-based sensors rely on redox reactions occurring at material surfaces. Oxidizing gases function as electron acceptors, inducing electron loss in surrounding molecules while themselves undergoing reduction. Conversely, reducing agents participate in reactions by surrendering electrons, thereby experiencing oxidation. When an N-type semiconductor is exposed to an oxidizing gas atmosphere, a significant electronic phenomenon occurs at the material interface. The surface electron concentration experiences a measurable depletion, resulting in increased electrical resistance throughout the semiconductor material. This resistance change forms the basis for detection mechanisms in many gas sensing applications. The interaction dynamics change substantially when reducing gases contact the metal oxide semiconductor surface. In this scenario, oxygen ions previously adsorbed onto the semiconductor surface engage in chemical reactions with the incoming reducing gas molecules. This interaction triggers the release of captured electrons back into the semiconductor's conduction band. The consequence of this electron replenishment is a characteristic decrease in the electrical resistance of the N-type semiconductor material. The incorporation of rare earth elements (REEs) as dopants introduces a notable modification to the electronic properties of N-type metal oxide semiconductors. Following the REE doping, these materials typically exhibit increased surface resistance when measured in ambient air conditions, as illustrated in Fig. 6.1. This resistance alteration stems from changes in the material's electronic structure and surface reactivity, which significantly influences gas detection performance parameters including sensitivity, selectivity, and response dynamics.

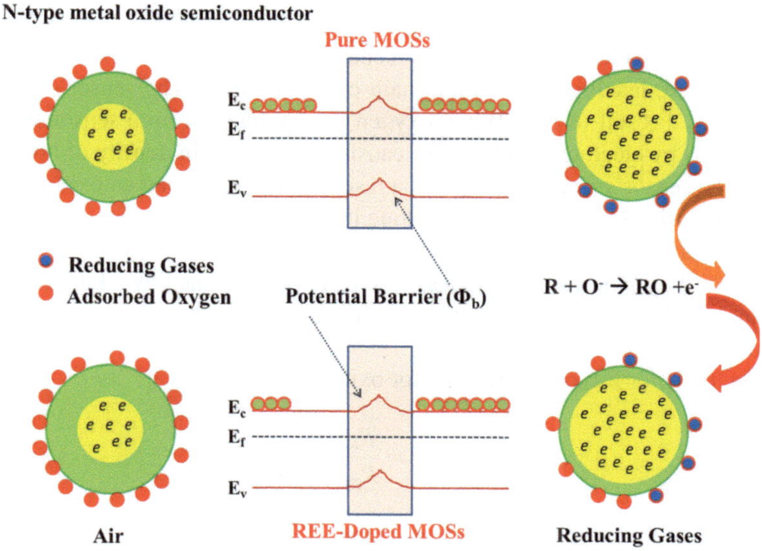

Fig. 6.1 Effect of rare earth element (REE) doping on surface resistance of N-type MOS under ambient conditions

5 Determinants of Gas Sensing Performance

Different factors, such as to be sensitive, selective, responsive, power consumption, reversibility, adsorption ability, and fabrication cost, are used to evaluate the gas sensor's effectiveness. A gas sensor's important features [45] are as follows.

5.1 *Sensitivity*

The ratio of resistance change caused by adding target gas into the air is characterized as the sensor response, response magnitude, or sensitivity. It refers to the ability to detect an analyte gas at a very low concentration or volume. The resistance of the chemoresistive sensor changes upon purging of the gas. The interplay of analyte gas with the sensor surface forces the electronic state of the sensor to change, as described in the receptor and transducer section. The n-type semiconductor metal oxides and the p-type semiconductor metal oxides are the two types of semiconductor metal oxides. Electrons are the primary charge carriers in an n-type semiconductor. When oxygen molecules adsorb on the sensor's surface, they absorb electrons and produce a depletion layer, which reduces the density of electron charge carriers. When n-type materials are exposed to a target gas (such as CO), it eliminates adsorbed oxygen and creates CO_2, which releases an electron back to the sensor. As a result, there is an increase in the density of electrons in the sensor, and simultaneously the resistance of the sensor starts reducing. The majority of charge carriers in p-type semiconductor oxides are holes. In this scenario, oxygen adsorption causes electrons to be removed, leaving holes as the main charge carriers. The sensor's resistance decreases as a result of this. When oxygen is eliminated from a sensor in case of reducing gas by ejecting the electrons, causing a decrease in density of holes in the sensor. The sensor's resistance rises as a result of all this. When an oxidizing gas is present, opposing reactions occur, resulting in an opposite trend in resistance. Porous structure, optimum operating temperature, dopants, thickness and size of crystal all affect sensitivity. The gas sensing response can be determined by taking the ratio of resistance of the metal oxide semiconducting gas sensor in air with respect to gas. The "*S*" gas reaction is written as.

In Eq. (6.7) for n-type semiconductor oxide

$$S = \frac{R_{\text{air}}}{R_{\text{gas}}}. \tag{6.7}$$

In Eq. (6.8) for p-type semiconductor oxide

$$S = \frac{R_{\text{gas}}}{R_{\text{air}}} \tag{6.8}$$

In some reports, it's also referred as in Eq. (6.9)

$$S = \frac{R_{\text{air}} - R_{\text{gas}}}{R_{\text{air}}} \tag{6.9}$$

The sensitivity of the sensor can be expressed as in Eq. (6.10)

$$\text{Sensitivity} = \frac{\text{Gas Response(s)}}{\text{Concentration of Analyte}} \tag{6.10}$$

5.2 *Selectivity*

Selectivity denotes to the sensor's ability to respond to a single analyte gas or a collection of them. At a particular operating environment, it should ideally have particularly elevated sensitivity of the analyte in the existence of inferring gases.

The term "selectivity" is stated as in Eq. (6.11),

$$\text{Sensitivity} = \frac{\text{Sensitivity of sensor for interferring gas}}{\text{Sensitivity of sensor for desired analyte gas}}. \tag{6.11}$$

Multiple strategies can be used to adjust or optimize selectivity in semiconductor oxides. These are divided into four categories, as listed below [46].

(a) Assessment in sensing characteristics including operating temperature, response time, type and concentration of the analyte.
(b) Differentiation of gases using chromatographic columns and filters depending on molecular size and other physical parameters.
(c) Fabrication process utilized for sensors
(d) Intrusions such as amplifiers, catalysts, and specialized surface dopants

According to Mizsei et al. [47], the physiochemical characteristics of semiconductors are influenced by temperature, and then modify the sensor's gas sensor response. The adsorption and desorption rate, Debye length, and carrier concentration are all influenced by the semiconductor's temperature. It is difficult to calibrate both the changes in sensor resistance in case of gas and along with the sensing temperature [48]. The developer of Figaro gas sensors claims that using a channel flow for gas sensors can affect sensor performance by lowering the sensor's temperature.

Besides that, temperature modification of sensors has several advantages, including:

(a) Cyclic variation in temperature can provide a particular response for each gas,
(b) Periodically shifting the sensor to a higher temperature can start cleaning up the sensor's surface,
(c) The variation in grain size and grain boundary due to thermal effect, can improve the selectivity of the gas, resulting in a maximum response for a given gas [49].

The key issue is the sensor's nonlinear gas response, which must be resolved. However, neural networks and soft-modeling techniques are being used to recognize the gas sensing patterns to solve this problem. As indicated earlier, a gas filter on the sensor's surface can be employed to enhance the sensor's selectivity. Coating SnO_2 film on the surface of a gas sensor via CVD of silicones improves sensitivity and selectivity toward hydrogen [50].

Dopants/additives can also be used to optimize or modify the acid–base characteristics of metal oxide. As a result, depending on the acidic or basic qualities of the analyte to be monitored, the acid–base characteristics of the sensor can indeed be adjusted. For instance, ethanol sensitivity improves in basic oxides but decreases in acidic oxides. Ethanol is catalytically oxidized by acidic or basic oxides in a number of ways. Selectivity of gas sensors can be enhanced by changing operating temperature, filter for selection of gas, and the use of dopants are all common ways to improve the selectivity of gas sensors.

5.3 Stability

The gas sensor must produce consistent and repeatable results throughout time, adhering to the norms established by the gas sensing industry. There should be no time-dependent drift in the gas sensor. Gas sensor stability is influenced by a number of things. Surface contaminants, thermal expansion effects, design flaws, structural alterations, phase shifts, toxicity from chemical interactions, and environmental changes are all possibilities. To address the aforementioned issues, the following procedures should be followed: the use of chemical and thermally stable components, particle size and composition optimization, and surface pre-treatment utilizing specialist techniques.

5.4 *Response and Recovery Time*

Response time is the time it takes for a sensor to achieve 90% of total signal response (resistance) after exposure to the target gas. A sensor's recovery time is when it takes to recover to 90% of its initial baseline resistance after the target gas has been removed. These variables are influenced by the material's properties such as doping atoms, crystal size, electrode material, and rate of diffusion etc.

5.5 *Operating Temperature*

The operational temperature of the sensor is the lowest temperature at which it responds to its maximum limit. The sensor's operating temperature is determined by the environmental target gas as well as the qualities of the chosen sensor material.

6 Gas Sensing Technologies

Almost all of the research in gas sensing technologies over the last fifty years has focused on following parameters. Different types of technologies used along with their Gas sensing principle is mentioned in Table 6.1.

(a) Developing and studying various types of sensors.
(b) Research on sensing mechanisms.
(c) Exploring different techniques for sensor fabrication.

Several new sensors as well as sensing methods have already been developed as a result of all of these years of scientific research. The important phenomenon of gas sensing is that the characteristics of the sensor devices change when they are exposed to a gaseous environment. The degree of the changes in attribute is proportional to the gas volumes. For gas sensors, variations in the following attributes acoustic, electrical, gas chromatography, optical are primarily used. But these sensing techniques have several fundamental limitations such as poor selectivity, poor responsiveness,

Table 6.1 Different Technology and Gas sensing Principle

S. No.	Gas sensing principle	Technology
1	Variation in frequency	Cantilever
2	Variation in capacitance	Chemo-capacitive
3	Variation in resistance	Chemo-resistive
4	Variation in emf/current	Electro-chemical
5	Variation in voltage	Field effect transistor
6	Separation of chemical components	Gas chromatography
7	Induces molecular vibration	IR spectroscopy
8	Variation in molecular weight	Mass spectroscopy
9	Variation in frequency	Piezoelectric
10	Variation in intensity of light/refractive index	Optical
11	Variation in resistance	Semiconductor
12	Variation in frequency	Surface acoustic wave
13	Oscillations of electrons	Surface plasmon resonance

difficulty miniaturization and bulky and expensive equipment [51]. The detecting technique is based on the change in sensor conductivity caused by gas molecules adhering to its active surface. Moreover, because of their high operating temperatures, the device's power consumption is very high, and selectivity is also a concern. In order to quantify and monitor the gases or vapors different kinds of techniques and approaches have been implemented as follows.

6.1 Gas Chromatography

Gas chromatography (GC) is an analytic method used to separate chemical components in a mixture and then detecting them to identify their occurrence, as well as how much of each component is there. Organic compounds or gases are the most commonly used chemical constituents. The Gas Chromatography (GC) approach, which was first developed in the early 1950s which is used to separate and analyze a mixture of chemicals, is among the most widely used techniques. Snow et al. [52] and Adahchour et al. [53] reported that this approach is now widely employed in applications like drug identification and assessment, polymer characterization, and gas/vapor sensing.

Gas chromatography is a type of laboratory analytical technology with high sensitivity, selectivity, and performance. These processes are expensive and require a large setup, which necessitates the use of experienced staff. The working phenomenon of gas chromatography says that whenever a mixture of chemicals is processed through a narrow tube (commonly termed as a column) at extremely high temperatures with a coat of a suitable material (called as the stationary phase); the chemicals present

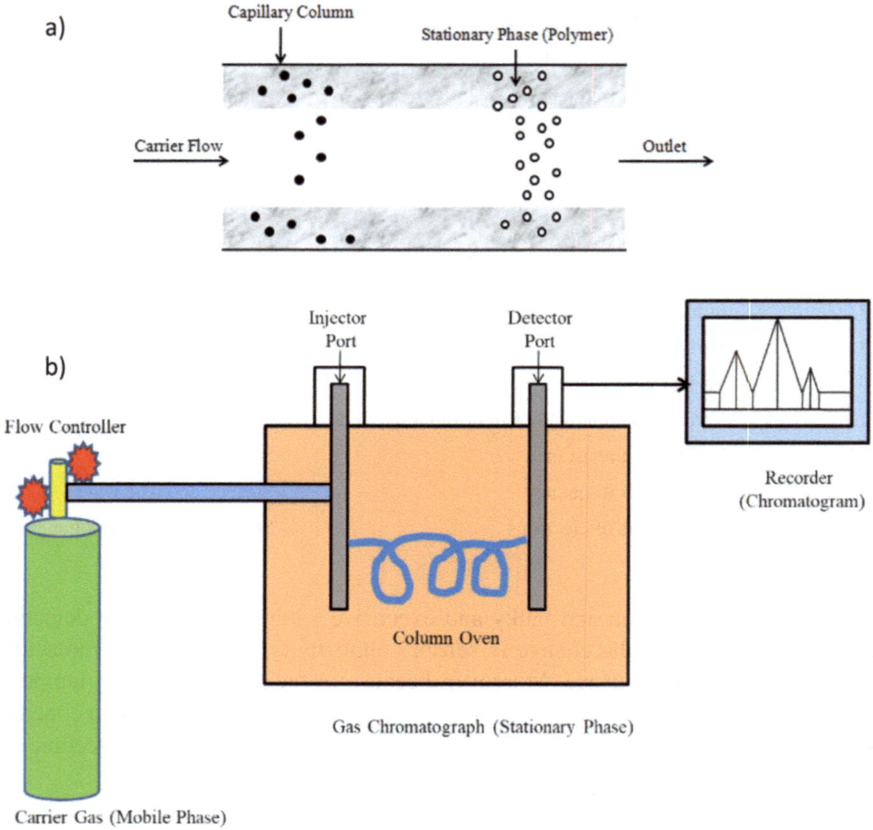

Fig. 6.2 **a** Retention of gas molecules in column, **b** schematic for sensing via gas chromatography

in the mixture are separated and drained at different time periods [54]. As illustrated in Fig. 6.2a the elution duration of various chemicals varies based on their affinities for specific gases/vapors. The output detector detects the separation of chemicals. Figure 6.2b depicts the GC schematic. The columns are primarily made of fused silica, and the substances used to separate compounds inside the columns are mostly inorganic salts and other minerals. This technology has a number of drawbacks, including its bulkiness, high energy consumption, and other transportation concerns.

6.2 *Electrochemical Sensor*

The working of electrodes in electrochemical sensors is usually based on the concept of change in electrochemical properties (such as reactive sites, conductivity/resistivity, and so on) during the interactions with gas/vapor molecules [55]. A three-electrode setup is used generally, including a counter electrode, a reference

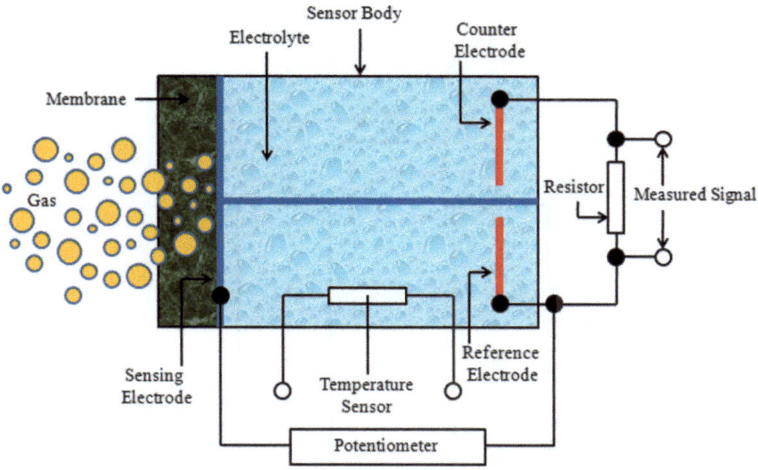

Fig. 6.3 Schematic for electrochemical sensing

electrode, and a measuring electrode. A hydrophobic permeable membrane diffuses the desired gas into the electrolyte enclosure. These sensors are used to measure the partial pressure of oxygen molecules in the ambient air, as well as the concentration of harmful toxins. With assistance of a potentiometer attached to the sensor, a consistent voltage is provided between both the reference electrode and the working electrode. For the electrochemical sensor, a schematic diagram is represented in Fig. 6.3.

The characteristics of the target gas (oxidizing/reducing) to be detected at the measurement electrode used to determine the electrode material, electrolyte, and operating potential. The current is sent through the counter electrode and allowed to pass across the sensor cell. The atmospheric oxygen would either be reduced or oxidized over the counter electrode, depending on the severity of the target gas (oxidizing/reducing), thus the resultant current that flows through the sensor is evaluated, which is directly proportional to the amount of target gas molecules. One of the most significant disadvantages of electrochemical sensors seems to be the instability of the sensor due to swelled electrolyte when humidity is encountered [56].

6.3 Infrared (IR) Spectroscopy

Infrared (IR) spectroscopy is another advanced technique utilized for the investigation and monitoring of gases. The IR waves inside the reference cell pass directly through the IR filters, whereas the IR waves in the sample cell are absorbed as they pass through the gas molecules inside and subsequently reach the IR filters. Figure 6.4 shows the schematic for the IR sensor. Every compound has a distinct characteristic spectrum with absorption bands that occur at specific frequencies, indicating the existence of gas molecules. Dobust et al. [57], presented NH_3 gas IR spectra and

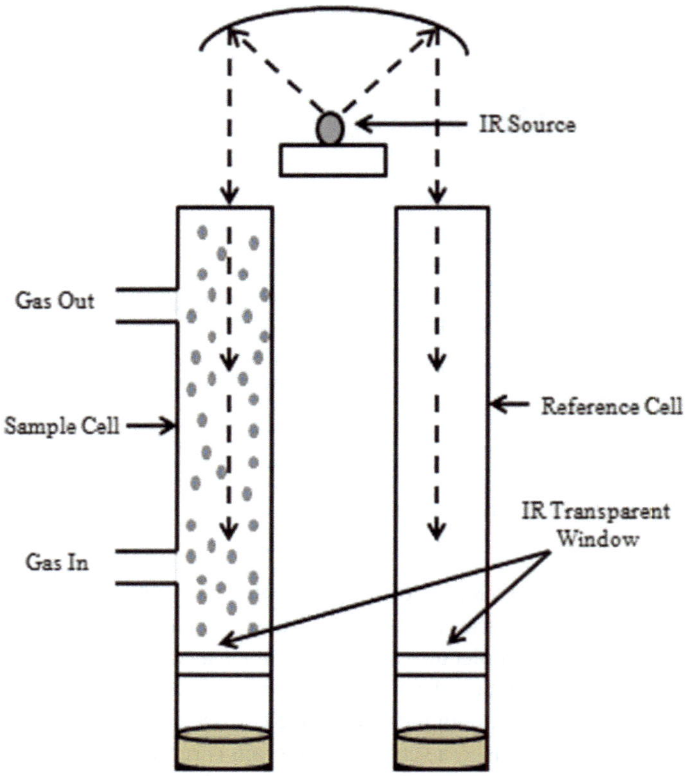

Fig. 6.4 Schematic for IR sensing

depicted the NH_3 characteristic bonds. However, the existing technique has a main drawback is its inability to monitor all gases, such as hydrogen, and the results might readily shift when different humidity levels are present.

6.4 Semiconductor Sensor

The semiconductor gas sensors are a type of sensor that works on the chemoresistive principle, in which the semiconducting sensing element's conductivity or resistance changes after it interacts with the target gas. According to Yamazoe et al. [58], any thin/thick film, pellet, or other capable nanostructured architecture can be used as a semiconducting sensing element. The surface potential of a semiconducting metal oxide changes when it is exposed to either an oxidizing or reducing gas due to gas molecules adsorbed on the surface. As a result, the conductivity of the sensing element changed, which was easily detectable by electronic instruments. These sensors are cost effective and simple for fabrication, and their robustness made them to be widely

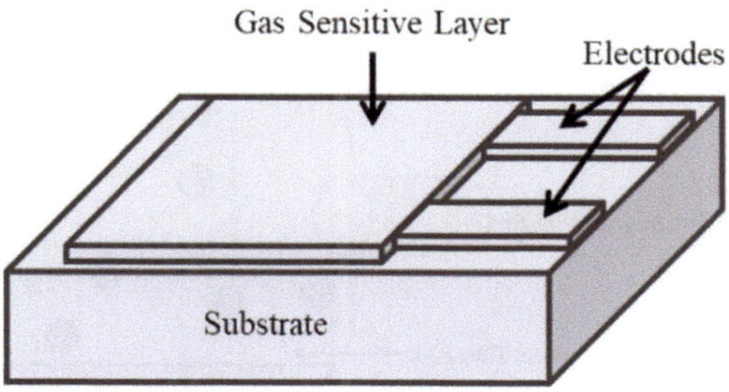

Fig. 6.5 Schematic for semiconductor-based sensing

used. The fundamental architecture of a bulk semiconductor gas/vapor sensor, which includes electrodes on both sides for monitoring the sensor's real-time resistance values. The same semiconducting materials must be heated, either by an external heater or an in-built microheater, in order to achieve optimal sensing at the right temperature for the sensor. By dipping a metal coil in a paste of the sensing material created with a colloidal solution of metal hydroxides as predecessors, a bulk semiconductor sensor can be made. The semiconductor thin film based gas sensor is deposited with semiconducting thin films on a substrate appropriately chosen for the sensor as base with pre-fabricated comb-like planar electrodes known as Inter-Digitated Electrodes/Transducers (IDE/ IDT's) with the under placed heater. Figure 6.5 shows schematic for Semiconductor based sensor and Thin Film. The main disadvantage of semiconducting metal oxide gas/vapor sensors is that they demand more power due to their higher operating temperatures. Efforts are still being made around the world to create either lower operating temperatures for manufactured gas sensors or a different technique for efficient gas sensors.

6.5 *Surface Plasmon Resonance*

Surface Plasmon Resonance (SPR) is the group oscillations of electrons in the conduction band in resonance with the incident light's oscillating electric field, which produces intense plasmon electrons by non-radiative excitation. Paliwal et al. [59], discuss Surface Plasmon Resonance (SPR) as a sensitive approach for gas/vapor and chemical sensing. Surface Plasmon's are charge density oscillations that can occur at the interface of two dielectric mediums with opposite charges. The SPs can be easily excited using the prism coupling approach with the Kretschmann configuration, which involves depositing a metal sheet of appropriate thickness on the face of the prism. When a monochromatic beam of light is incident on the prism surface, SPs mode activates, resulting in a resonance at a specific wavelength and angle of incidence, which is termed as SPR. SPR sensor concept in which gas is permitted to interact with the detecting (dielectric material) layer, and the change in reflected intensity is monitored by change in concentration of gas as shown in Fig. 6.6.

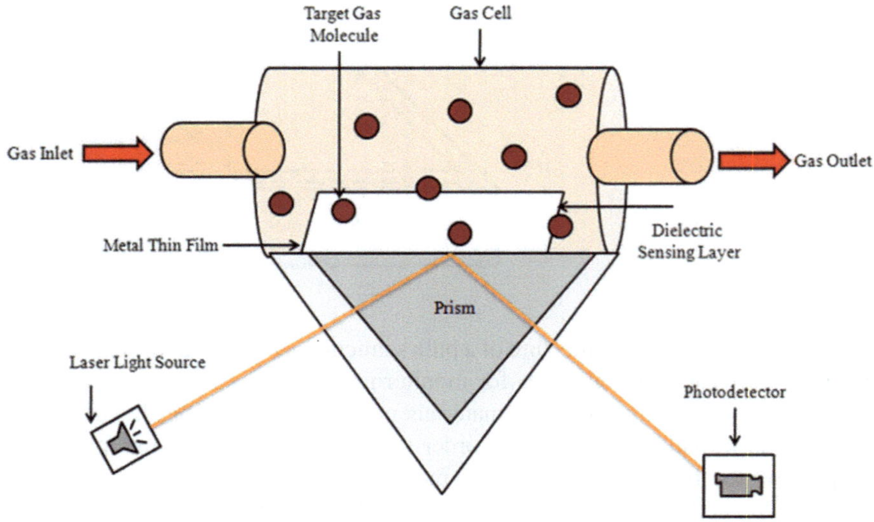

Fig. 6.6 Schematic for SPR-based gas sensor

The SPR mode changes at the interface due to a change in dielectric characteristics caused by the target gas interacting with the sensor layer. As a result, the change in SPR parameters of the sensing layer is proportional to gas concentrations, allowing the calibration curve to be used for finding the target gas concentration. The ability to work at room temperature is an advantage of this sensor, but there are a number of disadvantages to SPR-based sensors, including a complex and bulky experimental setup [60].

7 Rare Earth Element-Doped MOS for Gas Sensing Employments

Gas detection in semiconductor-based sensors operates on redox principles. When oxidizing gases interact with N-type semiconductors, they extract electrons from the material surface, increasing resistance. Conversely, when reducing gases encounter these semiconductors, by reacting with oxygen ions that have already been adsorbed, they lower the resistance and return electrons to the conduction band.

Rare earth element (REE) doping significantly modifies these electronic properties. After incorporation of REEs, N-type metal oxide semiconductors typically exhibit increased surface resistance in ambient conditions, altering their sensing characteristics.

7.1 Acetone Detection

Zhang et al., demonstrates In_2O_3 doped with various REEs (La, Pr, Sm, Ce, Yb) synthesized via solvothermal methods shows enhanced acetone detection. Yb-doping (6%) transforms compact nanosheets into porous structures (190 nm width, 23 nm thickness) with sensitivity of 15.3 toward 50 ppm acetone at 220 °C, exhibiting rapid response, recovery dynamics, and stability [20]. Research reflects that flower-like microstructures, synthesized hydrothermally, via In_2O_3 doped with Ce achieved sensitivity of 41.8 for acetone with a concentration of 200 ppm at 250 °C with 2 s of response time, 154 s of recovery time, and month-long stability Wei et al. [63]. $CoFe_{2-x}Ce_xO_4$ materials prepared by molten-salt methodology showed optimal performance at 4% Ce-doping with significantly improved sensitivity (177%) toward acetone at reduced operating temperature (225 °C) compared to undoped variants [64].

7.2 Ethanol Detection

CeO_2-doped ZnO nanostructures developed via assistance of microwave methods demonstrated enhanced sensing response toward ethanol, CO, and CH_4 as shown by Hamedani et al. At 5% ceria doping, recovery times decreased substantially across 200–400 °C, with sensitivity of 145 toward 500 ppm ethanol at 350 °C [65]. Rapid solution combustion technique has been implemented by Bhatia et al., who found that doping done by Dy to ZnO nanoparticles (4%) formed leaf-like structures (47–77 nm grain size), demonstrating 20–27% sensitivity toward ethanol even at low concentrations (25 ppm) at 150 °C of optimal temperature [12]. REE-doped ZnO nanowires synthesized solvothermally showed diameter increases after doping. At 300 °C operating temperature, ZnO doped with 1% of Ce reflected maximum sensitivity (25.3) toward 100 ppm ethanol at 25–30% relative humidity, with performance hierarchy (Ce > Eu > Er) corresponding to oxide basicity as per research given by Zhao et al. [11]. Gao et al. showed that nanorods of ZnO doped with Nd (2%) developed via solid-state reaction executed at low-heat demonstrated improved ethanol sensitivity (37.8 versus 20.3 for undoped) at 348 °C with response/recovery times of 5/10 s [27]. According to Li et al., hollow nanofibers of SnO_2 doped with Pr showed optimal performance at 0.6% doping with sensitivity of 64.33 toward 200 ppm ethanol at 300 °C, increasing to 212.7 at 500 ppm, attributed to increased oxygen vacancies and enhanced surface area [66]. Wang et al., found Yb-doped SnO_2 hollow nanofibers (1.0 wt%) showed excellent ethanol sensitivity (170) at 500 ppm with fast response-recovery (7/8 s) and superior selectivity against potential interferents [67]. Zhu et al. reported Ce-doped NiO microflowers demonstrated increased ethanol sensitivity (47.6 vs. 20.1 for undoped) at 300 °C due to enhanced active sites [36].

7.3 Other Gases

Guo et al., reported Y-doped ZnO nanospheres (4%) synthesized by template-free hydrothermal methods demonstrated improved formaldehyde detection (65.7 versus 47.4 for undoped) at reduced operating temperature (300 °C vs. 350 °C) with rapid response/recovery (4/6 s) [68]. According to Zhao et al., Y-doped In_2O_3 (5%) developed by electrospinning achieved formaldehyde sensitivity of 91.7 ± 7.9 at 100 ppm and 120 °C with extremely fast response (1 s), detection limit of 50 ppb, and 30-day stability [69]. Han et al., reported Ce-doped MoO_3 nanobelts demonstrated enhanced trimethylamine sensitivity (17.4 versus 4.6 for undoped) at reduced operating temperature (240 °C vs. 280 °C) with response/recovery under 10/20 s [18]. According to Li et al., Ce-doped NiO (0.5%) synthesized by sol–gel methods achieved 29% sensitivity toward 40 ppm NO_2 at 150 °C with selectivity against acetone, ammonia, and liquefied petroleum gas [22]. As per the research done by Mokoena et al., Eu^{3+} doped NiO (3%) demonstrated optimal toluene sensitivity (25% at 60 ppm, 100 °C) with 20 s of response time and 46 s of recovery time, increasing to 59% with extended exposure and higher bias voltage [70]. Research reported that ZnO thin films doped with Nd (3%) synthesized by SILAR method showed enhanced ammonia sensitivity with 43 s of response time, 12 s of recovery time, and 60-day stability at optimal room temperature Kasirajan et al. [37].

8 Conclusions and Future Perspectives

The past two decades have witnessed remarkable advancements in rare earth element-modified metal oxide semiconductor platforms for gas detection technologies. These materials have demonstrated exceptional capabilities in identifying environmentally and industrially significant gases with outstanding sensitivity, selectivity, and stability across diverse temperature ranges. Architectural considerations significantly influence performance characteristics, with zero-dimensional structures facilitating rapid response kinetics through enhanced surface interactions, while one-dimensional configurations improve sensitivity via optimized charge transport pathways. Several innovative approaches show considerable promise for future development, including heterojunction engineering, multi-component systems, and carbon-based GO, CNTs, and MWCNTs combined to rare earth elements which enable ambient temperature operation while maintaining high detection sensitivity. The integration of rare earth elements within polymer matrices offers additional advantages in temporal response and selectivity enhancement. Gas detection involves complex interdependent parameters, with rare earth element modifications providing superior performance through inherent chemical reactivity and surface basicity characteristics. Analysis reveals comparatively limited research in this field relative to noble metal alternatives, presenting significant opportunities for further investigation into these promising materials that form the foundation for next-generation environmental monitoring technologies, industrial safety systems, and advanced chemical detection architectures.

References

1. Anukunprasert, T., Saiwan, C., Traversa, E.: The development of gas sensor for carbon monoxide monitoring using nanostructure of Nb–TiO_2. Sci. Technol. Adv. Mater. **6**(3–4), 359–363 (2005)
2. Sberveglieri, G. (Ed.): Gas Sensors: Principles, Operation and Developments. Springer Science & Business Media (2012)
3. Brattain, W.H., Garrett, C.G.B.: Bell telephone system tech. Publs. Monogr. **2372**(1) (1955)
4. Liu, X., Cheng, S., Liu, H., Hu, S., Zhang, D., Ning, H.: A survey on gas sensing technology. Sensors **12**(7), 9635–9665 (2012). https://doi.org/10.3390/s120709635
5. Wagner, C., Hauffe, K.: Untersuchungen über den stationären Zustand von Katalysatoren bei homogenen Reaktionen I. Zeitschrift für Elektrochemie und Angewandte Physikalische Chemie **44**(3), 172–178 (1938). https://doi.org/10.1002/bbpc.19380440303
6. Seiyama, T., Kato, A., Fujiishi, K., Nagatani, M.: A new detector for gaseous components using semiconductive thin films. Anal. Chem. **34**(11), 1502–1503 (1962)
7. Zhao, S., Shen, Y., Zhou, P., Zhong, X., Han, C., Zhao, Q., Wei, D.: Design of Au@WO_3 core−shell structured nanospheres for ppb-level NO_2 sensing. Sens. Actuators, B Chem. **282**, 917–926 (2019). https://doi.org/10.1016/j.snb.2018.11.142
8. Saruhan, B., Lontio Fomekong, R., Nahirniak, S.: Influences of semiconductor metal oxide properties on gas sensing characteristics. Front. Sens. **2**, 657931 (2021)
9. Degler, D., Weimar, U., Barsan, N.: Current understanding of the fundamental mechanisms of doped and loaded semiconducting metal-oxide-based gas sensing materials. ACS Sens. **4**(9), 2228–2249 (2019)
10. Chen, X., Shen, Y., Zhou, P., Zhao, S., Zhong, X., Li, T., ... & Meng, D. (2019). NO_2 sensing properties of one-pot-synthesized ZnO nanowires with Pd functionalization. Sens. Actuators B: Chem. **280**, 151–161
11. Zhao, S., Shen, Y., Li, A., Chen, Y., Gao, S., Liu, W., Wei, D.: Effects of rare earth elements doping on gas sensing properties of ZnO nanowires. Ceram. Int. **47**(17), 24218–24226 (2021)
12. Bhatia, S., Verma, N., Kumar, R.: Morphologically-dependent photocatalytic and gas sensing application of Dy-doped ZnO nanoparticles. J. Alloy. Compd. **726**, 1274–1285 (2017)
13. Logamani, P., Rajeswari, R., Poongodi, G.: Enhanced photocatalytic activity of rare earth metal (Nd and Gd) doped ZnO nanostructures. Mech. Mater. Sci. Eng. J. **9**(2)
14. Ge, L., Huang, Q., Zhang, Y., Shen, Z.: Ring-opening polymerization of styrene oxide with rare earth coordination catalysts. Eur. Polymer J. **36**(12), 2699–2705 (2000)
15. Ge, C., Xie, C., Cai, S.: Preparation and gas-sensing properties of Ce-doped ZnO thin-film sensors by dip-coating. Mater. Sci. Eng. B **137**(1–3), 53–58 (2007)
16. Han, D., Yang, J., Gu, F., Wang, Z.: Effects of rare earth element doping on the ethanol gas-sensing performance of three-dimensionally ordered macroporous In_2O_3. RSC Adv. **6**(51), 45085–45092 (2016)
17. Du, N., Zhang, H., Chen, B.D., Ma, X.Y., Liu, Z.H., Wu, J.B., Yang, D.R.: Porous indium oxide nanotubes: layer-by-layer assembly on carbon-nanotube templates and application for room-temperature NH_3 gas sensors. Adv. Mater. **19**(12), 1641–1645 (2007)
18. Han, D., Song, P., Zhang, S., Zhang, H., Xu, Q., Wang, Q.: Enhanced methanol gas-sensing performance of Ce-doped In2O3 porous nanospheres prepared by hydrothermal method. Sens. Actuators, B Chem. **216**, 488–496 (2015)
19. Anand, K., Kaur, J., Singh, R.C., Thangaraj, R.: Effect of terbium doping on structural, optical and gas sensing properties of In_2O_3 nanoparticles. Mater. Sci. Semicond. Process. **39**, 476–483 (2015)

20. Liu, D., Lei, W., Qin, S., Hou, L., Liu, Z., Cui, Q., Chen, Y.: Large-scale synthesis of hexagonal corundum-type In_2O_3 by ball milling with enhanced lithium storage capabilities. J. Mater. Chem. A **1**(17), 5274–5278 (2013)
21. Zhang, W.H., Ding, S.J., Zhang, Q.S., et al.: Rare earth element-doped porous In_2O_3 nanosheets for enhanced gas-sensing performance. Rare Met. **40**, 1662–1668 (2021)
22. Yang, H., Zhang, X., Li, J., Li, W., Xi, G., Yan, Y., Bai, H.: Synthesis of mesostructured indium oxide doped with rare earth metals for gas detection. Microporous Mesoporous Mater. **200**, 140–144 (2014). https://doi.org/10.1016/j.micromeso.2014.08.021
23. Li, Z., Wang, W., Zhao, Z., Liu, X., Song, P.: One-step hydrothermal preparation of Ce-doped MoO_3 nanobelts with enhanced gas sensing properties. RSC Adv. **7**(45), 28366–28372 (2017)
24. Majhi, S.M., Mirzaei, A., Kim, H.W., Kim, S.S., Kim, T.W.: Recent advances in energy-saving chemiresistive gas sensors: a review. Nano Energy **79**, 105369 (2020). https://doi.org/10.1016/j.nanoen.2020.105369
25. Yang, F., Guo, J., Zhao, L., Shang, W., Gao, Y., Zhang, S., Gu, G., Zhang, B., Cui, P., Cheng, G., Du, Z.: Tuning oxygen vacancies and improving UV sensing of ZnO nanowire by micro-plasma powered by a triboelectric nanogenerator. Nano Energy **67**, 104210
26. Chu, Y.L., Ji, L.W., Xie, J.H., Chu, T.T.: High-sensitivity of self-powered gas sensors based on piezoelectric nanogenerators with Y-doped 1-D ZnO nanostructures. IEEE Sens. J. (2024)
27. Cao, Y., Pan, W., Zong, Y., Jia, D.: Preparation and gas-sensing properties of pure and Nd-doped ZnO nanorods by low-heating solid-state chemical reaction. Sens. Actuators, B Chem. **138**(2), 480–484 (2009). https://doi.org/10.1016/j.snb.2009.03.015
28. Qi, Q., Zhang, T., Zheng, X., Fan, H., Liu, L., Wang, R., Zeng, Y.: Electrical response of Sm_2O_3-doped SnO_2 to C_2H_2 and effect of humidity interference. Sens. Actuators, B Chem. **134**(1), 36–42 (2008). https://doi.org/10.1016/j.snb.2008.04.011
29. Qin, G., Gao, F., Jiang, Q., Li, Y., Liu, Y., Luo, L., Zhao, H.: Well-aligned Nd-doped SnO_2 nanorod layered arrays: preparation, characterization and enhanced alcohol-gas sensing performance. Phys. Chem. Chem. Phys. **18**(7), 5537–5549 (2016)
30. Sun, Y.F., Liu, S.B., Meng, F.L., Liu, J.Y., Jin, Z., Kong, L.T., Liu, J.H.: Metal oxide nanostructures and their gas sensing properties: a review. Sensors **12**(3), 2610–2631 (2012)
31. Sharma, N., Choudhury, S.P.: Gas sensing using metal oxide semiconductor doped with rare earth elements: a review. Mater. Sci. Eng. B **307**, 117505 (2024)
32. Selvakumar, D., Sonu, K., Ramadoss, G., Sivaramakrishnan, R., Jayavel, R., Eswaramoorthy, M., Venkateswara Rao, K., Pugazhendhi, A.: Heterostructures of polyaniline and Ce–ZnO nanomaterial coated flexible PET thin films for LPG gas sensing at standard environment. Chemosphere **314**, 137492 (2023). https://doi.org/10.1016/j.chemosphere.2022.137492
33. Bai, J., Luo, Y., Chen, C., Deng, Y., Cheng, X., An, B., Wang, Q., Li, J., Zhou, J., Wang, Y., Xie, E.: Functionalization of 1D In_2O_3 nanotubes with abundant oxygen vacancies by rare earth dopant for ultra-high sensitive ethanol detection. Sens. Actuators B Chem. **324**, 128755 (2020). https://doi.org/10.1016/j.snb.2020.128755
34. Bharathi, P., Krishna Mohan, M., Shalini, V., Harish, S., Navaneethan, M., Archana, J., Ganesh Kumar, M., Dhivya, P., Ponnusamy, S., Shimomura, M., Hayakawa, Y.: Growth and influence of Gd doping on ZnO nanostructures for enhanced optical, structural properties and gas sensing applications. Appl. Surf. Sci. **499**, 143857 (2019). https://doi.org/10.1016/j.apsusc.2019.143857
35. Zhu, L., Zeng, W., Yang, J., Li, Y.: Unique hierarchical Ce-doped NiO microflowers with enhanced gas sensing performance. Mater. Lett. **251**, 61–64 (2019). https://doi.org/10.1016/j.matlet.2019.05.055
36. Kasirajan, K., Bruno Chandrasekar, L., Maheswari, S., Karunakaran, M., Shunmuga Sundaram, P.: A comparative study of different rare-earth (Gd, Nd, and Sm) metals doped ZnO thin films and its room temperature ammonia gas sensor activity: Synthesis, characterization, and investigation on the impact of dopant. Opt. Mater. **121**, 111554 (2021). https://doi.org/10.1016/j.optmat.2021.111554
37. Morrison, S.R.: Selectivity in semiconductor gas sensors. Sens. Actuators **12**(4), 425–440 (1987). https://doi.org/10.1016/0250-6874(87)80061-6

38. Barsan, N., Weimar, U.: Conduction model of metal oxide gas sensors. J. Electroceramics **7**(3), 143–167 (2001). https://doi.org/10.1023/A:1014405811371
39. Kim, H.J., Lee, J.H.: Highly sensitive and selective gas sensors using p-type oxide semiconductors: overview. Sens. Actuators B Chem. **192**, 607–627 (2014). https://doi.org/10.1016/j.snb.2013.11.005
40. Ogawa, H., Nishikawa, M., Abe, A.: Hall measurement studies and an electrical conduction model of tin oxide ultrafine particle films. J. Appl. Phys. **53**(6), 4448–4455 (1982). https://doi.org/10.1063/1.331230
41. Schuetzle, D., Hammerle, R.: Fundamentals and applications of chemical sensors. In: ACS Symposium Series, vol. 163 (1986). https://doi.org/10.1016/s0003-2670(00)85041-2
42. International Union of Pure and Applied Chemistry: Commission on General Aspects of Analytical Chemistry. Chem. Sens. **63**(9), 1247–1250 (1991)
43. Szuber, J.: The effect of oxygen adsorption on the electronic properties of the polar GaAs(111) surface after thermal cleaning in ultrahigh vacuum. Thin Solid Films **170**(2), 219–226 (1989). https://doi.org/10.1016/0040-6090(89)90727-X
44. Mizsei, J., Lantto, V.: Simultaneous response of work function and resistivity of some SnO_2-based samples to H_2 and H_2S. Sens. Actuators B. Chem. **4**(1–2), 163–168 (1991). https://doi.org/10.1016/0925-4005(91)80193-N
45. Liu, X., Cheng, S., Liu, H., Hu, S., Zhang, D., Ning, H.: A survey on gas sensing technology. Sensors (Switzerland) **12**(7), 9635–9665 (2012). https://doi.org/10.3390/s120709635
46. Yamazoe, N., Miura, N.: Some Basic Aspects of Semiconductor Gas Sensor*s*. Kodansha Ltd. (1992)
47. Mizsei, J.: How can sensitive and selective semiconductor gas sensors be made? Sens. Actuators B. Chem. **23**(2–3), 173–176 (1995). https://doi.org/10.1016/0925-4005(94)01269-N
48. Sears, W.M., Colbow, K., Consadori, F.: General characteristics of thermally cycled tin oxide gas sensors. Semicond. Sci. Technol. **4**(5), 351–359 (1989). https://doi.org/10.1088/0268-1242/4/5/004
49. Sears, W.M., Colbow, K., Consadori, F.: Algorithms to improve the selectivity of thermally-cycled tin oxide gas sensors. Sens. Actuators **19**(4), 333–349 (1989). https://doi.org/10.1016/0250-6874(89)87084-2
50. Göpel, W., Schierbaum, K.D.: SnO_2 sensors: current status and future prospects. Sens. Actuators B: Chem. **26**(1–3), 1–12 (1995). https://doi.org/10.1016/0925-4005(94)01546-T
51. Ampuero, S., Bosset, J.O.: The electronic nose applied to dairy products: a review. Sens. Actuators B Chem. **94**(1), 1–12 (2003). https://doi.org/10.1016/S0925-4005(03)00321-6
52. Snow, N.H., Slack, G.C.: Head-space analysis in modern gas chromatography. TrAC—Trends Anal. Chem. **21**(9–10), 608–617 (2002). https://doi.org/10.1016/S0165-9936(02)00802-6
53. Adahchour, M., Beens, J., Vreuls, R.J.J., Brinkman, U.A.T.: Recent developments in comprehensive two-dimensional gas chromatography (GC $\times$ GC). II. Modulation and detection. TrAC—Trends Anal. Chem. **25**(6), 540–553 (2006). https://doi.org/10.1016/j.trac.2006.04.004
54. Cortes, H.J., Winniford, B., Luong, J., Pursch, M.: Comprehensive two dimensional gas chromatography review. J. Sep. Sci. **32**(5–6), 883–904 (2009). https://doi.org/10.1002/jssc.200800654
55. Dossi, N., et al.: An electrochemical gas sensor based on paper supported room temperature ionic liquids. Lab Chip **12**(1), 153–158 (2012). https://doi.org/10.1039/c1lc20663j
56. Weinberger, M., Hiner, S.L., Tierney, W.M.: Improving functional status in arthritis: the effect of social support. Soc Sci Med **23**(9), 899–904 (1986). https://doi.org/10.1016/0277-9536(86)90218-2
57. Dubost, H., Abouaf-Marguin, L.: Infrared spectra of carbon monoxide trapped in solid argon. Double-doping experiments with H_2O, NH_3 and N_2. Chem. Phys. Lett. **17**(2), 269–273 (1972). https://doi.org/10.1016/0009-2614(72)87071-4
58. Yamazoe, N.: Toward innovations of gas sensor technology. Sens. Actuators B Chem. **108**(1–2), 2–14 (2005). https://doi.org/10.1016/j.snb.2004.12.075
59. Paliwal, A., Sharma, A., Tomar, M., Gupta, V.: Surface plasmon resonance study on the optical sensing properties of tin oxide (SnO_2) films to NH_3 gas. J. Appl. Phys. **119**(16) (2016). https://doi.org/10.1063/1.4948332

60. Piliarik, M., Vaisocherová, H., Homola, J.: A new surface plasmon resonance sensor for high-throughput screening applications. Biosens. Bioelectron. **20**(10), 2104–2110 (2005). https://doi.org/10.1016/j.bios.2004.09.025
61. Rana, L., Gupta, R., Tomar, M., Gupta, V.: ZnO/ST-Quartz SAW resonator: an efficient NO_2 gas sensor. Sens. Actuators B Chem. **252**(2), 840–845 (2017). https://doi.org/10.1016/j.snb.2017.06.075
62. Jakubik, W.P.: Surface acoustic wave-based gas sensors. Thin Solid Films **520**(3), 986–993 (2011). https://doi.org/10.1016/j.tsf.2011.04.174
63. Wei, D., Huang, Z., Wang, L., Chuai, X., Zhang, S., Lu, G.: Hydrothermal synthesis of Ce-doped hierarchical flower-like In_2O_3 microspheres and their excellent gas-sensing properties. Sens. Actuators B Chem. **255**, 1211–1219 (2018). https://doi.org/10.1016/j.snb.2017.07.162
64. Khandekar, M., Tarwal, N., Mulla, I., Suryavanshi, S.: Nanocrystalline Ce doped $CoFe_2O_4$ as an acetone gas sensor. Ceram. Int. **40**(1), 447–452 (2013). https://doi.org/10.1016/j.ceramint.2013.06.021
65. Zhang, D., Fan, Y., Li, G., Ma, Z., Wang, X., Cheng, Z., Xu, J.: Highly sensitive BTEX sensors based on hexagonal WO_3 nanosheets. Sens. Actuators B Chem. **293**, 23–30 (2019). https://doi.org/10.1016/j.snb.2019.04.110
66. Li, W., Ma, S., Li, Y., Li, X., Wang, C., Yang, X., Cheng, L., Mao, Y., Luo, J., Gengzang, D., Wan, G., Xu, X.: Preparation of Pr-doped SnO_2 hollow nanofibers by electrospinning method and their gas sensing properties. J. Alloy. Compd. **605**, 80–88 (2014). https://doi.org/10.1016/j.jallcom.2014.03.182
67. Wang, T., Ma, S., Cheng, L., Luo, J., Jiang, X., Jin, W.: Preparation of Yb-doped SnO_2 hollow nanofibers with an enhanced ethanol–gas sensing performance by electrospinning. Sens. Actuators, B Chem. **216**, 212–220 (2015). https://doi.org/10.1016/j.snb.2015.04.040
68. Guo, W., Liu, T., Sun, R., Chen, Y., Zeng, W., Wang, Z.: Hollow, porous, and yttrium functionalized ZnO nanospheres with enhanced gas-sensing performances. Sens. Actuators, B Chem. **178**, 53–62 (2013). https://doi.org/10.1016/j.snb.2012.12.073
69. Zhao, Y., Zou, X., Chen, H., Chu, X., Li, G.D.: Tailoring energy level and surface basicity of metal oxide semiconductors by rare-earth incorporation for high-performance formaldehyde detection. Inorgan. Chem. Front. **6**(7), 1767–1774 (2019)
70. Mokoena, T.P., Swart, H.C., Hillie, K.T., Motaung, D.E.: Engineering of rare-earth Eu3+ ions doping on p-type NiO for selective detection of toluene gas sensing and luminescence properties. Sens. Actuators B Chem. **347**, 130530 (2021). https://doi.org/10.1016/j.snb.2021.130530

Computational Modelling of Rare Earth-Doped Nanostructures: Applications in Energy Harvesting and Environmental Sensing

Shivani Singla and Naveen Bansal

Abstract Rare earth (RE)-doped nanostructures are advanced functional materials with exceptional optical, electronic, and catalytic properties arising from their unique 4f orbital configurations. At the nanoscale, RE ion incorporation into host matrices tunes band structures, defect states, and light–matter interactions, driving innovations in energy harvesting and environmental sensing. Computational modelling plays a vital role in understanding these phenomena and guiding material design. Methods such as density functional theory, time-dependent density functional theory, molecular dynamics, Monte Carlo simulations, and finite-difference time-domain techniques offer multiscale insights into dopant incorporation, defect chemistry, charge transfer, and optical transitions. The theoretical foundations and computational strategies employed to investigate RE-doped nanostructures, emphasizing their applications in photovoltaics, photocatalysis, thermoelectrics, gas and pollutant sensing, and bio/chemical detection are reviewed. Case studies demonstrate how simulations complement experiments by predicting luminescence modulation, adsorption energetics, and spectral conversion. This chapter underscores the pivotal role of computational modelling as both an explanatory and predictive tool, accelerating the development of next-generation RE-doped nanomaterials for sustainable energy and environmental technologies.

Keywords 4f-orbitals · Photocatalysis · Thermoelectric · Gas sensing · Computational modelling

S. Singla (✉)
Department of Physics, University Institute of Sciences, Chandigarh University, Mohali 140413, Punjab, India
e-mail: shivanisingla88@gmail.com

N. Bansal
School of Computer Applications, Lovely Professional University, Jalandhar 144411, Punjab, India

© The Author(s), under exclusive license to Springer Nature Singapore Pte Ltd. 2026

N. Kumar et al. (eds.), *Rare Earth-Doped Metal Oxide Nanostructures*, Engineering Materials, https://doi.org/10.1007/978-981-95-6192-6_7

1 Introduction

The rapid global demand for sustainable energy and environmental remediation has stimulated intensive research into advanced functional nanomaterials. Among them, rare earth (RE)-doped nanostructures have emerged as a unique class of materials owing to their exceptional optical and electronic properties, which originate primarily from the partially filled 4f orbitals of RE ions [1, 2]. These properties make RE-doped nanostructures highly attractive for applications in energy harvesting, light conversion, sensing, and photonic devices [3–5].

At the nanoscale, however, material properties deviate significantly from their bulk counterparts due to effects such as quantum confinement, increased surface-to-volume ratio, and defect-dominated dynamics [5–7]. Incorporating RE ions into nanostructures not only enables tuneability of their electronic band structure but also enhances light–matter interactions, opening opportunities for designing materials that can absorb, convert, and detect photons with unprecedented efficiency. Such characteristics are crucial for photovoltaics (where spectral conversion can overcome the Shockley–Queisser limit), photocatalysis (where visible-light absorption is enhanced), and environmental sensing (where luminescence modulation can indicate pollutant presence) [5, 8, 9].

Despite these advantages, the complexity of RE-doped systems poses challenges for experimental optimization. Doping often leads to competing effects. For instance, while RE ions can introduce desirable energy levels within the host bandgap, they can also act as recombination centres if poorly distributed [10]. Furthermore, the interplay between dopant concentration, host matrix composition, defect chemistry, and nanoscale morphology determines the final performance [11]. Relying exclusively on trial-and-error experimental approaches is time-consuming and costly.

Here, computational modelling emerges as a transformative tool. By employing first-principles calculations, atomistic simulations, and optical response modelling, researchers can predict the structural, electronic, and optical behaviour of RE-doped nanostructures prior to synthesis. Using computational approaches such as density functional theory (DFT), time-dependent DFT (TD-DFT), molecular dynamics (MD), and finite-difference time-domain (FDTD) simulations, researchers can explore how RE dopants integrate within the host lattice, identify the electronic states introduced by doping and their interactions with the host conduction and valence bands, and examine how structural defects and local environments influence luminescence efficiency. Furthermore, advanced approaches such as DFT, TD-DFT, MD, and FDTD simulations provide a powerful framework to investigate energy transfer mechanisms including upconversion and cross-relaxation in co-doped systems. These methods also enable optimization of nanoscale morphologies to enhance light absorption and emission. By answering these questions, computational modelling significantly accelerates rational material design, helping to narrow down promising dopant-host combinations before experimental validation.

This chapter aims to provide a comprehensive overview of computational modelling strategies applied to RE-doped nanostructures, emphasizing their role in energy harvesting (solar photovoltaics, photocatalysis, thermoelectrics) and environmental sensing (gas detection, pollutant monitoring, chemical and biological sensing). Starting with a theoretical background on RE doping and computational methods, the chapter will then explore structural, electronic, and optical modelling, followed by detailed application-driven discussions. Finally, the chapter will address current challenges, limitations, and future perspectives, before drawing overall conclusions. This chapter bridges the gap between fundamental modelling concepts and real-world applications, highlighting how computation can serve as a design compass for next-generation RE-doped nanomaterials tailored for sustainable technologies.

2 Theoretical Background

2.1 Rare Earth Doping in Nanostructures

RE ions (such as Eu^{3+}, Er^{3+}, Nd^{3+}, Tb^{3+}, and Yb^{3+}) have long been recognized for their exceptional optical behaviour, which arises from their 4f–4f transitions. Unlike transition metal ions, whose optical transitions are highly sensitive to the surrounding crystal field due to their partially filled d-orbitals, the 4f orbitals in RE ions are shielded by outer 5s and 5p orbitals. This shielding results in narrow, atom-like emission lines, long excited-state lifetimes, and a relative insensitivity to host lattice distortions [12, 13].

When RE ions are incorporated into nanostructured host matrices such as oxides (TiO_2, ZnO, SiO_2), fluorides ($NaYF_4$, CaF_2), perovskites (ABO_3 oxides, halide perovskites), or semiconductors (Si, GaN, g-C_3N_4), their optical response is profoundly influenced by both the local symmetry and long-range lattice environment. The modifications imposed by RE doping to the host are summarised in Fig 1. Specifically, their relatively large ionic radii and strong electro positivity allow them to distort local lattice environments, modify defect chemistry, and induce lattice strain. Oxygen vacancies generated through RE doping improve charge separation efficiency, thereby enhancing photocatalytic and energy storage performance. For example, La-doped TiO_2 nanoparticles show that La doping induces oxygen vacancies and Ti^{3+} formation, which act as charge traps that prevent electron–hole recombination which is an important mechanism for enhanced photodegradation of methylene blue dye [14].

Importantly, RE doping introduces localized energy levels within the bandgap of host material, thereby extending light absorption into the visible or near-infrared regions [1]. Ce^{3+} doping in TiO_2 introduces 4f levels, enabling visible-light-driven photocatalysis [15]. Similarly, Gd^{3+}-doped ZnO nanorods demonstrated enhanced charge transport and photocatalytic dye degradation efficiency [16].

While RE ions predominantly exist in the trivalent state (RE^{3+}), certain elements such as cerium (Ce^{3+}/Ce^{4+}) and europium (Eu^{2+}/Eu^{3+}) exhibit multiple oxidation

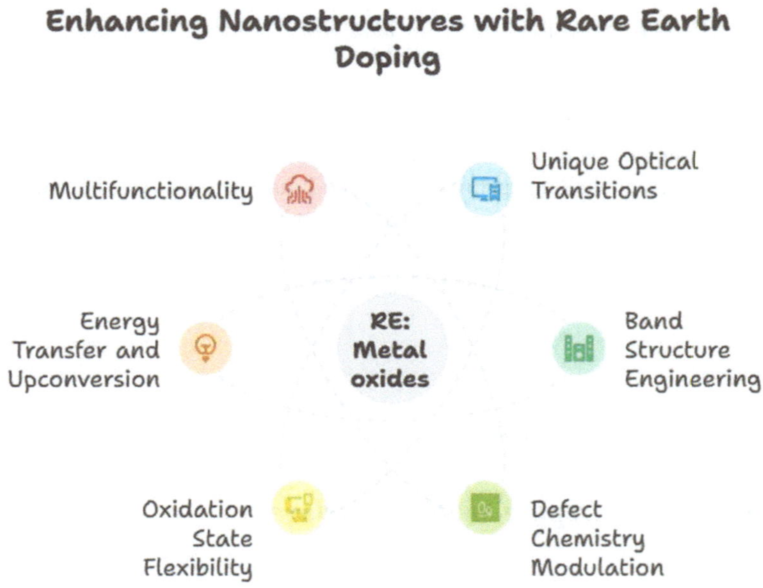

Fig. 1 Key advantages of rare earth doping in nanostructures

states, thereby providing additional flexibility for tuning charge transfer and defect dynamics in host materials. For example, Eu^{3+} doping typically yields strong red luminescence due to the $^5D_o \rightarrow {}^7F_2$ transition, while Er^{3+} doping introduces near-infrared emission (~ 1.54 μm), making it critical for optical communication and photovoltaic enhancement [17–19].

Furthermore, the most striking property of RE ions is their narrow, atom-like f–f transitions, which result in distinct luminescence signatures. For instance, Eu^{3+} produces sharp red emission, Tb^{3+} yields green, and Tm^{3+} enables blue emission, facilitating multicolour tuning within a single host lattice [19–21].

RE ions act as luminescent centres capable of receiving energy from excitons or surface plasmons. For instance, Eu^{3+}-doped $CsPbBr_3$ perovskite quantum dots have demonstrated significant improvements in photoluminescence quantum yield and structural stability, attributed to preserved crystalline integrity and effective energy transfer pathways. These enhancements indicate their increasing potential for light-emitting applications, particularly when co-doped with ions like Tb^{3+} to achieve tuneable emission colours [22]. Er^{3+}/Yb^{3+} co-doped $NaYF_4$ nanostructures efficiently convert NIR photons into visible light, offering applications in photovoltaics and bioimaging. For instance, $NaYF_4:Yb^{3+}/Er^{3+}$ nanoparticles, especially when structured as core–shell or coated with silver, exhibit strong green and red upconversion luminescence under 980 nm excitation, and serve in bioimaging applications [23].

Beyond optical properties, RE doping introduces multifunctionality. Gd^{3+}-doped ceria nanoparticles provided dual catalytic activity for reactive oxygen species (ROS) scavenging and magnetic resonance imaging (MRI) contrast enhancement, bridging energy materials with biomedical diagnostics [24].

At the nanoscale, quantum confinement and surface effects introduce additional complexity. The high surface-to-volume ratio of nanoparticles leads to increased defect densities, which can act as non-radiative recombination centres. Therefore, it is crucial to capture not only bulk substitutional doping but also surface adsorption, interstitial incorporation, and defect-dopant interactions [25]. Furthermore, dopant clustering at high concentrations can quench luminescence due to cross-relaxation, a phenomenon that is crucial to study when designing co-doped systems [26].

2.2 Computational Approaches

Computational modelling has emerged as a foundation in understanding and predicting the behaviour of RE-doped nanostructures, offering atomistic to device-level insights that are often inaccessible through experiments alone. Techniques such as DFT, MD, and FDTD provide complementary perspectives ranging from electronic-level phenomena to thermal stability and macroscopic performance. The key features of these techniques are listed in Table 1. The most widely employed approaches are summarized below.

2.2.1 Density Functional Theory

DFT has become the basis of computational modelling for RE-doped nanostructures, owing to its balance between accuracy and computational cost. It provides critical insights into the ground-state electronic structure, allowing calculations of band structure, density of states, charge density distribution, and defect formation energies. In RE-doped systems, DFT is particularly useful in describing how the

Table 1 Summary of computational methods used for RE-doped nanostructures

Method	Key features	Properties studied	References
DFT	Ground-state electronic structure	Band structure, defect states, charge density	[47]
TD-DFT	Excited states, optical response	Absorption/emission spectra, charge transfer states	[33]
MD	Finite-temperature atomic dynamics	Dopant diffusion, thermal stability, phonon behaviour	[48]
MC	Stochastic sampling of dopant/energy transfer	Förster/dexter energy transfer, clustering, luminescence quenching	[49]
FDTD	Light–matter interaction via Maxwell's equations	Absorption, scattering, plasmonic enhancement, photonic crystals	[43]

partially filled 4f orbitals interact with host lattice states, often introducing mid-gap defect levels that influence optical and catalytic properties. For instance, Guoxi and Yongxin used first-principles DFT to study Sn, Fe, and Ce co-doped anatase TiO_2 and showed that co-doping narrows the bandgap more effectively than single doping [27]. While Fe introduces mid-gap states that may act as recombination centres, Sn and Ce primarily shift the conduction band edge, enhancing visible-light absorption. Optical simulations revealed a red-shifted absorption spectrum and improved photoconductivity, with Sn–Ce co-doping offering the most significant improvement. This study highlights co-doping as a powerful band gap engineering strategy for boosting TiO_2 visible-light photocatalytic efficiency. Similarly, DFT studies on Y-doped hafnia provided atomistic insight into the stabilization of the ferroelectric orthorhombic phase at elevated temperatures, aligning with experimental trends in non-volatile memory devices [28].

However, standard exchange–correlation functionals such as the local density approximation (LDA) and generalized gradient approximation (GGA) often struggle to capture the localized nature of RE 4f electrons, leading to well-known issues such as bandgap underestimation and misaligned RE impurity states. To address this limitation, corrective approaches like DFT + U (which applies a Hubbard U correction to localized orbitals) and hybrid functionals (such as HSE06, which incorporate a fraction of exact Hartree–Fock exchange) are widely employed [29]. These methods improve the accuracy of f-electron localization, enabling better predictions of optical transitions, charge transfer states, and redox chemistry. For example, DFT + U has been applied to CeO_2 nanoparticles to capture the stabilization of Ce^{3+} oxidation states via oxygen vacancies [30].

Recent advances combine machine learning-assisted DFT with high-throughput screening, accelerating the discovery of RE-doped materials with tailored optoelectronic properties. For example, neural network-accelerated DFT has been used to predict catalytic activity of RE-doped graphene [31]. These developments point towards a future where DFT not only explains experimental results but also guides the rational design of RE-doped nanostructures for applications in photocatalysis, solid-state lighting, and quantum information technologies.

2.2.2 Time-Dependent DFT (TD-DFT)

While DFT is well-established for investigating ground-state properties such as crystal structure, electronic density distribution, and total energy, it cannot directly capture the excited-state phenomena that are central to optical and photonic applications of RE-doped nanomaterials. TD-DFT extends the capabilities of conventional DFT by treating the system under time-dependent perturbations, thereby enabling accurate modelling of excited states, optical transitions, and charge transfer processes. This makes TD-DFT particularly powerful for predicting and interpreting the absorption and emission properties of RE-doped nanostructures, where intrashell transitions, charge transfer states, and host-to-dopant interactions play critical roles [32].

One of the primary applications of TD-DFT in RE-doped systems is the simulation of absorption spectra. By calculating excitation energies and oscillator strengths, TD-DFT allows researchers to reproduce and assign spectral features observed experimentally [33]. Moreover, TD-DFT can probe how changes in the dielectric environment of the host influence the emission intensity, making it a valuable tool for designing host–dopant combinations tailored for photonic applications.

Another important aspect where TD-DFT excels is the modelling of charge transfer states, particularly host-to-dopant and dopant-to-host transitions. In RE-doped oxides, charge transfer from O^{2-} 2p orbitals to RE^{3+} ions often competes with intra-4f excitations and significantly modifies the absorption edge [34]. TD-DFT captures such interactions by explicitly describing the coupling between localized 4f states of the RE ion and the extended states of the host conduction or valence band.

2.2.3 Molecular Dynamics (MD)

MD simulations, both classical and ab initio (AIMD), play a pivotal role in understanding the finite-temperature behaviour of RE-doped nanostructures. While DFT and TD-DFT primarily capture ground- and excited-state electronic properties, MD extends this insight to dynamic processes by solving Newton equations of motion for atoms under realistic temperature and pressure conditions. This capability makes MD an indispensable tool for investigating dopant diffusion, thermal stability, phonon transport, and structural relaxation in doped materials. Classical MD, using empirical or semi-empirical interatomic potentials, enables long time-scale simulations (nanoseconds to microseconds) of large systems, making it suitable for studying grain boundaries, extended defects, and large nanoparticles. In contrast, AIMD, based on first-principles forces (typically from DFT), offers higher accuracy and is particularly valuable for capturing localized dopant-lattice interactions, short-range defect dynamics, and temperature-dependent band-edge fluctuations.

A key application of MD in RE-doped systems is the study of annealing processes, which critically influence dopant incorporation, clustering, and defect healing [35]. For instance, AIMD simulations of Eu-doped perovskites have demonstrated enhanced lattice rigidity and reduced halide ion migration, directly explaining the improved operational stability of perovskite solar cells [8]. Another emerging direction is the use of reactive force fields (ReaxFF), which allow bond formation and breaking, thereby extending MD to catalytic and redox processes in RE-doped oxides. For example, ReaxFF simulations of a high-quality parameterization of Ce/O/H ReaxFF for ceria which accurately describes oxygen vacancy formation energies and surface reactivity lays a strong foundation for such future applications [36].

MD bridges the gap between static electronic structure calculations and real experimental processing conditions, enabling the prediction of dopant diffusion pathways, lattice relaxation under heat treatment, phonon scattering behaviour, and catalytic activity. With ongoing advancements in machine learning-based force fields and GPU-accelerated MD, simulations are expected to capture even longer time- and length-scales, thereby providing a comprehensive picture of RE-doped nanostructure dynamics in functional applications.

2.2.4 Monte Carlo (MC) Simulations

MC simulations are powerful stochastic methods that provide insights into processes governed by probability and randomness, making them particularly well-suited for investigating dopant distribution, exciton migration, and energy transfer dynamics in RE-doped nanostructures [37]. Unlike deterministic approaches such as DFT or MD, which solve explicit equations of motion or electronic structures, MC relies on random sampling of possible configurations to statistically approximate system behaviour. This makes MC an indispensable tool for exploring large configurational spaces and processes involving photon–ion interactions, dopant clustering, and defect distributions.

A major application of MC in RE-doped systems lies in modelling energy transfer mechanisms, such as Förster resonant energy transfer (FRET) and Dexter exchange interactions, which dictate luminescence quenching and cooperative upconversion efficiency [38, 39]. For example, MC simulations of Yb^{3+}/Er^{3+} co-doped nanophosphors have revealed how varying Yb^{3+} concentration influences the probability of exciton hopping and subsequent green/red emission intensity, closely matching experimental upconversion spectra [40].

Another important use of MC is in statistical modelling of defect distributions in nanostructures, where random dopant incorporation leads to non-uniform electronic and optical behaviour. For instance, MC approaches have been employed to simulate oxygen vacancy clustering in RE-doped oxides, providing insights into the origin of non-radiative recombination channels that degrade luminescence quantum yield [41].

Beyond optical modelling, MC is increasingly being integrated with multiscale simulations, where output from DFT or MD is used as input parameters for large-scale stochastic modelling. This hybrid approach has been applied in RE-doped fluoride nanocrystals to predict concentration-dependent upconversion quantum yields, where MC captured the stochastic nature of ion–ion interactions beyond the reach of deterministic models [42].

2.2.5 Finite-Difference Time-Domain (FDTD) Simulations

FDTD simulations are among the most widely used numerical methods for investigating light–matter interactions in nanostructured materials. Based on the direct solution of Maxwell's equations in the time domain, FDTD provides both spatial and temporal evolution of electromagnetic fields in complex nanostructures. This makes it a particularly versatile approach for modelling optical absorption, scattering, plasmonic enhancement, and near-field interactions in RE-doped nanomaterials, where experimental characterization at nanoscale is often challenging.

FDTD simulations have been instrumental in exploring local field enhancement effects that strongly influence luminescence intensity and spectral distribution of RE-doped nanostructures. For example, the placement of RE ions such as Eu^{3+} or Er^{3+} near metallic nanostructures (Ag, Au, Al) has been shown to induce significant plasmonic enhancement of emission through local electromagnetic field confinement [6]. FDTD enables quantitative predictions of how parameters such as metallic nanoparticle size, shape, and ion–metal separation distance affect the Purcell factor, emission branching ratios, and radiative lifetimes.

Another major application of FDTD is in photonic crystal engineering, where periodic dielectric structures are designed to manipulate the density of optical states. FDTD simulations reveal that periodic dielectric structures (photonic crystals) can generate localized field intensity peaks several times higher than in homogeneous media, especially when the incident wavelength aligns with the photonic bandgap [43]. Similarly, designs incorporating metallic nanoparticles (like Au) within a photonic crystal matrix, simulated via FDTD, show nearly 2 × enhancement in peak electric field amplitude compared to standalone nanoparticles [44]. This field enhancement is pivotal for boosting light–matter interactions in RE-doped environments. However, not specific to RE doping, FDTD-related methods have been successfully utilized to the best of our knowledge.

FDTD is also a crucial tool in solar energy applications, particularly in designing RE-doped upconversion and downconversion layers for photovoltaics. By simulating optical absorption across broad spectral ranges, FDTD helps optimize light trapping and spectral conversion mechanisms. For example, FDTD simulations of Yb^{3+}/Er^{3+}-doped upconversion layers in Si solar cells have shown enhanced near-infrared light harvesting due to resonant coupling with plasmonic nanostructures [45]. Furthermore, incorporation of RE-doped phosphors into plasmonic back-reflector designs has been modelled to improve both absorption efficiency and carrier generation rates.

Recent advances have also combined FDTD with atomistic methods (DFT/TD-DFT), where electronic structure information from quantum mechanical calculations is used as input to FDTD optical models. This hybrid multiscale approach allows for realistic simulations of emission spectra, nonlinear optical response, and spectral conversion efficiency in nanostructures [46].

3 Modelling of Rare Earth-Doped Nanostructures

The design of RE-doped nanostructures requires a comprehensive understanding of how dopant ions incorporate into host lattices, how they modify the electronic band structure, and how these changes translate into optical properties. Computational modelling offers critical insights into each of these aspects, enabling a predictive approach towards optimizing material performance. Table 2 displays a comparison between the simulated results and obtained experimental results.

Table 2 Computational predictions and corresponding experimental validations

System	Computational prediction	Experimental observation	References
Ce-doped TiO_2	DFT: bandgap narrowing and lattice distortion; oxygen vacancy formation improves charge separation	Enhanced methylene blue degradation efficiency	[9]
Ho^{3+}-doped g-C_3N_4	DFT: reduced HOMO–LUMO gap, increased electrophilicity, enhanced visible-light absorption	Nearly doubled photocatalytic degradation efficiency of methylene blue	[50]
Ce-doped CeO_2 nanoparticles	DFT + U: stabilization of Ce^{3+} oxidation states via oxygen vacancies	Enhanced catalytic activity and redox performance	[30]
Y-doped HfO_2(epitaxial thin films)	DFT: predicted intrinsic ferroelectric polarization ~ 64 μC cm^{-2}; stabilization of Pca21 orthorhombic phase with rhombohedral distortion	Stable and enhanced out-of-plane polarization of 50 μC cm^{-2} at RT; confirmation of orthorhombic ferroelectric phase	[28]
Yb^{3+}–Er^{3+} U CNPs	MC versus ordinary differential equation (ODE) simulations of population evolution, lifetime, spectral purity, and saturation curves; MC predicted more accurate ET dynamics than ODE	Experimental validation shows MC-based results better match observed lifetimes and saturation behaviour	[40]
Eu-doped TiO_2	DFT + U predicted red-shift in absorption edge and enhanced optical activity	UV–vis absorption shows strong red-shift, enhanced visible-light absorption	[51]
$NaYF_4$ upconversion nanoparticle growth	DFT predicted facet-dependent ligand coverage driving shape evolution in UCNPs	Hydrothermal syntheses confirmed directional growth consistent with modelling	[52]
Na^+-co-doped $ZnAl_2O_4$:Ho^{3+},Yb^{3+} UCNPs	DFT and PALS studies showed Na^+ co-doping reduces zinc vacancies, lowering non-radiative quenching	DFT and PALS studies showed Na^+ co-doping reduces zinc vacancies, lowering non-radiative quenching	[53]

3.1 Structural Modelling

3.1.1 Dopant Incorporation

Structural modelling of RE-doped nanostructures provides atomistic insight into how dopant ions are accommodated within host lattices and how these modifications influence thermodynamic stability and functional performance. The first consideration in such modelling is identifying the most favourable dopant incorporation sites. RE ions may substitute host cations (substitutional doping), occupy interstitial voids, or adsorb onto surfaces. Substitutional incorporation is generally favoured in dense oxide and fluoride hosts due to charge and size compatibility, whereas interstitial incorporation becomes possible in layered or porous frameworks. For instance, Eu^{3+} in TiO_2 preferentially replaces Ti^{4+} ions rather than occupying interstitial sites, a behaviour confirmed by defect formation energy calculations using DFT [51].

Computationally, defect formation energy (E_f) is calculated using DFT to determine the stability of different dopant configurations:

$$E_f = E_{\text{doped}} - E_{\text{host}} - \mu_{\text{dopant}} + \mu_{\text{host}}, \tag{1}$$

where E_{doped} and E_{host} are the total energies of the doped and pristine systems, while μ_{dopant} and μ_{host} are the chemical potentials of the dopant and substituted host atom. Lower values of E_f indicate more favourable incorporation [47]. Ce^{3+} incorporation in ZnO demonstrates that substitutional doping not only minimizes formation energy but also induces desirable optical modifications [54].

3.1.2 Size and Charge Effects

Another crucial factor is the size and charge effect associated with RE doping. RE ions generally possess larger ionic radii compared to the host cations they replace. For example, the ionic radius of Eu^{3+} (0.95 Å, CN = 6) is significantly larger than Ti^{4+} (0.61 Å, CN = 6), leading to pronounced lattice strain and distortions around the dopant site [55]. These distortions are captured through structural relaxation in DFT calculations and are directly correlated with changes in phonon spectra and band structure. Moreover, the charge mismatch resulting from heterovalent substitution (e.g. Eu^{3+} replacing Ti^{4+}) requires charge compensation through intrinsic defects, most commonly oxygen vacancies. Such defect–dopant associations strongly affect optical behaviour. For instance, Sm^{3+}-doped SnO_2 has been shown to form stable Sm^{3+}–VO complexes, which act as sensitizers for visible-light emission [56]. Similarly, Ce^{3+}/Ce^{4+} coexistence in Ce-doped TiO_2 is stabilized through oxygen vacancy formation, improving electrocatalytic activity by enhancing charge separation [57].

3.1.3 Surface and Interface Effects

At the nanoscale, surface and interface effects dominate dopant behaviour. Dopants located near surfaces experience under-coordinated environments, altered electrostatic potentials, and greater structural flexibility compared to bulk lattice sites. Recent MC simulations of lanthanide-doped nanocrystals indicate that thermal diffusion can cause dopant redistribution. This process leads to dopant clustering or surface segregation, which in turn influences upconversion efficiency by modifying local energy transfer pathways and non-radiative decay channels [49]. Experimental studies corroborate these predictions; Eu^{3+} in ZnO has been observed to segregate towards grain boundaries and surfaces, altering emission intensity and thermal stability [58]. Furthermore, interfaces between RE-doped nanoparticles and external matrices such as polymers, silica coatings, or aqueous media play a pivotal role in determining dopant stability and energy transfer. For example, $NaYF_4{:}Yb^{3+}$, Er^{3+} UCNPs encapsulated in a nonporous silica shell demonstrate sustained luminescence intensity and structural integrity after two years of storage [59]. These findings underscore the importance of incorporating interfacial effects into computational models to more accurately reflect real-world operating environments of doped nanostructures.

3.2 Electronic Properties

3.2.1 Band Structure Modifications

Depending on the nature of the host lattice and dopant oxidation state, RE incorporation can either narrow or widen the effective bandgap. Hybrid functional calculations have confirmed that these dopant-induced states reduce the bandgap from 3.2 eV (pristine TiO_2) to values as low as 2.6–2.8 eV depending on dopant concentration and configuration [60]. Such modifications enable band engineering for energy harvesting and sensing applications, where bandgap reduction enhances visible-light absorption while controlled widening can suppress unwanted recombination. For instance, first-principles calculations demonstrate that Eu^{3+} substitution in anatase TiO_2 introduces impurity states that overlap with the valence band, yielding visible-light absorption absent in pristine TiO_2. This alters the optical absorption edge and improves photocatalytic response under visible illumination [51]. In addition, RE ions often introduce trap states that can play dual roles: they can facilitate sub-bandgap photon absorption, which is advantageous for photovoltaics, but they may also act as non-radiative recombination centres that degrade luminescence performance. Similarly, Ce^{3+} doping introduces occupied 4f states just below the conduction band of ZnO, resulting in a narrowed effective bandgap and enhanced photocatalytic efficiency [54]. Complexes involving dopants and intrinsic defects create mid-gap states that strongly influence optical and catalytic responses. As an example, Er^{3+} doping in Si nanocrystals enables efficient infrared emission by introducing defect-assisted radiative pathways, whereas Nd^{3+} doping in TiO_2 reduces the bandgap and promotes enhanced visible-light absorption, making it a promising modification for photocatalysis and solar energy harvesting [61, 62].

3.2.2 Charge Transfer Dynamics

The interaction between dopant states and the host conduction/valence bands also determines charge transfer dynamics, which are central to luminescence and photocatalytic processes. Computational studies have demonstrated that host-to-dopant charge transfer transitions often dominate excitation pathways in RE-doped oxides. For instance, in Eu^{3+}-doped oxide systems, charge transfer from O^{2-} 2p orbitals to the Eu^{3+} ion significantly contributes to absorption spectra, often more strongly than direct f–f transitions [63]. This interplay between localized dopant levels and extended band states enables tuneable excitation mechanisms, allowing materials to be optimized for specific functionalities such as wavelength-selective luminescence or photocatalytic charge separation.

3.2.3 Defect Interactions

Another important feature is the charge compensation mechanism associated with RE doping, which influences electronic properties through defect formation. When trivalent RE ions (e.g. Eu^{3+}, Nd^{3+}, Sm^{3+}) substitute tetravalent cations in oxides (e.g. Ti^{4+} in TiO_2, Sn^{4+} in SnO_2), oxygen vacancies are created to maintain charge neutrality. These oxygen vacancies act as shallow donor states, introducing additional electronic levels close to the conduction band minimum (CBM). Modelling defects–dopant complexes provides insights into how they stabilize certain dopant oxidation states (e.g. Ce^{3+} vs. Ce^{4+}) and influence optical activity [64]. For example, DFT + U studies on doped SnO_2 have shown that the presence of oxygen vacancies increases electron density at the CBM, thereby improving conductivity and visible-light absorption [65]. In this way, defect engineering, when coupled with RE doping, becomes an essential strategy for achieving a balance between improved absorption and minimized non-radiative losses.

3.2.4 Others

The 4f–band coupling also plays a vital role in determining optical and transport behaviour. In certain hosts, hybridization between RE 4f/5d orbitals and host conduction/valence bands alters effective mass, mobility, and carrier lifetimes. For instance, Er^{3+} doping in GaN introduces discrete 4f transitions that support green luminescence at ~ 540 nm [66]. Similarly, Nd^{3+}-doped TiO_2 shows strong f–d coupling, which accelerates visible-light absorption but can increase recombination unless compensated by additional defect engineering [67].

At the nanoscale, quantum confinement effects further amplify the role of RE dopants in modifying electronic properties. In smaller nanoparticles, the host's conduction and valence bands shift, altering the relative position of dopant-induced states. Computational studies have shown that in RE-doped ZnO quantum dots, the confinement-induced widening of the bandgap increases the relative importance of localized 4f levels, leading to discrete, sharp absorption features that are absent in bulk counterparts [68, 69]. Such quantum-size effects explain why RE-doped nanostructures often display richer optical spectra and stronger luminescence than bulk-doped crystals.

3.3 Optical Properties

3.3.1 Absorption and Emission Spectra

The optical behaviour of RE-doped nanostructures is primarily governed by the unique nature of RE 4f orbitals, which are shielded by outer 5s and 5p electrons. This shielding results in sharp intra-4f transitions that are relatively insensitive to the local crystal field, thereby producing narrow emission bands that are advantageous for sensing and photonic applications. In addition to these intra-4f transitions, charge transfer states (CTS) involving host–dopant interactions, and band-to-dopant transitions significantly contribute to absorption and emission processes. On the luminescence side, configuration coordinate diagrams are employed to map radiative versus non-radiative decay pathways, offering insights into emission efficiencies and quenching mechanisms. For example, DFT + U simulations of Eu^{3+}-doped TiO_2 have successfully reproduced the characteristic red emission associated with the $^5D_o \rightarrow {}^7F_2$ transition [51]. Similarly, Er^{3+}/Yb^{3+} co-doped Y_2O_3 systems exhibit stepwise infrared absorption leading to prominent green upconversion emission—consistent with the upconversion mechanisms that TD-DFT can help model [70]. These computational insights are invaluable for tailoring dopant–host combinations toward specific optical functionalities such as efficient light emitters or broad-spectrum solar absorbers.

3.3.2 Energy Transfer Mechanisms

Energy transfer processes in RE-doped nanostructures play a pivotal role in determining luminescence performance, particularly in co-doped or sensitizer–activator systems. Computational modelling enables the quantification of transfer mechanisms such as FRET, which operates via long-range dipole–dipole interactions, and Dexter electron exchange, which requires short-range wavefunction overlap [38]. MC simulations have proven especially powerful in this context, as they allow researchers to model dopant distribution, critical transfer distances, and concentration quenching effects under realistic conditions [42]. For instance, co-doping with Yb^{3+} as a sensitizer in $NaYF_4$ nanocrystals has been modelled to optimize Er^{3+}-based upconversion processes, thereby enabling efficient infrared-to-visible light conversion [42]. Such insights not only provide a mechanistic understanding of luminescence but also guide experimental efforts towards rational dopant engineering for energy harvesting and sensing applications.

3.3.3 Nonlinear Optical Properties

Beyond linear absorption and emission, nonlinear optical (NLO) properties of RE-doped nanostructures are gaining increasing attention due to their potential applications in bioimaging, optical switching, and environmental sensing. Processes such as two-photon absorption (TPA), third-harmonic generation (THG), and optical Kerr effects depend strongly on the electronic structure and local field enhancements around RE dopants. Computational modelling of these effects typically employs perturbation theory applied to electronic wavefunctions, enabling calculation of second- and third-order susceptibilities. Recent TD-DFT and wavefunction-based studies have shown that co-doping strategies significantly enhance nonlinear responses by introducing intermediate states that facilitate multiphoton transitions. For example, Yb^{3+}/Er^{3+} co-doped $NaYF_4$ nanoparticles exhibit pronounced two-photon absorption, consistent with theoretical predictions of intermediate-assisted upconversion pathways [71]. Similarly, simulations have indicated strong third-harmonic generation, which could be leveraged for nanoscale photonic devices [72]. The integration of computational modelling with experimental spectroscopy thus provides a powerful framework for designing RE-doped nanomaterials with tailored NLO responses.

3.4 Vibrational and Thermodynamic Properties

3.4.1 Phonon Spectra and Lattice Dynamics

The incorporation of RE dopants into nanostructures significantly influences lattice vibrations, which in turn affect optical activity, thermal stability, and carrier relaxation pathways. Computational studies, particularly those using density functional perturbation theory (DFPT), provide insights into phonon dispersion relations and vibrational density of states (VDOS). These calculations reveal how RE ions, due to their larger ionic radii and localized 4f electrons, perturb the lattice and introduce localized vibrational modes. For instance, phonon calculations of Eu^{3+}-doped TiO_2 have shown shifts in optical phonon frequencies, which strongly couple with electronic transitions and influence luminescence lifetimes [73]. Similarly, phonon spectra of Ce^{3+}-doped oxides indicate enhanced electron–phonon coupling that can either facilitate multiphonon relaxation or stabilize defect-related emission pathways

[74]. Such modelling is essential for interpreting experimental Raman and infrared spectra and for correlating lattice vibrations with optical efficiency.

3.4.2 Electron–Phonon Coupling and Non-radiative Processes

A key factor in determining the efficiency of RE-doped nanomaterials is the competition between radiative and non-radiative relaxation. Strong electron–phonon coupling often promotes multiphonon relaxation processes, which quench luminescence by dissipating excitation energy as heat. First-principles modelling ties the optics of RE-doped hosts directly to electron–phonon coupling and multiphonon relaxation. In practice, one maps the ground- and excited-state potential energy surfaces with a configuration-coordinate diagram, then uses those displacements and vibrational frequencies to estimate non-radiative rates, an approach consistent with the energy-gap law for RE ions (larger gaps $\rightarrow$ exponentially lower multiphonon decay) and widely used to rationalize quenching trends [75]. In silicate glasses, high-energy Si–O phonons ($\approx$1000–1100 cm^{-1}) efficiently bridge 4f-level gaps of Er^{3+}, accelerating non-radiative decay and depressing the 1.5 μm quantum efficiency; by contrast, low-phonon-energy fluoride hosts suppress multiphonon losses and therefore support brighter RE emission and upconversion [76]. In fluoride lattices exhibit maximum phonon energies only in the few hundred cm^{-1} range (e.g. $LiYF_4$ modes above ~ 360 cm^{-1}), which helps explain why Yb^{3+}/Er^{3+} systems in fluorides (such as $NaYF_4$) remain benchmark upconverters [77]. Modern ab initio workflows (DFT/DFPT) now compute these vibrational spectra and electron–phonon couplings directly, enabling quantitative, materials-specific predictions of radiative vs non-radiative pathways and identification of "killer" modes that dominate quenching. Such computational insights are critical for guiding the choice of host lattices that balance structural stability with minimal phonon-induced quenching.

3.4.3 Thermodynamic Stability and Dopant Solubility

Thermodynamic modelling provides another crucial perspective by assessing the formation energies of dopant incorporation and defect–dopant complexes. DFT calculations of formation energies, combined with phase diagram modelling, allow predictions of dopant solubility limits, charge compensation mechanisms, and phase stability under varying synthesis conditions. For instance, DFT studies of Nd^{3+}-doped perovskites have shown that oxygen vacancy formation is energetically favourable, leading to stable dopant–defect complexes that simultaneously influence both luminescence and conductivity [78]. Similarly, thermodynamic stability analysis of Ce^{3+} versus Ce^{4+} oxidation states in oxide hosts reveals that local charge compensation by defects plays a decisive role in determining dopant activity [79]. Such predictions not only rationalize experimental doping limits but also provide guidelines for synthesis strategies such as co-doping and defect engineering to enhance RE incorporation.

3.4.4 Thermal Transport and Phonon Engineering

The vibrational and thermodynamic properties of RE-doped nanostructures also directly impact thermal transport, which is particularly relevant for applications in thermophotonics, lasers, and energy conversion. Computational methods such as MD and Boltzmann transport equation (BTE) approaches are used to calculate thermal conductivity and assess how dopant-induced phonon scattering reduces heat transport [80, 81]. For example, simulations of Yb^{3+}/Er^{3+} co-doped fluoride nanocrystals show that dopant-induced lattice strain and mass disorder suppress phonon propagation, leading to reduced thermal conductivity that stabilizes optical emission by minimizing thermal quenching [80]. Such findings highlight the dual role of RE dopants: while they enhance optical performance, they also act as phonon scatterers, enabling thermal management in nanophotonic devices.

4 Applications in Energy Harvesting

RE-doped nanostructures have become an integral component of modern energy harvesting strategies due to their ability to manipulate photon energy, extend spectral absorption, modulate charge dynamics, and scatter phonons, all of which contribute to improved performance in photovoltaics, photocatalysis, and thermoelectrics. Computational modelling plays a central role in predicting these effects, offering atomistic insights into how RE doping modifies band structures, introduces defect states, and enhances light–matter interactions, thereby enabling rational design of nanostructures for optimized device performance.

4.1 Photovoltaics

One of the primary limitations of single-junction solar cells is the Shockley–Queisser limit (~ 33%), which arises from spectral mismatch, i.e. photons with energies below the bandgap cannot be absorbed, while those with much higher energies lose their excess energy as heat. RE doping provides two promising solutions to this problem through spectral conversion processes. *Upconversion* (UC) enables the absorption of two or more low-energy photons to generate a higher-energy photon that can be absorbed by the solar cell, as demonstrated in Yb^{3+}/Er^{3+} co-doped inorganic phosphors that upconvert near-infrared photons into the visible range, thereby improving Si solar cell efficiency [82]. Conversely, *downconversion* (DC) splits a single high-energy photon into two lower-energy photons, a process that Tb^{3+}/Yb^{3+} co-doping has been computationally modelled to exploit in Si and GaAs photovoltaics [83]. Computational approaches such as DFT and TD-DFT have been widely applied to model dopant-induced electronic transitions and emission spectra, while MC simulations have enabled estimation of UC efficiencies under varying dopant concentrations [42]. Furthermore, FDTD modelling has guided the design of plasmon-enhanced light-trapping structures [84]. A notable study is Er^{3+}-doped Si nanocrystals, where first-principles simulations revealed the role of defect-assisted transitions in achieving efficient 1.54 μm emission, a wavelength matching the telecommunication band that also enhances infrared response in Si photovoltaics [85].

4.2 *Photocatalysis*

RE doping provides solutions to the intrinsic limitations of conventional photocatalysts such as TiO_2, which has a wide bandgap (~ 3.2 eV) and therefore absorbs only a small fraction of the solar spectrum. RE dopants can introduce mid-gap states or shift absorption edges, extending activity into the visible region. For example, RE doping in TiO_2 was predicted by DFT + U to introduce localized 4f states and defect-mediated charge transfer, enhancing visible-light absorption [86]. Similarly, Ho^{3+}-doped g-C_3N_4 demonstrated that holmium incorporation modifies both structural and electronic properties, as confirmed by extensive experimental characterization and DFT calculations. Theoretical results indicated that Ho doping decreases the HOMO–LUMO gap and chemical hardness while increasing electrophilicity, thereby enhancing reactivity and visible-light absorption. Experimentally, the 0.2 mmol Ho-doped sample achieved nearly double the photocatalytic degradation efficiency of methylene blue compared to pristine g-C_3N_4, underscoring the synergistic role of RE doping and electronic structure modulation in boosting photocatalytic performance [50].

Beyond absorption, RE ions also improve charge separation by acting as trapping centres. Ce doping has been shown to enhance the photocatalytic activity of TiO_2 by inducing lattice distortions, charge redistribution, and orbital modifications, which increase the number of excited electrons and improve light absorption. First-principles calculations confirmed these structural and electronic changes, correlating with experimental methylene blue degradation tests that revealed optimal activity at moderate Ce concentrations. Furthermore, when applied to asphalt pavements, Ce–TiO_2 coatings, particularly those prepared by surface spraying exhibited superior NO removal efficiency, demonstrating the practical potential of RE-doped TiO_2 in environmental remediation [9]. Computational modelling is vital in explaining these effects.

4.3 *Thermoelectrics*

Thermoelectric (TE) materials convert temperature gradients into electrical voltage, quantified by the figure of merit (ZT) [87]:

$$\mathrm{ZT} = \frac{S^2 \sigma T}{\kappa}, \tag{2}$$

where S is the Seebeck coefficient, σ is electrical conductivity, T is absolute temperature, and κ is thermal conductivity. Enhancing ZT requires simultaneously maximizing the power factor ($S^2\sigma$) while minimizing thermal conductivity [88]. RE doping contributes in two ways. Primarily, by inducing strong phonon scattering due to ionic size mismatch and mass differences, which reduces lattice thermal conductivity, and secondly, by modulating band structures to tune carrier concentrations and effective mass, thereby optimizing the power factor [89, 90]. Computationally, these effects are probed using Boltzmann transport theory in combination with DFT to predict Seebeck coefficients, conductivity, and carrier mobility, while DFPT allows phonon dispersion and thermal transport simulations. Reported literature highlight the promise of this approach. La-doped $SrTiO_3$ was predicted to exhibit enhanced Seebeck coefficients via band convergence effects, while RE-filled skutterudites, such as Yb-filled $CoSb_3$, demonstrated reduced lattice thermal conductivity due to localized rattling modes of the dopant [91]. Yb-filled $CoSb_3$ skutterudites, DFT revealed emergent conduction band minima and increased effective masses due to Yb inclusion, indicative of band structure modification that boosts carrier transport. Simultaneously, DFPT-based studies and inelastic neutron data show that RE rattler atoms like Yb introduce localized vibrational modes ("Einstein modes") that strongly scatter heat-carrying phonons, reducing lattice thermal conductivity [92]. These findings highlight the central role of modelling in identifying dopant–host combinations that balance charge carrier mobility with low thermal conductivity, a key requirement for high-efficiency thermoelectrics.

5 Applications in Environmental Sensing

RE-doped nanostructures have gained increasing attention in environmental sensing due to their luminescent properties, high chemical stability, and sensitivity to surface adsorption phenomena. By leveraging computational modelling, researchers can understand the interaction of gas molecules, pollutants, and biomolecules with RE-doped surfaces, predict sensing mechanisms, and optimize material design for real-world applications. Key areas of application include gas sensing, luminescence-based environmental monitoring, and biosensing platforms.

5.1 Gas Sensing

Gas sensors are critical for monitoring air quality, industrial emissions, and hazardous gases. RE-doped nanostructures function primarily via resistance modulation and luminescence modulation. In resistance-based sensing, the adsorption of gas molecules alters the carrier concentration in semiconductor hosts such as RE-doped ZnO, thereby changing conductivity [93]. In luminescence-based sensing, RE ions serve as optical probes, where variations in emission either through quenching or enhancement in signal with the presence of specific gases [94]. Computational studies reveal that RE dopants modify the surface chemistry by introducing oxygen vacancies and altering adsorption energetics. For example, mesoporous La-doped SnO_2 spheres have shown excellent sensitivity and selectivity towards glutaraldehyde, with DFT calculations indicating enhanced adsorption energies compared to pure SnO_2—though not specifically involving oxygen vacancies or NO_2 adsorption [95]. MD simulations further provide insights into diffusion kinetics of gases across doped surfaces, while ab initio thermodynamics helps evaluate sensor performance under varying temperature and pressure conditions [96].

5.2 Heavy Metal Sensing

RE luminescence is highly sensitive to interactions with heavy metal ions (e.g. Pb^{2+}, Hg^{2+}, Cd^{2+}). Computational studies reveal that adsorption of these ions on RE-doped nanostructures leads to spectral shifts or luminescence quenching, which can be exploited for detection. For instance, DFT studies of As(III) and other pollutant adsorption on doped TiO_2 surfaces show how surface adsorption alters local electronic structure and activity, providing a conceptual basis for potential luminescence modulation [97]. Complementarily, TD-DFT optical simulations predict spectral changes induced by pollutant binding, thereby guiding the rational design of luminescent probes [98].

RE doping emerges as an effective strategy for enhancing g-C_3N_4 performance. Ce^{3+}-doped g-CN narrows its bandgap, improves light absorption, and suppresses electron–hole recombination, thereby enhancing H_2 evolution under visible light [50, 99]. Computational studies elucidate how organic molecules adsorb and react on doped surfaces. For example, DFT modelling of methylene blue adsorption on Ce^{3+}-doped TiO_2 demonstrated reduced bandgap and enhanced charge transfer pathways, which explained the experimentally observed improvement in photocatalytic degradation efficiency [3].

5.3 Bio and Chemical Sensing

RE-doped nanostructures are equally important in bio and chemical sensing applications. Their sharp, long-lived emission lines make them excellent fluorescent probes, and computational modelling provides predictive insights into how biomolecule adsorption influences emission properties. TD-DFT studies on carbon-based quantum dots show that surface functional groups like –NH_2 and –C = O introduce new NIR absorption bands, demonstrating how local chemistry can reshape electronic transitions [100]. In RE-doped nanoparticles, analogous functionalization—such as –COOH, $–NH_2$ or –OH proximal to RE ions—can perturb the local crystal field, shifting emission spectra, altering branching ratios, and modifying luminescence efficiency. In addition, molecular docking combined with MD simulations is increasingly applied to explore the anchoring mechanisms of biomolecules on RE-doped surfaces, providing information on binding stability and preferred interaction sites.

For chemical sensing, RE-doped nanostructures are utilized for the detection of explosives, volatile organic compounds (VOCs), and environmental toxins [101]. However, not studies on DFT adsorption studies predicting selective binding of VOCs such as ethanol and formaldehyde on RE-doped surfaces are reported to the best of our knowledge. Meanwhile, optical simulations using TD-DFT reveal spectral fingerprints that enable selective chemical recognition. A representative example is the computational study of Er^{3+}-doped Y_2O_3 nanoparticles, which demonstrated selective fluorescence quenching upon interaction with nitroaromatic molecules, thereby supporting their application in explosive detection [102].

6 Conclusion

RE-doped nanostructures are a versatile class of materials with significant potential in energy harvesting and environmental sensing. Their unique optical and electronic characteristics, derived from the shielding of 4f orbitals, enable sharp emission, long-lived excited states, and efficient upconversion. At the nanoscale, these properties are further tuned by size confinement, surface effects, and defect chemistry, enabling highly adaptable functional materials.

Experimental approaches alone are insufficient to fully exploit these systems, making computational modelling indispensable. Techniques such as density functional theory (DFT), molecular dynamics, and multiscale simulations reveal insights into structural stability, electronic modifications, optical transitions, and energy transfer mechanisms. Modelling not only complements experiments but also accelerates discovery by predicting material behaviour, reducing costs, and guiding rational design.

Applications in photovoltaics, photocatalysis, thermoelectrics, and pollutant sensing illustrate the impact of computational predictions on optimizing performance. Case studies further demonstrate the synergy between simulations and experiments in advancing RE-doped nanomaterials.

Key challenges remain in modelling strongly correlated 4f electrons, capturing multiscale effects, and achieving experimental validation. Future progress will integrate advanced quantum methods, machine learning, and collaborative data platforms, bridging quantum-level insight with device-level performance to enable sustainable next-generation technologies.

References

1. Daksh, D., Agrawal, Y.K: Rare earth-doped zinc oxide nanostructures: a review. Rev. Nanosci. Nanotechnol. **5**, 1–27
2. Zheng, B., Fan, J., Chen, B., Qin, X., Wang, J., Wang, F., Deng, R., Liu, X.: Rare-earth doping in nanostructured inorganic materials. Chem. Rev. **122**, 5519–5603 (2022)
3. Alberoni, C., Barroso-Martin, I., Infantes-Molina, A., Rodriguez-Castellon, E., Talon, A., Zhao, H., You, S., Vomiero, A., Moretti, E.: Ceria doping boosts methylene blue photodegradation in titania nanostructures. Mater. Chem. Front. **5**, 4138–4152 (2021)
4. Thor, W., Kai, H.Y., Yeung, Y.H., Wu, Y., Cheung, T.L., Tam, L.K.B., Zhang, Y., Charbonnière, L.J., Tanner, P.A., Wong, K.L.: Unearthing the real-time excited state dynamics from antenna to rare earth ions using ultrafast transient absorption. JACS Au **4**, 3813–3822 (2024)
5. Crawford, S.E., Ohodnicki, P.R., Baltrus, J.P.: Materials for the photoluminescent sensing of rare earth elements: challenges and opportunities. J. Mater. Chem. C **8**, 7975–8006
6. Feng, A.L., You, M.L., Tian, L., Singamaneni, S., Liu, M., Duan, Z., Lu, T.J., Xu, F., Lin, M.: Distance-dependent plasmon-enhanced fluorescence of upconversion nanoparticles using polyelectrolyte multilayers as tunable spacers. Sci. Rep. **5**, 1–10 (2015)
7. Zheng, J., Meng, D., Guo, J., Liu, X., Zhou, L., Wang, Z.: Defect Engineering for enhanced electrocatalytic oxygen reaction on transition metal oxides: the role of metal defects. Adv. Mater. **36**, 2405129 (2024)
8. Liang, Y., Li, F., Cui, X., Stampfl, C., Ringer, S.P., Yang, X., Huang, J., Zheng, R.: Multiple B-site doping suppresses ion migration in halide perovskites. Sci. Adv. **11**, 1–11 (2025)
9. Cao, X., Yang, X., Li, H., Huang, W., Liu, X.: Investigation of Ce-TiO_2 photocatalyst and its application in asphalt- based specimens for NO degradation. Constr. Build. Mater. **148**, 824–832 (2017)
10. Sood, S., Kumar, P., Raina, I., Misra, M., Kaushal, S., Gaur, J., Kumar, S., Singh, G.: Enhancing optoelectronic performance through rare-earth-doped ZnO: insights and applications. Photonics **12** (2025)
11. Makkar, M., Viswanatha, R.: Frontier challenges in doping quantum dots: synthesis and characterization. RSC Adv. **8**, 22103–22112 (2018)
12. Bünzli, J.C.G.: Lanthanide luminescence for biomedical analyses and imaging. Chem. Rev. **110**, 2729–2755 (2010)
13. Auzel, F.: Upconversion and anti-stokes processes with f and d ions in solids. Chem. Rev. **104**, 139–173 (2004)
14. Verma, V., Singh, S.V.: La-doped TiO_2 nanoparticles for photocatalysis: synthesis, activity in terms of degradation of methylene blue dye and regeneration of used nanoparticles. Arab. J. Sci. Eng. **48**, 16431–16443 (2023)
15. Lin, B.L., Chen, R., Zhu, M.L., She, A.S., Chen, W., Niu, B.T., Chen, Y.X., Lin, X.M.: Enhanced photoelectrochemical water splitting performance of Ce-doped TiO_2 nanorod array photoanodes for efficient hydrogen production. Catalysts **14** (2024)
16. Akhtar, J., Tahir, M.B., Sagir, M., Bamufleh, H.S.: Improved photocatalytic performance of Gd and Nd co-doped ZnO nanorods for the degradation of methylene blue. Ceram. Int. **46**, 11955–11961 (2020)
17. Edgar, A.: Luminescent materials, Springer handbook of electronic and photonic materials. Lumin. Mater. **38**, 999–1011
18. Zhang, H., Gao, H., Sun, Y., Zhao, J., You, W., Pan, G., Liu, Y., Zhang, H., Zhang, Z., Mao, Y.: Boosting the 1540 nm near-infrared luminescence of Er^{3+} doped double perovskites via cross-relaxation and energy transfer. Ceram. Int. **49**, 18068–18075 (2023)
19. De, M., Sharma, S., Jana, S.: Enhancement of $^5D_o \rightarrow {}^7F_2$ red emission of Eu^{3+} incorporated in lead sodium phosphate glass matrix. Phys. B Condens. Matter **556**, 131–135 (2019)
20. Yang, J., Zhang, C., Peng, C., Li, C., Wang, L., Chai, R., Lin, J.: Controllable red, green, blue (RGB) and bright white upconversion luminescence of Lu_2O_3:Yb^{3+}/Er^{3+}/Tm^{3+} nanocrystals through single laser excitation at 980 nm. Chem.—A Eur. J. **15**, 4649–4655 (2009)

21. Jin, Y., Lü, W., Zhang, J., Hao, Z., Zhang, X.: Luminescence properties of Tb^{3+}, Eu^{3+}, Tm^{3+} Co-doped $Na_5La(MoO_4)_4$ for white light-emitting diode. J. Nanosci. Nanotechnol. **14**, 3683–3686 (2014)
22. Cheng, Y., Shen, C., Shen, L., Xiang, W., Liang, X.: Tb^{3+}, Eu^{3+} co-doped CsPbBr 3 QDs glass with highly stable and luminous adjustable for white LEDs. ACS Appl. Mater. Interfaces **10**, 21434–21444 (2018)
23. Dong, B., Xu, S., Sun, J., Bi, S., Li, D., Bai, X., Wang, Y., Wang, L., Song, H.: Multifunctional $NaYF_4$:Yb^{3+}, E^{r3}+@Ag core/shell nanocomposites: integration of upconversion imaging and photothermal therapy. J. Mater. Chem. **21**, 6193–6200 (2011)
24. Eriksson, P., Truong, A.H.T., Brommesson, C., Du Rietz, A., Kokil, G.R., Boyd, R.D., Hu, Z., Dang, T.T., Persson, P.O.A., Uvdal, K.: Cerium oxide nanoparticles with entrapped gadolinium for high T1 relaxivity and ROS-scavenging purposes. ACS Omega **7**, 21337–21345
25. Wang, F., Liu, X.: Recent advances in the chemistry of lanthanide-doped upconversion nanocrystals. Chem. Soc. Rev. **38**, 976–989 (2009)
26. Chen, G., Qiu, H., Prasad, P.N., Chen, X.: Upconversion nanoparticles: design, nanochemistry, and applications in theranostics. Chem. Rev. **114**, 5161–5214 (2014)
27. Guoxi, R., Yongxin, G.: Regulation of photoelectric properties of multi-element (Sn, Fe, Ce) co-doped anatase TiO_2: the first principles study. Chem. Phys. Lett. **844**, 141265 (2024)
28. Yun, Y., Buragohain, P., Li, M., Ahmadi, Z., Zhang, Y., Li, X., Wang, H., Li, J., Lu, P., Tao, L., Wang, H., Shield, J.E., Tsymbal, E.Y., Gruverman, A., Xu, X.: Intrinsic ferroelectricity in Y-doped HfO_2 thin films. Nat. Mater. **21**, 903–909 (2022)
29. Beridze, G., Birnie, A., Koniski, S., Ji, Y., Kowalski, P.M.: DFT + U as a reliable method for efficient ab initio calculations of nuclear materials. Prog. Nucl. Energy **92**, 142–146 (2016)
30. Han, Z.K., Liu, W., Gao, Y.: Advancing the understanding of oxygen vacancies in ceria: insights into their formation, behaviour, and cataytic roles. JACS Au **5**, 1549–1569 (2025)
31. Fu, Q., Xu, T., Wang, D., Liu, C.: Rare earth modified carbon-based catalysts for oxygen electrode reactions: a machine learning assisted density functional theory investigation. Carbon N. Y. **223**, 119045 (2024)
32. Casida, M.E., Huix-Rotllant, M.: Progress in time-dependent density-functional theory. Annu. Rev. Phys. Chem. **63**, 287–323 (2012)
33. Yang, Y., Zhang, L., Xiao, C., Huang, Z., Zhao, F., Yin, J.: Highly efficient upconversion photodynamic performance of rare-earth-coupled dual-photosensitizers: ultrafast experiments and excited-state calculations. Nanophotonics **13**, 443–455 (2024)
34. Zhao, Z., Zhao, X., Zhang, M., Sun, X.: Charge-transfer process in surface-enhanced raman scattering based on energy level locations of rare-earth Nd^{3+}-doped TiO_2 nanoparticles. Nanomaterials **11** (2021)
35. Chaussedent, S., Teboul, V., Monteil, A.: Molecular dynamics simulations of rare-earth-doped glasses. Curr. Opin. Solid State Mater. Sci. **7**, 111–116 (2003)
36. Brugnoli, L., Menziani, M.C., Urata, S., Pedone, A.: Development and application of a ReaxFF reactive force field for cerium oxide/water interfaces. J. Phys. Chem. A **125**, 5693–5708 (2021)
37. Grieshammer, S., Grope, B.O.H., Koettgen, J., Martin, M.: A combined DFT + U and Monte Carlo study on rare earth doped ceria. Phys. Chem. Chem. Phys. **16**, 9974–9986 (2014)
38. Skourtis, S.S., Liu, C., Antoniou, P., Virshup, A.M., Beratan, D.N.: Dexter energy transfer pathways. Natl. Acad. Sci. **113**, 8115–8120 (2016)
39. Sahoo, H.: Förster resonance energy transfer—a spectroscopic nanoruler: principle and applications. J. Photochem. Photobiol. C Photochem. Rev. **12**, 20–30 (2011)
40. Zheng, Y.: Investigation of the Energy Transfer Network in Upconverting Nanoparticles (Massachusetts Institute of Technology) (2025)
41. Mulwa, M.: A Combined ab Initio and Experimental Study of Lanthanides and/or Transition Metal Doped Oxides (2017)
42. Pawlik, G., Niczyj, J., Noculak, A., Radosz, W., Podhorodecki, A.: Multiband Monte Carlo modelling of upconversion emission in sub 10 nm β-$NaGdF_4$:Yb^{3+}, Er^{3+} nanocrystals-effect of Yb^{3+} content. J. Chem. Phys. **146**, 244111 (2017)

43. Zaytsev, K.I., Gorelik, V.S., Khorokhorov, A.M., Yurchenko, S.O.: FDTD simulation of the electromagnetic field surface states in 2D photonic crystals. J. Phys. Conf. Ser. **486** (2014)
44. Mu, Z., Gu, H., Zhang, B., Zheng, J., Zhai, Z., He, X., Zhao, Y.: FDTD modelling of photonic crystal-incorporated gold nanoparticles for enhancing the localized electric field. J. Mater. Chem. C **5**, 9540–9544 (2017)
45. Paudel, H.P., May, P.S.: Design, fabrication, and characterization of a plasmonic upconversion enhancer and its prospects for photovoltaics. J. Photonics Energy **3**, 035598 (2014)
46. You, J.W., Panoiu, N.C.: Analysis of the interaction between classical and quantum plasmons via FDTD-TDDFT method. IEEE J. Multiscale Multiphys. Comput. Tech. **4**, 111–118 (2019)
47. Kuganathan, N., Srikaran, R., Fossati, P.C.M., Chroneos, A.: Theoretical modelling of defects, dopants, and diffusion in the mineral ilmenite. Minerals **9**, 1–13 (2019)
48. Kotliar, G., Savrasov, S.Y., Haule, K., Oudovenko, V.S., Parcollet, O., Marianetti, C.A.: Electronic structure calculations with dynamical mean-field theory. Rev. Mod. Phys. **78**, 865–951 (2006)
49. Pawlik, G., Mitus, A.C.: Anomalous diffusion and decay of clusters of dopants in lanthanide-doped nanocrystals. Mater. (Basel) **18**, 1–15 (2025)
50. Yavuz, A., Aydin, D., Disli, B., Ozturk, T., Gul, B., Gubbuk, I.H., Ersoz, M.: Enhancing visible light photocatalytic activity of holmium doped g-C_3N_4 and DFT theoretical insights. Environ. Sci. Pollut. Res. **31**, 44828–44847 (2024)
51. Lin, Y., Jiang, Z., Hu, X., Zhang, X., Fan, J.: The electronic and optical properties of Eu/Si-codoped anatase TiO_2. Appl. Phys. Lett. **100**, 102105 (2012)
52. Tessitore, G., Mandl, G.A., Brik, M.G., Park, W., Capobianco, J.A.: Recent insights into upconverting nanoparticles: spectroscopy, modelling, and routes to improved luminescence. Nanoscale **11**, 12015–12029 (2019)
53. Balhara, A., Gupta, S.K., Aggarwal, N., Srivastava, S., Panda, J.J., Patra, S., Chakraborty, A., Rakshit, S., Chakravarty, R.: Remarkably enhanced upconversion luminescence in Na^+ codoped spinel nanoparticles for photothermal cancer therapy and SPECT imaging. Mater. Adv. **4**, 5338–5352 (2023)
54. Khuili, M., Fazouan, N., El Makarim, H.A., Atmani, E.H., Rai, D.P., Houmad, M.: DFT study of rare earth (Tm, Yb, Ce) doped ZnO: structural, optoelectronic and electrical properties. arXiv 1–15
55. Shannon, R.D., Baur, N.H., Gibbs, O.H., Eu, M., Cu, V.: Revised effective ionic radii in halides and chaleogenides. Acta Crystallogr. **32**, 751–767 (1976)
56. Cojocaru, B., Avram, D., Kessler, V., Parvulescu, V., Seisenbaeva, G., Tiseanu, C.: Nanoscale insights into doping behavior, particle size and surface effects in trivalent metal doped. Sci. Rep. **7**, 4–12 (2017)
57. Chu, K., Cheng, Y., Li, Q., Liu, Y., Tian, Y.: Fe-doping induced morphological changes, oxygen vacancies and Ce^{3+}–Ce^{3+} pairs in CeO_2 for promoting electrocatalytic nitrogen fixation. J. Mater. Chem. A **8**, 5865–5873 (2020)
58. Feng, S., Naim, S., Katea, Ek, M., Westin, G., Tai, C.W.: Atomistic structure investigation of Eu-doped ZnO nanosponges. Inorg. Chem. **64**, 232–241 (2025)
59. Iglesias-Mejuto, A., Lamy-Mendes, A., Pina, J., Costa, B.F.O., García-González, C. A., Durães, L.: Synthesis of highly luminescent silica-coated upconversion nanoparticles from lanthanide oxides or nitrates using co-precipitation and sol–gel methods. Gels **10** (2024)
60. Yang, K., Dai, Y., Huang, B.: Understanding photocatalytic activity of S- and P-doped TiO_2 under visible light from first-principles. J. Phys. Chem. C **111**, 18985–18994 (2007)
61. Nassoko, D, Li, Y.-F., Li, J.-L., Li, X., Yu, Y.: Neodymium-doped TiO_2 with anatase and brookite two phases.pdf. Int. J. Photoenergy **1**, 716087 (2012)
62. Izeddin, I., Timmerman, D., Gregorkiewicz, T., Moskalenko, A.S., Prokofiev, A.A., Yassievich, I.N., Fujii, M.: Energy transfer in Er-doped SiO_2 sensitized with Si nanocrystals. Phys. Rev. B **78**, 35327 (2008)
63. Liu, X., Li, L., Noh, H., Moon, B., Choi, B., Jeong, J.: Chemical bond properties and charge transfer bands of O^{2-}-Eu^{3+}, O^{2-}-Mo^{6+} and O^{2-}-W^{6+} in Eu^{3+}-doped garnet hosts $Ln_3M_5O_{12}$ and ABO_4 molybdate and tungstate phosphors. Dalton Trans. **43** (2014)

64. Trovarelli, A.: Catalysis by ceria and related materials. In: Trovarelli, A., Fornasiero, P. (eds.). Imperial College Press, London (2002)
65. Carey, J.J., McKenna, K.P.: Screening doping strategies to mitigate electron trapping at anatase TiO_2 surfaces. J. Phys. Chem. C **123**, 22358–22367 (2019)
66. Ruiz-caridad, A.: Rare-Earth Doped Crystalline Oxide on Silicon Photonics Platform (2021)
67. Alam, M. W., Khalid, N.R., Naeem, S., Niaz, N.A., Ahmad Mir, T., Nahvi, I., Souayeh, B., Zaidi, N.: Novel Nd-N/TiO_2 nanoparticles for photocatalytic and antioxidant applications using hydrothermal approach. Materials (Basel) **15** (2022)
68. Toma, M., Selyshchev, O., Havryliuk, Y., Pop, A., Zahn, D.R.T.: Optical and structural characteristics of rare earth-doped ZnO nanocrystals prepared in colloidal solution. Photochem **2**, 515–527 (2022)
69. Singh, P., Singh, R. K., Kumar, R.: Journey of ZnO quantum dots from undoped to rare-earth and transition metal-doped and their applications. RSC Adv. **11**, 2512–2545 (2021)
70. Lojpur, V.M., Ahrenkiel, P.S., Dramićanin, M.D.: Color-tunable up-conversion emission in Y_2O_3:Yb^{3+}, Er^{3+} nanoparticles prepared by polymer complex solution method. Nanoscale Res. Lett. **8**, 1–6 (2013)
71. Ghazyani, N., Majles Ara, M.H., Raoufi, M.: Nonlinear photoresponse of $NaYF_4$:Yb,Er@$NaYF_4$ nanocrystals under green CW excitation: a comprehensive study RSC Adv. **10**, 25696–25702 (2020)
72. Yu, C., Hansen, K.K., Madsen, L.B.: Enhanced high-order harmonic generation in donor-doped band-gap materials. Phys. Rev. A **99**, 1–8 (2019)
73. Mendez, V., Fabre, M., Cornier, T., Bosselet, F., Loridant, S., Asaad, S., Daniele, S.: Enhanced photoluminescence of europium-doped TiO_2 nanoparticles using a single-source precursor strategy. Molecules **29** (2024)
74. Lin, Y.-C., Erhart, P., Bettinelli, M., George, N.C., Parker, S.F., Karlsson, M.: Understanding the interactions between vibrational modes and excited state relaxation in $Y_{3-x}Ce_xAl_5O_{12}$: design principles for phosphors based on 5d–4f transitions. Chem. Mater. **30**, 1865–1877 (2018)
75. Yu, D., Ballato, J., Riman, R.E.: The temperature-dependence of multiphonon relaxation of rare-earth ions in solid-state hosts. J. Phys. Chem. C **176**, 139–148 (2017)
76. Boetti, N.G., Negro, D., Scarpignato, G.C., Perrone, G., Abrate, S., Lousteau, J., Mura, E., Milanese, D.: Photonic glasses for IR and mid-IR spectral range. Proc. SPIE **10564**, 74 (2017)
77. Minisini, B., Wang, Q.A., Tsobnang, F.: Ab initio investigation of the lattice dynamics of fluoride scheelite LiYF4. J. Phys. Condens. Matter **17**, 4953 (2005)
78. Sundell, P.G., Björketun, M.E., Wahnström, G.: Thermodynamics of doping and vacancy formation in $BaZrO_3$ perovskite oxide from density functional calculations. Phys. Rev. B **73**, 104112 (2006)
79. Nolan, M.: Charge compensation and Ce^{3+} formation in trivalent doping of the CeO_2 (110) surface: the key role of dopant ionic radius. J. Phys. Chem. C **115**, 6671–6681 (2011)
80. Liu, K., Bian, G., Zhang, Z., Ma, F., Su, L.: Modelling and analyzing the glass-like heat transfer behavior of rare-earth doped alkaline earth fluoride crystals. CrystEngComm **24**, 6468–6476 (2022)
81. Lan, G., Ouyang, B., Song, J.: The role of low-lying optical phonons in lattice thermal conductance of rare-earth pyrochlores: a first-principle study. Acta Mater. **91**, 304–317 (2015)
82. Shalav, A., Richards, B.S., Green, M.A.: Luminescent layers for enhanced silicon solar cell performance: up-conversion. Sol. Energy Mater. Sol. Cells **91**, 829–842 (2007)
83. Trupke, T., Green, M.A., Würfel, P.: Improving solar cell efficiencies by down-conversion of high-energy photons. J. Appl. Phys. **92**, 1668–1674 (2002)
84. Liu, X., Lei, D.Y.: Simultaneous excitation and emission enhancements in upconversion luminescence using plasmonic double-resonant gold nanorods. Sci. Rep. **5**, 1–13 (2015)
85. Kenyon, A.J.: Erbium in silicon. Semicond. Sci. Technol. **20**, 65–84 (2005)
86. Chen, W., Yuan, P., Zhang, S., Sun, Q., Liang, E.: Electronic properties of anatase TiO_2 doped by lanthanides: a DFT+U study. Phys. B Condens. Matter **407**, 1038–1043 (2012)
87. Tuley, R., Simpson, K.: ZT optimization: an application focus. Materials (Basel) **10** (2017)

88. Yatim, N.M., Sallehin, N.Z.I.M., Suhaimi, S., Hashim, M.A.: A review of ZT measurement for bulk thermoelectric material. AIP Conf. Proc. **1972** (2018)
89. Lyu, W.Y., Liu, W.D., Li, M., Hong, M., Guo, K., Luo, J., Xing, J., Sun, Q., Xu, S., Zou, J., Chen, Z.G.: The effect of rare earth element doping on thermoelectric properties of GeTe. Chem. Eng. J. **446** (2022)
90. Lyu, W.Y., Hong, M., Liu, W.D., Li, M., Sun, Q., Xu, S.D., Zou, J., Chen, Z.G.: Rare-earth Nd inducing record-high thermoelectric performance of $(GeTe)_{85}(AgSbTe_2)_{15}$. Energy Mater. Adv. (2021)
91. Sikam, P., Thirayatorn, R., Kaewmaraya, T., Thongbai, P., Moontragoon, P., Ikonic, Z.: Improved thermoelectric properties of $SrTiO_3$ via (La, Dy and N) co-doping: DFT approach. Molecules **27** (2022)
92. Isaacs, E.B., Wolverton, C.: Electronic structure and phase stability of Yb-filled $CoSb_3$ skutterudite thermoelectrics from first-principles. Chem. Mater. **31**, 6154–6162 (2019)
93. Sayago, I., Santos, J.P., Sánchez-Vicente, C.: The effect of rare earths on the response of photo UV-activate ZnO gas sensors. Sensors **22** (2022)
94. Chu, H., Yao, D., Chen, J., Yu, M., Su, L.: Double-emission ratiometric fluorescent sensors composed of rare-earth-doped ZnS quantum dots for Hg^{2+}detection. ACS Omega **5**, 9558–9565 (2020)
95. Li, P., Feng, B., Feng, Y., Song, G., Cheng, X., Deng, Y., Wei, J.: Synthesis of mesoporous lanthanum-doped SnO_2 spheres for sensitive and selective detection of the glutaraldehyde disinfectant. ACS Sens. **8**, 3723–3732 (2023)
96. Hong, S.-N., Kye, Y.-H., Yu, C.-J., Jong, U.-G., Ri, G.-C., Choe, C.-S., Kim, K.-H., Han, J.-M.: Ab initio thermodynamic study of the SnO_2 (110) surface in an O_2 and NO environment: a fundamental understanding of the gas sensing mechanism for NO and NO_2. Phys. Chem. Chem. Phys. **18**, 31566–31578 (2016)
97. Huang, X., Wu, M., Huang, R., Yang, G.: How doping regulates As(III) adsorption at TiO_2 surfaces: a DFT + U study. Molecules **29** (2024)
98. Al-Thani, A.F.A.A., Ba-Busail, F.E.A., Fadaly, M.K.S., Shraim, A.M., Salih, K.S.M.: Spectral, TD-DFT, and metal sensing investigations of four chromone-based compounds. J. Mol. Struct. **1325**, 141029 (2025)
99. Jin, R., Hu, S., Gui, J., Liu, D.: A convenient method to prepare novel rare earth metal ce-doped carbon nitride with enhanced photocatalytic activity under visible light. Bull. Korean Chem. Soc. **36**, 17–23 (2015)
100. Umami, R., Permatasari, F.A., Sundari, C.D.D., Muttaqien, F., Iskandar, F.: Surface functional groups effect on the absorption spectrum of carbon dots: initial TD-DFT study. J. Phys. Conf. Ser. **2243** (2022)
101. Benamara, M., Ly, A., Soltani, S., Essid, M., Dahman, H., Dhahri, R., El Mir, L., Debliquy, M., Lahem, D.: Enhanced detection of low concentration volatile organic compounds using advanced doped zinc oxide sensors. RSC Adv. **13**, 30230–30242 (2023)
102. Park, C.W., Park, D.J.: Development of Er^{3+}, Yb^{3+} co-doped Y_2O_3 NPs according to Yb^{3+} concentration by LPPLA method: potential further biosensor. Biosensors **11** (2021)

Thermoelectric Applications of Rare Earth-Doped Metal Oxides

Indu Gupta, Rahul Jain, and Kaushlendra Pandey

Abstract Thermoelectric (TE) materials play a crucial role in waste heat recovery and energy harvesting, converting thermal gradients directly into electricity. Among the many materials studied, oxide-based thermoelectrics have attracted significant attention due to their thermal stability, environmental friendliness, and low cost. Yet they have historically functioned less well than traditional chalcogenide-based counterparts. The existing works have reported various RE doping as new tactics to improve TE performance by adjusting carrier concentration, enhancing Seebeck coefficient, and diminishing thermal conductivity. This chapter thoroughly examines the function of rare earth dopants in diverse oxide systems, including ZnO, $CaMnO_3$, $SrTiO_3$, and $NaCo_2O_4$. We emphasize the doping mechanisms, structural alterations, charge transport phenomena, and phonon scattering techniques introduced by rare earth (RE) elements.

1 Introduction

The growing energy needs of today's society, which are caused by technological progress and global industrialization, have made it very important to find sustainable and eco-friendly energy sources. Not only are conventional fossil fuels limited, but they also release greenhouse gases that cause climate change and damage the environment. In this situation, the search for other ways to convert energy has become more urgent. Thermoelectric (TE) materials stand out as a promising option because they can directly turn waste heat into electrical energy without using moving parts or chemical reactions. This is possible because of the Seebeck and Peltier effects [1–3].

I. Gupta (✉)
Department of Physics, UIS, Chandigarh University, Mohali, Punjab, India
e-mail: indugupta17887@gmail.com

R. Jain
Department of Applied Sciences, CGC University, Mohali, India

K. Pandey
Department of Electricity, Metropolitan University of Technology, Santiago, Chile

© The Author(s), under exclusive license to Springer Nature Singapore Pte Ltd. 2026
N. Kumar et al. (eds.), *Rare Earth-Doped Metal Oxide Nanostructures*, Engineering Materials, https://doi.org/10.1007/978-981-95-6192-6_8

Thermoelectric energy conversion is based on the principles of solid-state physics. The Seebeck effect happens when charge carriers (electrons or holes) move from the hot side of a TE material to the cold side, creating a voltage difference. On the other hand, passing an electric current through the material can create a temperature gradient (the Peltier effect), which makes thermoelectrics useful for both generating electricity and cooling solid objects, as shown in Fig. 1. A dimensionless figure of merit is used to show how well such a material works:

$$ZT = \frac{S^2 \sigma T}{\kappa},$$

where

S is the Seebeck coefficient (μV/K),
σ is the electrical conductivity (S/cm),
κ is the thermal conductivity (W/mK),
T is the absolute temperature (K).

Increased ZT is a measure of improved thermoelectric efficiency [4]. But it is difficult to optimize ZT because its parameters are coupled. For instance, an increase in carrier concentration increases σ but reduces S, whereas decrease in κ is typically limited by the inherent phonon transport characteristics of the material. Conventional high-performance thermoelectric materials are chalcogenides such as bismuth telluride (Bi_2Te_3), lead telluride (PbTe), and silicon–germanium (SiGe) alloys [5–7]. They have outstanding ZT values at various temperature ranges and were commercialized for specialized applications in aerospace, consumer electronics, and medical

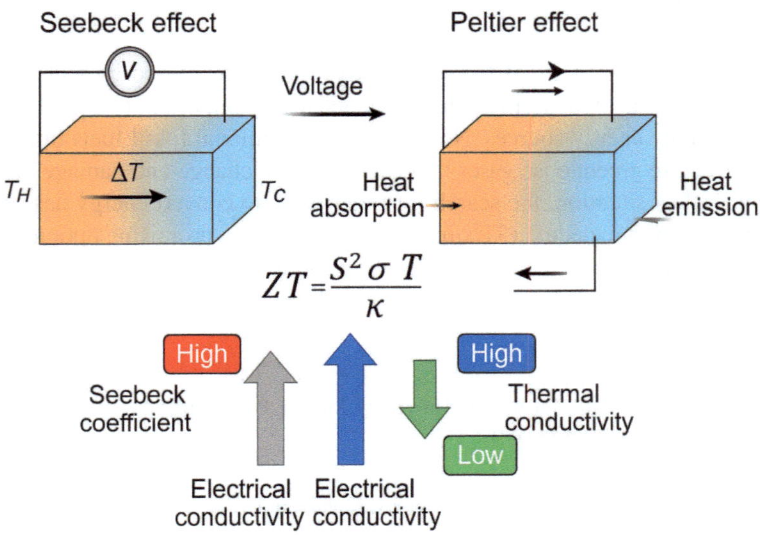

Fig. 1 Thermoelectric principles and ZT optimization

areas. But there are some inherent limitations that prevent them from universal application:

- Toxicity (e.g., Pb-based materials),
- Poor oxidation resistance,
- Lack of constituent elements (e.g., Te),
- Degradation at high temperatures.

These limitations have necessitated a revolutionary shift toward less environmentally harmful thermoelectric oxides, which are chemically stable at high temperatures and are made up of elements readily available in the Earth's crust [8–10]. Oxide ceramics such as ZnO, $SrTiO_3$, $CaMnO_3$, and $NaCo_2O_4$ have emerged as potential candidates, especially for high-temperature (500–1000 °C) thermoelectric devices. Although superior in most regards, the materials are susceptible to relatively low electrical conductivity and high lattice thermal conductivity, leading to moderate ZT values (typically less than 0.5), thus limiting their usability [11, 12].

One of the most promising approaches to enhance the thermoelectric performance of these oxides is through rare earth (RE) element doping. Rare earth elements (La–Lu, Sc, Y) possess distinctive electronic configurations (partially filled 4f orbitals), high atomic mass, and a wide range of ionic radii, making them suitable for modifying the physical and electronic structure of host oxide materials [13]. Doping with rare earth elements can improve thermoelectric performance through several mechanisms:

- Carrier concentration tuning: RE ions with higher valence states can act as electron donors, thereby increasing electrical conductivity without significantly compromising the Seebeck coefficient.
- Lattice distortion and strain engineering: The size mismatch between the dopant and host cation introduces strain, which scatters phonons and reduces thermal conductivity.
- Mass fluctuation scattering: The large atomic mass and localized 4f electrons of RE ions enhance phonon scattering, further decreasing the lattice thermal conductivity.
- Band structure modification: Certain RE elements introduce shallow donor or acceptor levels that can enhance power factor ($S^2\sigma$) by promoting beneficial charge transport pathways [14–16].

In ZnO, rare earth dopants such as La^{3+} and Y^{3+} are known to substitute Zn^{2+} sites, effectively increasing the carrier concentration and reducing grain boundary resistance. In $SrTiO_3$, La or Dy substitution at the Sr site can increase electron mobility, while in $CaMnO_3$, RE doping at the Ca site enhances mixed-valent Mn^{3+}/Mn^{4+} conduction pathways [17–19]. Layered cobaltites like $NaCo_2O_4$ also benefit from RE doping in terms of structural stability and phonon scattering. Moreover, the synergistic effects of RE doping with nanostructuring, defect engineering, and composite formation open new avenues for engineering oxide-based thermoelectrics with ZT values approaching or exceeding 1.0. Such values make them viable candidates for scalable applications, including industrial waste heat recovery, automotive thermoelectric generators (TEGs), and off-grid energy harvesting systems [20].

This chapter aims to present a comprehensive overview of the current state-of-the-art in rare earth-doped metal oxide thermoelectrics. We explore fundamental concepts, structural chemistry, doping strategies, charge and heat transport mechanisms, and performance trends in various oxide systems. Special emphasis is given to ZnO, $SrTiO_3$, $CaMnO_3$, and $NaCo_2O_4$, the most widely studied RE-doped oxides. We also discuss recent breakthroughs, challenges in doping efficiency and material stability, and potential pathways for the design of next-generation thermoelectric devices.

2 Fundamentals of Thermoelectricity in Oxides

2.1 Thermoelectric Parameters

The efficiency of thermoelectric (TE) materials is governed by three primary and interdependent physical properties: the Seebeck coefficient (S), electrical conductivity (σ), and thermal conductivity (κ). These factors are combined in the dimensionless figure of merit:

$$ZT = \frac{S^2 \sigma T}{\kappa},$$

where

S is the Seebeck coefficient (μV/K),
σ is the electrical conductivity (S/cm),
\kappa is the thermal conductivity (W/mK),
T is the absolute temperature (K).

An ideal thermoelectric material should possess a high Seebeck coefficient, high electrical conductivity, and low thermal conductivity. However, optimizing these parameters simultaneously is challenging due to their inherent coupling.

2.1.1 Seebeck Coefficient (S)

The Seebeck coefficient measures the voltage generated per unit temperature difference and is described by:

$$S = \Delta V / \Delta T.$$

It is sensitive to the carrier concentration and the density of states (DOS) near the Fermi level. In general, materials with low carrier concentration exhibit high Seebeck coefficients due to the steep energy gradient in the DOS [2, 21]. However, this also results in low electrical conductivity, highlighting the trade-off in optimizing ZT. In oxide semiconductors, the wide bandgaps typically yield low intrinsic carrier concentrations, but this can be tuned via extrinsic doping. Rare earth (RE) doping, in particular, enables manipulation of the Fermi level and band structure. For instance, La^{3+} or Y^{3+} doping introduces donor levels and adjusts carrier energy distribution to enhance the Seebeck coefficient [22, 23].

2.1.2 Electrical Conductivity (Σ)

Electrical conductivity is defined as:

$$\sigma = ne\mu,$$

Where n is the carrier concentration, e is the elementary charge, and μ is the carrier mobility [24]. High electrical conductivity is desirable for efficient thermoelectric performance, but increasing n tends to reduce the Seebeck coefficient, again underscoring the need for a balanced design approach. In RE-doped oxides, electrical conductivity is typically improved through substitutional doping, where a higher valence RE ion replaces a host cation and contributes free electrons. For example, La^{3+} replacing Sr^{2+} in $SrTiO_3$ results in additional n-type carriers, thereby boosting σ [25]. However, increased doping can lead to ionized impurity scattering and carrier localization, reducing the mobility and offsetting gains in conductivity [26].

2.1.3 Thermal Conductivity (K)

Thermal conductivity comprises two components:

$$\kappa = \kappa_e + \kappa_l$$

- κ_e is the electronic thermal conductivity, associated with the flow of charge carriers and typically related to σ via the Wiedemann–Franz law.
- κ_l is the lattice thermal conductivity, arising from phonon transport within the crystal lattice [27].

In oxide semiconductors, κ_l dominates due to the relatively low carrier concentration. To improve ZT, reducing κ_l is crucial. Techniques such as nanostructuring, alloying, and especially rare earth doping introduce mass fluctuation and strain fields that scatter phonons and reduce lattice thermal conductivity [15]. The large atomic mass and ionic radius mismatch of RE elements create point defects and localized lattice distortions that effectively disrupt phonon propagation. For instance, doping $CaMnO_3$ with Sm^{3+} or Gd^{3+} results in significant phonon scattering and a marked reduction in thermal conductivity [22, 25]. In summary, optimizing thermoelectric performance demands a fine-tuned balance among Seebeck coefficient, electrical conductivity, and thermal conductivity. While these parameters are interdependent, rare earth doping has proven an effective strategy to simultaneously manipulate all three. This multidimensional control makes RE-doped oxides strong candidates for high-temperature and environmentally sustainable thermoelectric applications.

2.2 *Mechanisms in Oxide Semiconductors*

Transition metal oxides (TMOs) have emerged as promising thermoelectric (TE) materials due to their unique electronic configurations, chemical stability, and thermal robustness at elevated temperatures. However, the mechanisms that govern charge transport in oxides are more complex than in traditional semiconductors like Bi_2Te_3 or PbTe, primarily due to the localized nature of d-electrons and strong electron–phonon interactions [28, 29]. Understanding these conduction mechanisms is critical for tailoring and optimizing their thermoelectric performance. In oxide semiconductors, electronic transport can occur via multiple mechanisms, including:

2.2.1 Band Conduction

In well-ordered crystalline oxides with delocalized electronic states, conduction occurs through the extended states in the conduction band. This is typical for *n*-type ZnO and doped $SrTiO_3$, where donor levels introduced via doping (e.g., Al or La) provide additional free electrons that occupy the conduction band [30]. The conduction mechanism in these materials resembles that of classical semiconductors, governed by band theory and characterized by relatively high carrier mobility and moderate Seebeck coefficients. The effective mass of the charge carriers and the density of states (DOS) near the Fermi level are key parameters influencing band conduction. In $SrTiO_3$, the large degeneracy of the Ti 3d t_2g orbitals leads to a high DOS near the Fermi level, enhancing the Seebeck coefficient even at high carrier concentrations [31].

2.2.2 Small Polaron Hopping

Many transition metal oxides—especially $CaMnO_3$ and $NaCo_2O_4$—exhibit significant localization of charge carriers due to strong electron-lattice coupling. In these systems, charge carriers can become trapped in local potential wells, forming polarons, which are quasi-particles consisting of an electron or hole coupled to a lattice distortion. Conduction in this case occurs through thermally activated hopping between localized states, known as small polaron hopping [32]. This mechanism is characterized by an Arrhenius-type temperature dependence of electrical conductivity:

$$\sigma(T) = \sigma_0 \exp(-E_a k_B T),$$

where E_a is the activation energy for hopping, and k_B is the Boltzmann constant. In materials like $CaMnO_3$, the mixed-valent nature of Mn^{3+}/Mn^{4+} facilitates polaron hopping and contributes to electrical conduction [33]. However, mobility is often low, which limits the electrical conductivity and thus the power factor.

2.2.3 Spin and Orbital Degrees of Freedom

Layered cobaltites like $NaCo_2O_4$ demonstrate another unique conduction behavior resulting from strong correlation between spin, charge, and orbital degrees of freedom. The CoO_2 layers in $NaCo_2O_4$ form a two-dimensional triangular lattice where the coexistence of high-spin and low-spin Co ions leads to spin entropy transport—a mechanism where entropy carried by spin degeneracy contributes significantly to the Seebeck coefficient [34]. This explains the unusually high thermopower (~ 100 μV/K) observed in these materials despite their metallic conductivity [35].

2.2.4 Role of Oxygen Vacancies and Defect Chemistry

Oxide semiconductors are highly sensitive to oxygen stoichiometry, which can significantly alter their conduction mechanisms. Oxygen vacancies act as donor defects in n-type materials like ZnO and $SrTiO_3$, enhancing carrier concentration and modifying the conduction pathway [36]. In reduced $SrTiO_3$, Ti^{4+} is partially converted to Ti^{3+}, facilitating conduction via mixed-valent hopping, similar to polaronic behavior [37].

Overall, oxide semiconductors exhibit diverse and often complex conduction mechanisms that are strongly influenced by crystal structure, defect chemistry, orbital overlap, and temperature. A comprehensive understanding of these mechanisms—especially the balance between band-like and hopping conduction—is essential for the rational design of high-performance thermoelectric oxides. Importantly, rare earth doping can modulate these mechanisms by tuning the carrier concentration, modifying the band structure, or introducing beneficial disorder to enhance phonon scattering while maintaining favorable charge transport pathways.

3 Role of Rare Earth Elements in Oxide Thermoelectrics

Rare earth (RE) elements—comprising the lanthanide series (La to Lu), along with Scandium (Sc) and Yttrium (Y)—play a critical role in advancing the performance of oxide-based thermoelectric (TE) materials. Their unique electronic configurations, characterized by partially filled and localized 4f orbitals, enable them to effectively alter the electronic, structural, and phononic properties of host lattices in a controlled manner. In TE oxides, RE elements primarily function as dopants, substituting either A-site or B-site cations in perovskite, spinel, or layered structures, thereby improving carrier transport and suppressing lattice thermal conductivity.

3.1 Site-Specific Substitution and Structural Considerations

Oxide thermoelectrics such as $SrTiO_3$, $CaMnO_3$, ZnO, and $NaCo_2O_4$ typically have perovskite, wurtzite, or layered crystal structures with well-defined cation sites. In these structures:

- **A-site substitution** involves replacing large, divalent cations like Sr^{2+} or Ca^{2+} with trivalent RE^{3+} ions (e.g., La^{3+}, Sm^{3+}), introducing extra electrons for n-type conduction (Fig. 2) [38, 39].
- **B-site substitution**, though less common, may involve RE ions replacing smaller transition metal cations (e.g., Ti^{4+}, Mn^{4+}), affecting local magnetic and electronic states (Fig. 2) [40].

In $SrTiO_3$, La^{3+} typically substitutes Sr^{2+} on the A-site, contributing one extra electron per La ion and increasing n-type carrier density [20]. Similarly, in $CaMnO_3$, RE^{3+}

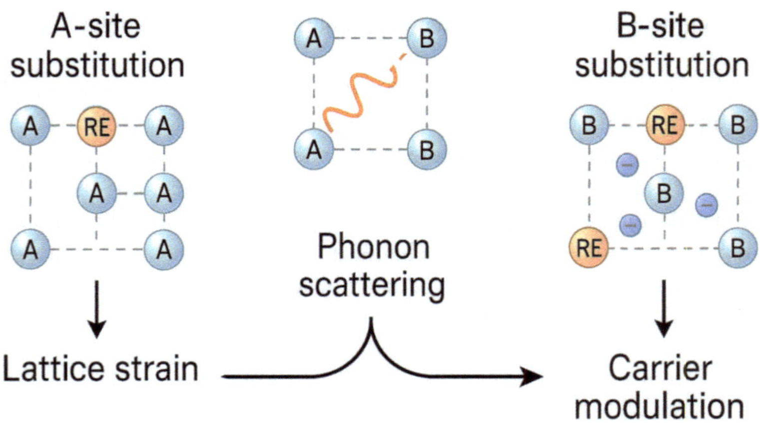

Fig. 2 Schematic showing RE substitution at A- or B-sites, inducing lattice strain, phonon scattering, and carrier modulation

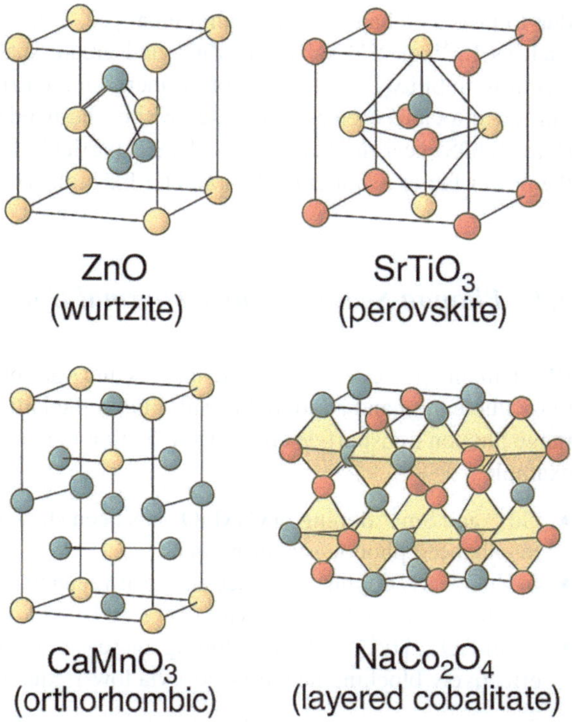

Fig. 3 Crystal structures of ZnO (wurtzite), $SrTiO_3$ (perovskite), $CaMnO_3$ (orthorhombic), and $NaCo_2O_4$ (layered cobalitate)

(e.g., Sm^{3+}, Y^{3+}) replaces Ca^{2+}, resulting in increased Mn^{3+}/Mn^{4+} mixed valency that enhances polaron hopping conduction [33]. The ionic radius mismatch between RE dopants and host cations induces local lattice distortions that influence both electronic structure and phonon scattering behavior [15]. The crystal structures are shown in Fig. 3.

3.2 *Carrier Concentration Control and Electronic Structure Modulation*

One of the primary thermoelectric benefits of RE doping is its ability to tailor carrier concentration through valence mismatch. For instance, La^{3+} and Nd^{3+}, when substituted at Sr^{2+} or Ca^{2+} sites, act as donor dopants, increasing the number of conduction electrons:

$$\mathrm{RE}^{3+} \rightarrow M^{2+} + e^{-}.$$

This mechanism has been effectively applied in $SrTiO_3$, where La doping increases electrical conductivity without significantly compromising the Seebeck coefficient due to favorable band structure alignment [20, 28]. Similarly, in ZnO, RE elements such as La^{3+} or Y^{3+} substitute Zn^{2+} and create shallow donor levels, enhancing electron mobility [36]. In addition to increasing carrier density, some RE ions can introduce localized electronic states within the bandgap, modifying the density of states (DOS) near the Fermi level. This is particularly advantageous for optimizing the power factor $S^2\sigma$ a crucial parameter for high ZT [7].

3.3 Phonon Scattering and Thermal Conductivity Reduction

RE elements possess large atomic masses and variable ionic radii, which introduce mass fluctuation and strain field scattering in the lattice. These effects disrupt long-range phonon transport, leading to a reduction in lattice thermal conductivity κ_l. For example:

- Gd^{3+} and Sm^{3+} doping in $CaMnO_3$ has been shown to reduce κ by over 30% due to enhanced phonon scattering [41].
- In ZnO, La^{3+} doping introduces oxygen vacancies and grain boundary scattering that lower thermal conductivity [42].
- In $SrTiO_3$, Dy^{3+} and Pr^{3+} doping increases disorder in the Sr–O sublattice, effectively blocking heat transport via low-frequency phonons [43].

These effects are especially beneficial in oxides where κ_l typically dominates over electronic thermal conductivity κ_e, especially in wide-bandgap systems.

3.4 Influence on Defect Chemistry and Microstructure

RE doping also significantly affects the defect chemistry, grain growth, and microstructural evolution of oxide thermoelectrics:

- In ZnO, RE doping can suppress grain coarsening and enhance density during sintering, contributing to better charge transport across grain boundaries [44].
- In $NaCo_2O_4$, RE substitution stabilizes Na-deficient structures and maintains the integrity of CoO_2 layers, which are essential for high Seebeck coefficients [45].

The nature of the rare earth ion also influences the formation energy of oxygen vacancies and point defects, indirectly affecting the balance between electrical conductivity and thermal conductivity.

Table 1 Summary of effects across oxide systems

Oxide system	Common RE dopants	Primary effects	References
ZnO	La^{3+}, Y^{3+}, Nd^{3+}	Carrier concentration ↑, κ ↓	[36, 42, 44]
$SrTiO_3$	La^{3+}, Dy^{3+}, Pr^{3+}	σ ↑, band modulation, κ ↓	[20, 43]
$CaMnO_3$	Sm^{3+}, Gd^{3+}, Y^{3+}	Mixed-valence hopping ↑, κ ↓	[33, 35, 41]
$NaCo_2O_4$	Ce^{3+}, Pr^{3+}, Sm^{3+}	Thermopower ↑, structural stability ↑	[25, 30, 45]

3.5 *Multiband and Entropy Effects*

Certain RE-doped oxides exhibit multiband conduction or enhanced electronic entropy transport, especially in cobalt-based oxides. For instance, RE-doped $NaCo_2O_4$ exhibits unusually high thermopower due to entropy-driven mechanisms and spin degeneracy effects in Co^{3+}/Co^{4+} systems [25, 30]. Furthermore, advanced first-principles studies suggest that some RE ions may contribute to band convergence or band flattening, leading to higher DOS near the Fermi level, enhancing Seebeck coefficient without excessive sacrifice in conductivity [35, 46, 47]. Rare earth doping has become an indispensable tool for enhancing the thermoelectric performance of metal oxides. By enabling precise control over carrier concentration, inducing phonon scattering through mass and strain fluctuations, and modifying electronic structures, RE elements provide a versatile platform for tuning oxide TE materials. Moreover, the structural and chemical stability imparted by RE doping makes these materials promising for high-temperature, large-scale energy recovery systems. Future efforts may explore multi-RE doping, high-entropy oxides, and RE–transition metal co-doping as new directions to break existing performance barriers. Effect of doping on various oxides are summarized in Table 1.

4 ZnO-Based Thermoelectrics and RE Doping

4.1 *Overview of ZnO*

Zinc oxide (ZnO) has garnered significant interest as a potential thermoelectric (TE) material due to its wide bandgap (~ 3.3 eV), high chemical and thermal stability, non-toxicity, and abundance in nature [48]. As an *n*-type oxide semiconductor, ZnO primarily conducts through electrons and possesses relatively high carrier mobility, making it suitable for high-temperature TE applications. However, the major drawback of pristine ZnO lies in its high lattice thermal conductivity ($\kappa \approx 50$ W/m·K), which significantly lowers its thermoelectric figure of merit (ZT). In addition, the intrinsic Seebeck coefficient of ZnO is moderate (~ 300 μV/K at 800 K), and the electrical conductivity is limited by grain boundary scattering and impurity levels [33, 49]. To overcome these limitations and enhance the thermoelectric performance of ZnO, researchers have employed various strategies, including cation substitution (doping), nanostructuring, and grain boundary engineering. Among these, doping with rare earth (RE) elements has emerged as an effective approach to tailor the electronic and phonon transport properties of ZnO [50, 51].

4.2 *Doping with La, Y, and Nd*

4.2.1 Lanthanum (La) Doping

La^{3+} ions can substitute for Zn^{2+} in the ZnO lattice due to their relatively close ionic radii (La^{3+}: ~ 1.03 Å; Zn^{2+}: ~ 0.74 Å) [33]. This substitution introduces extra electrons, thereby enhancing the *n*-type carrier concentration and increasing electrical conductivity (σ). Furthermore, the introduction of La alters the electronic band structure, narrowing the bandgap slightly and improving charge carrier mobility [52]. Studies have demonstrated that La doping can lead to a two-fold increase in power factor and reduce the thermal conductivity of ZnO, with ZT values reaching up to 0.3 at 1000 K [53, 54]. Additionally, La-doped ZnO displays better thermal stability, a key requirement for high-temperature TE devices.

4.2.2 Yttrium (Y) Doping

Yttrium ions (Y^{3+}), which are heavier and smaller than La^{3+}, are known to improve thermoelectric performance by two primary mechanisms: (1) modifying the carrier density and (2) enhancing phonon scattering due to mass fluctuation and strain field effects [55]. The mismatch between the ionic radius of Y^{3+} and Zn^{2+} creates local lattice distortions, promoting stronger phonon scattering and leading to a decrease in lattice thermal conductivity [56]. Experimental results have shown that Y doping in ZnO can reduce κ by over 30% and moderately improve the Seebeck coefficient while maintaining acceptable electrical conductivity, resulting in enhanced ZT values at high temperatures (~ 0.25 at 973 K) [57].

4.2.3 Neodymium (Nd) Doping

Nd^{3+} ions also substitute at the Zn-site and contribute additional charge carriers, enhancing electrical conductivity. Similar to La and Y, Nd introduces lattice distortion and increases phonon scattering. Compared to La, however, Nd tends to show a slightly lower solubility limit in ZnO, which can influence phase stability and microstructure [7, 56]. Nevertheless, optimized Nd-doping has shown potential in enhancing both the Seebeck coefficient and power factor at elevated temperatures.

4.3 Microstructure and Grain Boundary Effects

Rare earth doping does more than modify carrier concentration—it also significantly alters the microstructure of ZnO, particularly grain size, grain boundary density, and defect distribution. When ZnO is doped with RE elements and subjected to controlled nanostructuring, the increased density of grain boundaries serves as scattering centers for mid- to long-wavelength phonons, thereby reducing lattice thermal conductivity (κ_l) [58]. Grain boundary engineering, especially in combination with RE doping, creates energy barriers that filter low-energy carriers while preserving high-mobility carriers, which enhances the Seebeck coefficient (energy filtering effect). Additionally, nano-sized grains and dopant segregation at the grain boundaries impede phonon propagation, improving the ZT value without compromising electrical performance as shown in Fig. 4 [7, 56].

The synergy between dopant-induced phonon scattering and grain boundary-driven thermal resistance leads to a balanced enhancement in TE properties. Recent reports on La and Y co-doped ZnO indicate that this approach can reduce thermal conductivity to below 5 W/m·K while maintaining a power factor greater than 300 μW/mK^2 [55].

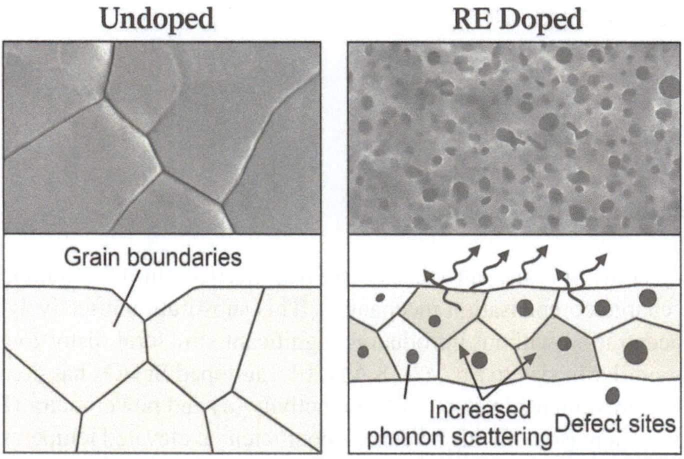

Fig. 4 Microstructural modifications from RE doping

5 $SrTiO_3$-Based Thermoelectrics and RE Doping

5.1 *Properties of $SrTiO_3$*

Strontium titanate ($SrTiO_3$) is a cubic perovskite oxide semiconductor that has been extensively studied for thermoelectric applications in the high-temperature regime. With a bandgap of ~ 3.2 eV and high-temperature chemical stability, it is especially promising for power generation in harsh environments [59]. $SrTiO_3$ exhibits an inherently large Seebeck coefficient (S ~ 400–600 μV/K) at elevated temperatures (800–1000 K), making it a favorable candidate despite its relatively high thermal conductivity and moderate electrical conductivity [60, 61]. The conduction in $SrTiO_3$ is dominated by electron transport in the Ti 3d t_2g conduction band, and this behavior can be substantially tuned by chemical doping at either the A-site (Sr^{2+}) or B-site (Ti^{4+}) in the ABO_3 structure [62]. However, pristine $SrTiO_3$ is highly insulating; thus, extrinsic doping is necessary to induce and optimize *n*-type conductivity [4]. In its undoped state, $SrTiO_3$ exhibits low carrier concentrations (< 10^{17} cm^{-3}), which are unsuitable for thermoelectric applications. Therefore, controlled doping with suitable cations like rare earths (RE^{3+}) has emerged as a critical strategy to enhance both electrical conductivity and phonon scattering, thereby improving the figure of merit (ZT) [23, 63].

5.2 *La and Dy Doping*

5.2.1 Lanthanum (La) Doping

La^{3+} doping at the Sr^{2+} site introduces extra free electrons into the conduction band due to the charge compensation mechanism. This substitution effectively increases carrier concentration without introducing significant structural distortions, as La^{3+} (1.03 Å) is similar in size to Sr^{2+} (1.18 Å) [20]. La-doped $SrTiO_3$ has demonstrated significant improvements in electrical conductivity (σ) and power factor ($PF = S^2\sigma$) while retaining a reasonably high Seebeck coefficient at elevated temperatures [64]. For instance, Terasaki et al. reported that La doping ($Sr_{1-x}La_xTiO_3$) achieved a ZT value of ~ 0.35 at 1000 K due to optimized carrier concentration and improved mobility [65]. The La-doped $SrTiO_3$ system also retains its stability under oxidizing conditions, making it highly suitable for practical thermoelectric modules.

5.2.2 Dysprosium (Dy) Doping

Dy^{3+} doping in $SrTiO_3$ serves dual roles: it substitutes at the A-site (Sr^{2+}), thereby increasing carrier concentration, and introduces significant mass and strain field fluctuations due to its heavier mass and smaller ionic radius ($Dy^{3+} = 1.027$ Å) [66].

This induces enhanced phonon scattering, resulting in a considerable reduction in lattice thermal conductivity (κ_l). Saitoh et al. demonstrated that Dy-doped $SrTiO_3$ samples showed reduced κ_l by more than 30%, with a slight retention in power factor, leading to a net increase in ZT values up to 0.4 at 973 K [33]. Moreover, Dy doping has been reported to enhance the density of states near the Fermi level, contributing positively to the Seebeck coefficient [34].

5.3 *Double Doping Strategy*

To further boost thermoelectric performance, researchers have explored double-doping strategies, where two dopants are introduced simultaneously at different lattice sites to tailor both electronic and thermal transport properties. A notable example is La/Nb co-doping, where La^{3+} substitutes Sr^{2+} at the A-site and Nb^{5+} substitutes Ti^{4+} at the B-site. This co-doping provides a synergistic enhancement in carrier concentration and mobility due to optimized band filling and reduced scattering. Nb-doping contributes additional electrons and modifies the conduction band curvature, while La provides a more homogeneous charge distribution [67]. Double-doped $SrTiO_3$ systems have demonstrated ZT values as high as 0.5–0.6 at 1000 K, far surpassing those of singly doped analogs [68]. Other successful combinations include La/Pr and La/Sm co-doping, where RE elements not only provide charge carriers but also introduce beneficial lattice distortions that suppress thermal conductivity via mass disorder scattering and strain-induced phonon localization [23, 34].

6 $CaMnO_3$-Based Thermoelectrics

6.1 *Orthorhombic Structure and Transport*

Calcium manganite ($CaMnO_3$) is a promising n-type thermoelectric oxide material that crystallizes in an orthorhombic distorted perovskite structure (Pnma space group). This structure is stabilized by the corner-sharing MnO_6 octahedra, where Mn exists in mixed-valence states (Mn^{3+}/Mn^{4+}), essential for electrical conduction via the small polaron hopping mechanism [69, 70]. In undoped $CaMnO_3$, Mn predominantly exists in the 4 + state (Mn^{4+}), leading to relatively poor electronic transport properties due to insufficient charge carriers. The intrinsic Seebeck coefficient of $CaMnO_3$ is high (> -300 μV/K), but the electrical conductivity is limited (~ 10^{-3} S/cm) because of the low carrier concentration [71, 72]. The thermal conductivity (κ) of undoped $CaMnO_3$ lies in the range of 4–5 W/mK at room temperature, which is moderate but can be reduced through appropriate doping strategies [73]. The conduction mechanism in $CaMnO_3$ is governed by thermally activated hopping between

Mn^{3+} and Mn^{4+} ions. As such, manipulating the Mn valence state through aliovalent doping at the A-site (Ca^{2+}) with trivalent rare earth (RE^{3+}) ions increases Mn^{3+} content, thereby improving charge carrier density and electrical conductivity [74, 75]. Moreover, the orthorhombic distortion and the size mismatch between the host and dopant ions introduce lattice strain and increase phonon scattering. This makes $CaMnO_3$ particularly suitable for rare earth doping aimed at reducing κ while maintaining or enhancing electrical performance [76, 77].

6.2 *Sm, Yb, and Gd Doping*

6.2.1 Samarium (Sm) Doping

Sm^{3+} ions (ionic radius ~ 1.08 Å) closely match the size of Ca^{2+} (1.00 Å), enabling successful substitution with minimal lattice distortion. Sm^{3+} doping in $Ca_{1-x}Sm_xMnO_3$ has been reported to increase the Mn^{3+}/Mn^{4+} ratio, thereby enhancing small polaron hopping and electrical conductivity [78]. Several studies have reported optimized power factors in Sm-doped compositions, particularly in the doping range $x = 0.05$–0.1 [79]. Additionally, Sm^{3+} introduces mass disorder and strain field fluctuations, leading to a suppression of the lattice thermal conductivity to values as low as ~ 2 W/mK, thereby improving the thermoelectric figure of merit (ZT) [80]. For instance, ZT ~ 0.12 at 1000 K was reported for $Ca_{0.95}Sm_{0.05}MnO_3$ [81].

6.2.2 Ytterbium (Yb) Doping

Yb^{3+}, being a heavy rare earth ion with a relatively smaller ionic radius (~ 0.985 Å), induces significant mass and strain field scattering, making it especially effective in reducing κ [82]. Yb-doped $CaMnO_3$ also promotes increased electron concentration and local lattice distortion, leading to improved thermoelectric performance at high temperatures. Experiments have shown that $Ca_{1-x}Yb_xMnO_3$ exhibits lower κ than Sm or Gd-doped analogs, with reported thermal conductivities dropping below 1.8 W/mK for $x = 0.1$ [81]. This makes Yb doping an attractive route for enhancing ZT, especially in high-temperature applications (> 800 K).

6.2.3 Gadolinium (Gd) Doping

Gd^{3+} (ionic radius ~ 1.053 Å) has been successfully used to substitute Ca^{2+} in $CaMnO_3$ to create $Ca_{1-x}Gd_xMnO_3$ solid solutions. Gd^{3+} doping results in an increase in Mn^{3+} concentration and thus promotes small polaron hopping conduction [74].

Moreover, Gd's relatively high atomic mass contributes to enhanced phonon scattering, leading to reduced thermal conductivity [75]. Gd doping is known to maintain a balance between electrical conductivity and thermal conductivity reduction, which helps achieve improved ZT values. For example, $Ca0_{.95}Gd_{0.05}MnO_3$ has been reported to show ZT values of ~ 0.13 at 1000 K [76].

7 $NaCo_2O_4$ and Layered Cobaltates

7.1 Thermoelectric Features

Sodium cobaltate ($NaCo_2O_4$) and related layered cobaltites are among the most promising p-type oxide thermoelectric materials, particularly suited for high-temperature applications. The structure of $NaCo_2O_4$ comprises alternating layers of conductive CoO_2 sheets and insulating Na layers, forming a misfit-layered structure similar to that of high-temperature superconductors [24, 83]. This structural anisotropy plays a crucial role in the unique transport behavior of the material. $NaCo_2O_4$ exhibits a large Seebeck coefficient (S) of approximately 100 μV/K at 300 K, which can exceed 150 μV/K at elevated temperatures [84]. The material also shows relatively low thermal conductivity (~ 2–3 W/mK), owing to strong phonon scattering in the Na layers and structural complexity [85, 86]. Its electrical conductivity is comparatively high for oxides (10^3–10^4 S/m), resulting in a high power factor ($S^2\sigma$) and decent thermoelectric figure of merit (ZT $\approx$ 0.1–0.2) at around 800–1000 K [34]. The favorable thermoelectric performance of $NaCo_2O_4$ stems from the unique strong electronic correlations and spin entropy flow within the CoO_2 layers, which enhance the Seebeck coefficient without significantly compromising electrical conductivity [87]. The average Co oxidation state in $NaCo_2O_4$ is close to + 3.5, leading to a mixed-valence state of Co^{3+}/Co^{4+} that supports hole-type conduction via small polaron hopping [34, 88]. Another benefit of $NaCo_2O_4$ is its structural flexibility, allowing for various dopants (particularly rare earth elements) to be incorporated into the Na or Co sites or to stabilize Na-deficient compositions that often arise due to high-temperature volatilization or instability of Na [34].

7.2 Substitution with Ce, Sm, and Pr

7.2.1 Ce Doping

Cerium (Ce^{3+}/Ce^{4+}) substitution in $NaCo_2O_4$ significantly alters the electronic structure and carrier concentration due to its multivalent nature. When Ce is introduced at the Na site or interstitial sites, it can stabilize Na-deficient phases and promote charge compensation through the formation of oxygen vacancies [24, 34]. Ce doping often leads to a slight reduction in Seebeck coefficient due to increased carrier concentration but compensates with improved electrical conductivity [89]. Additionally, Ce acts as a phonon scattering center due to its relatively large atomic mass, thereby contributing to reduced lattice thermal conductivity [90]. Ce-doped $NaCo_2O_4$ has shown enhanced thermoelectric performance with ZT values reaching 0.2 at 900 K [91].

7.2.2 Sm Doping

Samarium (Sm^{3+}) doping also leads to charge carrier tuning and lattice distortion in the $NaCo_2O_4$ matrix. Sm ions can substitute into the Na layer or occupy interstitial positions, affecting the ionic transport pathways and introducing structural strain, which scatters phonons and lowers κ [92]. Sm doping also suppresses Na volatilization at high temperatures, thereby enhancing the thermal stability of the material [93]. Reports show that $Na_{1-x}Sm_xCo_2O_4$ exhibits improved thermoelectric performance, particularly at doping levels $x \approx 0.05$–0.10, where a balance is achieved between S and σ [94].

7.2.3 Pr Doping

Praseodymium (Pr^{3+}) ions introduce additional structural and electronic modifications, similar to Sm and Ce. Due to its size and valence, Pr^{3+} substitution influences both the Na layer arrangement and the CoO_2 electronic states, which can affect carrier effective mass and mobility [24]. Moreover, Pr doping in $NaCo_2O_4$ leads to increased phonon scattering due to mass and strain field fluctuations, lowering the lattice thermal conductivity without severely compromising the power factor [95]. Some studies have also suggested a stronger localization of charge carriers in Pr-doped samples, contributing to a higher S at the cost of slightly reduced conductivity [96].

7.3 Stabilization of Na-Deficient Phases

Rare earth doping is particularly beneficial in stabilizing Na-deficient or non-stoichiometric $Na_xCo_2O_4$ ($x < 1$) compositions, which often show superior thermoelectric properties due to increased carrier localization and enhanced phonon scattering [96]. The presence of RE ions acts as a sintering aid, improving grain connectivity and microstructural homogeneity, further enhancing electrical transport [97, 98]. These doping strategies also contribute to microstructural refinement, which enhances grain boundary scattering and helps reduce total thermal conductivity. This synergy makes rare earth–modified $NaCo_2O_4$ a strong candidate for commercial oxide thermoelectrics, especially for intermediate to high-temperature waste heat recovery applications [99].

8 Comparative Analysis of RE-Doped Oxides

Rare earth (RE) doping in oxide-based thermoelectrics has emerged as a powerful approach to optimize the thermoelectric (TE) performance of various metal oxides by tailoring their electrical, thermal, and structural properties. This strategy has been extensively investigated in several promising oxides, including ZnO, $SrTiO_3$, $CaMnO_3$, and $NaCo_2O_4$, each of which offers distinct mechanisms for thermoelectric enhancement and exhibits material-specific advantages and limitations which has been tabulated in Table 2. The comparative study of RE-doped oxides is graphically represented in Fig. 5.

Table 2 Thermoelectric performance of rare earth-doped oxide materials

Oxide Material	RE Dopants	S (μV/K)	σ (S/m)	κ (W/m·K)	Max ZT (at K)	Mechanism	References
ZnO	La, Gd, Nd	~ 100	~ 10^4	~ 2	~ 0.3 (1000 K)	Shallow donor doping, bandgap narrowing, phonon scattering	[100–104]
$SrTiO_3$	La, Dy, Pr	~ 200–300	~ 10^3	~ 3	~ 0.4–0.6 (1000 K)	Degenerate semiconductor, co-doping, phonon scattering	[24, 105–107]
$CaMnO_3$	Sm, Yb, Gd	~ 150	~ 10^3	~ 2	~ 0.2–0.3 (900 K)	Mixed-valence conduction, polaron hopping, lattice distortion	[36, 42, 108]
$NaCo_2O_4$	Ce, Sm, Pr	~ 150–200	~ 10^3	~ 2	~ 0.2–0.3 (800 K)	Spin entropy, 2D CoO_2 layers, RE-induced lattice stabilization	[34, 109, 110]

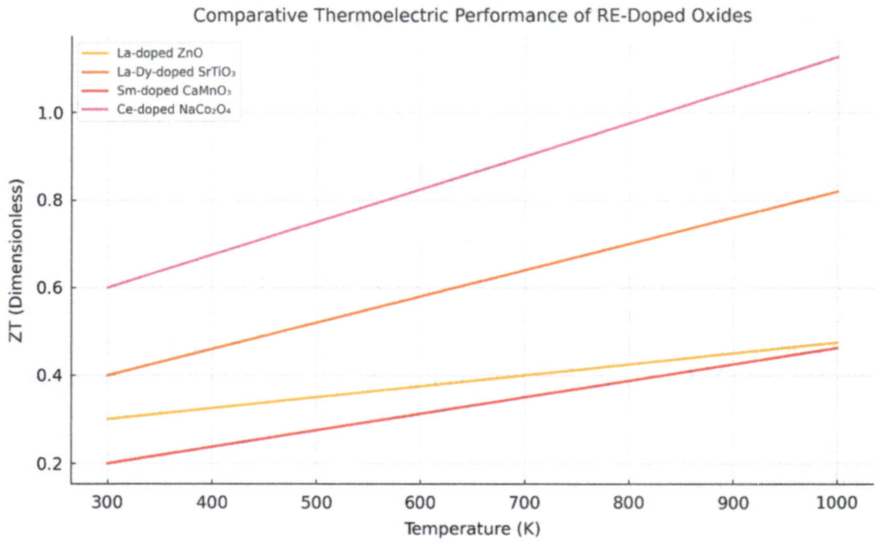

Fig. 5 Comparative analysis of RE-doped oxides

9 Challenges and Opportunities

Despite the significant advancements in rare earth (RE)-doped oxide thermoelectrics, several persistent challenges continue to hinder their widespread adoption in thermoelectric devices. A major issue is the limited solubility of RE dopants in host oxide matrices due to disparities in ionic radii and oxidation states. Most RE ions (e.g., La^{3+}, Nd^{3+}, Gd^{3+}) possess larger ionic sizes than the transition metal ions they replace (such as Zn^{2+} or Ti^{4+}), leading to structural strain and defect formation when their concentration exceeds the solubility limit. In ZnO, for instance, La^{3+} doping is typically effective below 2–3 at%. Beyond this, phase separation into La_2O_3 or $La(OH)_3$ phases occurs, which interrupts carrier transport and reduces electronic conductivity, counteracting any gains from doping [107, 111]. Similarly, in $SrTiO_3$ and $CaMnO_3$, excessive RE doping leads to the emergence of impurity phases like $RE_2Ti_2O_7$ or RE_2MnO_4, which scatter charge carriers, increase electrical resistivity, and degrade overall thermoelectric performance [112, 113]. Moreover, such phase segregation can induce unwanted grain boundary effects, which exacerbate phonon scattering and reduce carrier mobility. To mitigate these effects, precise control over processing parameters—including sintering temperature, dopant concentration, and atmosphere—is essential. Techniques such as spark plasma sintering (SPS) allow densification at relatively lower temperatures and reduced dwell times, limiting unwanted phase formation [48]. Additionally, the adoption of wet chemical synthesis and sol–gel methods can help achieve better dopant dispersion at the nanoscale, avoiding macro-phase separation and improving homogeneity [114].

Another critical issue lies in interdependent trade-offs among thermoelectric parameters. According to the equation $ZT = S^2\sigma T/\kappa$, any enhancement in one parameter can adversely affect others, complicating material optimization. For example, increasing electrical conductivity (σ) through doping raises carrier concentration, which can lower the Seebeck coefficient (S) due to a downward shift in the Fermi level, thus reducing the energy differential of carriers [70]. This trade-off limits the maximum achievable power factor ($PF = S^2\sigma$). Conversely, attempts to preserve a high S by minimizing carrier density often result in poor σ, making the material inefficient as a thermoelectric generator [12]. Additionally, while RE doping often improves phonon scattering and lowers lattice thermal conductivity (κ), it may also introduce ionized impurity scattering that hinders carrier mobility. Overcoming these trade-offs requires advanced materials engineering strategies such as band structure manipulation via band convergence (merging multiple valence or conduction bands to enhance S without compromising σ), energy filtering (preferential scattering of low-energy carriers), and co-doping (using two dopants to balance the carrier concentration and lattice strain) [21, 115]. These complex relationships necessitate a highly tailored approach to doping, often requiring computational modeling and experimental validation to identify the optimal dopant-host combinations for maximum ZT.

10 Conclusion

In the pursuit of sustainable and efficient energy solutions, thermoelectric (TE) materials represent a compelling class of functional materials capable of directly converting heat into electricity and vice versa. Among various TE candidates, metal oxides have emerged as attractive alternatives to conventional telluride- and antimonide-based compounds due to their inherent advantages: high thermal stability, low toxicity, earth abundance, and long-term durability at elevated temperatures. However, oxide materials have traditionally been hindered by low thermoelectric efficiency, primarily due to high thermal conductivity and limited electrical conductivity. A promising strategy to overcome these limitations lies in rare earth (RE) doping, which introduces a new dimension of tuneability to the structural, electronic, and phononic properties of oxide thermoelectrics. Rare earth elements, with their unique 4f electron configurations and wide range of ionic radii and oxidation states, offer versatile routes to engineer the performance of oxides such as ZnO, $SrTiO_3$, $CaMnO_3$, and $NaCo_2O_4$.

RE doping serves multiple roles across oxide systems. First, it modifies carrier concentration through aliovalent substitution (e.g., La^{3+} for Sr^{2+} in $SrTiO_3$ or Ca^{2+} in $CaMnO_3$), thus enhancing the electrical conductivity without severely compromising the Seebeck coefficient. Second, the mass and size mismatch between RE dopants and host cations introduces lattice strain, leading to enhanced phonon scattering and significant reductions in lattice thermal conductivity. This dual improvement in the power factor ($S^2\sigma$) and the suppression of thermal conductivity contributes to higher

ZT values across RE-doped oxide materials. The chapter highlights several system-specific advancements. For instance, La and Y doping in ZnO improves electrical transport and disrupts heat conduction pathways via nanostructuring. $SrTiO_3$ benefits from La and Dy doping strategies, and the implementation of double-doping schemes such as La/Nb or La/Pr has achieved higher ZT values by balancing carrier concentration and lattice thermal conductivity. In $CaMnO_3$, doping with Sm, Yb, or Gd increases Mn^{3+} content, optimizing the mixed-valence conduction pathway and simultaneously reducing κ through RE-induced phonon scattering. Layered cobaltates such as $NaCo_2O_4$, with inherently high Seebeck coefficients, gain further benefits from RE substitution (Ce, Sm, Pr) in their misfit-layered structures, contributing to improved electronic and thermal transport.

Despite this progress, challenges persist. The solubility limits of RE elements in host matrices can result in the formation of undesirable secondary phases or defect clusters, which may degrade electrical properties. Achieving a synergistic optimization of Seebeck coefficient, electrical conductivity, and thermal conductivity remains difficult, as improvements in one property often compromise others. Moreover, synthesis reproducibility, dopant distribution control, and scalability remain non-trivial issues for transitioning these materials from laboratory demonstrations to industrial-scale thermoelectric modules. Looking forward, emerging trends offer exciting opportunities. The concept of high-entropy oxides (HEOs), where multiple RE dopants are introduced simultaneously, is promising for further enhancement of thermoelectric performance through enhanced entropy-driven stabilization and lattice disorder. Additionally, the use of machine learning and high-throughput computational screening provides a powerful toolset to accelerate the discovery and optimization of RE-doped oxides. Furthermore, nanostructuring and composite formation, particularly through bottom-up approaches, could further tune thermal and electrical transport by enhancing grain boundary effects and promoting selective carrier pathways.

In conclusion, rare earth doping has established itself as a transformative approach for tailoring the thermoelectric properties of oxide materials. As materials scientists and engineers continue to unlock the full potential of these strategies, RE-doped metal oxides are well-positioned to make meaningful contributions to waste heat recovery, renewable energy harvesting, and green energy conversion. The synergy between fundamental materials research, predictive modeling, and scalable fabrication methods will be critical in transitioning RE-doped oxide thermoelectrics from concept to commercially viable technologies in the coming decades.

References

1. Snyder, G.J., Toberer, E.S.: Complex thermoelectric materials. Nat. Mater. **7**(2), 105–114 (2008)
2. Rowe, D.M. (ed.): Thermoelectrics Handbook: Macro to Nano. CRC Press (2006)
3. Bell, L.E.: Cooling, heating, generating power, and recovering waste heat with thermoelectric systems. Science **321**(5895), 1457–1461 (2008)

4. He, J., Tritt, T.M.: Advances in thermoelectric materials research: looking back and moving forward. Science **357**(6358), eaak9997 (2017
5. Heremans, J.P., Jovovic, V., Toberer, E.S., Saramat, A., Kurosaki, K., Charoenphakdee, A., Yamanaka, S., Snyder, G.J.: Enhancement of thermoelectric efficiency in PbTe by distortion of the electronic density of states. Science **321**(5888), 554–557 (2008)
6. Goldsmid, H.J.: Introduction to Thermoelectricity. Springer (2010)
7. Pei, Y., Shi, X., LaLonde, A., Wang, H., Chen, L., Snyder, G.J.: Convergence of electronic bands for high performance bulk thermoelectrics. Nature **473**(7345), 66–69 (2011)
8. Vineis, C.J., Shakouri, A., Majumdar, A., Kanatzidis, M.G.: Nanostructured thermoelectrics: big efficiency gains from small features. Adv. Mater. **22**(36), 3970–3980 (2010)
9. Liu, W., Jie, Q., Kim, H.S., Ren, Z.: Current progress and future challenges in thermoelectric power generation: from materials to devices. Acta Mater. **87**, 357–376 (2015)
10. Zhao, L.D., Lo, S.H., Zhang, Y., Sun, H., Tan, G., Uher, C., Dravid, V.P., Kanatzidis, M.G.: Ultralow thermal conductivity and high thermoelectric figure of merit in SnSe crystals. Nature **508**(7496), 373–377 (2014)
11. Koumoto, K., Mori, T. (eds.): Thermoelectric Nanomaterials: Materials Design and Applications. Springer
12. Ohtaki, M.: Recent aspects of oxide thermoelectric materials for power generation from mid-to-high temperature heat source. J. Ceram. Soc. Jpn. **119**(1391), 770–775 (2011)
13. Funahashi, R., Shikano, M.: $Bi_2Sr_2Co_2Oy$ whiskers with high thermoelectric figure of merit. Appl. Phys. Lett. **76**(17), 2385–2387 (2000)
14. Ito, M., Nagira, T., Hara, S.: Thermoelectric properties of $Na_xCo_2O_4$ with rare-earth metals doping prepared by polymerized complex method. J. Alloy. Compd. **408**, 1217–1221 (2006)
15. Fergus, J.W.: Oxide materials for high temperature thermoelectric energy conversion. J. Eur. Ceram. Soc. **32**(3), 525–540 (2012)
16. Lan, Y., Minnich, A.J., Chen, G., Ren, Z.: Enhancement of thermoelectric figure-of-merit by a bulk nanostructuring approach. Adv. Func. Mater. **20**(3), 357–376 (2010)
17. Hu, L., Gao, H., Liu, X., Xie, H., Shen, J., Zhu, T., Zhao, X.: Enhancement in thermoelectric performance of bismuth telluride based alloys by multi-scale microstructural effects. J. Mater. Chem. **22**(32), 16484–16490 (2012)
18. Raghavan, S., Zhang, J.Y., Stemmer, S.: Two-dimensional electron liquid at the (111) $SmTiO_3$/ $SrTiO_3$ interface. Appl. Phys. Lett. **106**(13), 132104 (2015)
19. Zhao, L. D., Tan, G., Hao, S., He, J., Pei, Y., Chi, H., Wang, H., Gong, S., Xu, H., Dravid, V.P., Kanatzidis, M. G.: Ultrahigh power factor and thermoelectric performance in hole-doped single-crystal SnSe. Science **351**(6269), 141–144 (2016)
20. Ohta, H., Kim, S.W., Mune, Y., Mizoguchi, T., Nomura, K., Ohta, S., Nomura, T., Nakanishi, Y., Ikuhara, Y., Hirano, M., Hosono, H., Koumoto, K.: Giant thermoelectric Seebeck coefficient of a two-dimensional electron gas in $SrTiO_3$. Nat. Mater. **6**(2), 129–134 (2007)
21. Tritt, T.M.: Thermoelectric phenomena, materials, and applications. Annu. Rev. Mater. Res. **41**, 433–448 (2011)
22. Hébert, S., Daou, R., Maignan, A., Martin, C.: Thermoelectric oxides: a review. J. Phys.: Condens. Matter **28**(1), 013001 (2016)
23. Koumoto, K., Funahashi, R., Guilmeau, E., Miyazaki, Y., Weidenkaff, A., Wang, Y., Wan, C.: Thermoelectric ceramics for energy harvesting. J. Am. Ceram. Soc. **96**(1), 1–23 (2013)
24. Wang, Y., Rogado, N.S., Cava, R.J., Ong, N.P.: Spin entropy as the likely source of enhanced thermopower in $NaxCo_2O_4$. Nature **423**(6938), 425–428 (2003)
25. Mune, Y., Ohta, H., Koumoto, K., Mizoguchi, T., Ikuhara, Y.: Enhanced seebeck coefficient of quantum-confined electrons in $SrTiO_3$/ $SrTi_{0.8}Nb_{0.2}O_3$ superlattices. Appl. Phys. Lett. **91**(19), 192105 (2007)
26. Zhu, Y., Wang, C., Su, W., Liu, J., Li, J., Zhang, X., Mei, L.: Influence of rare-earth elements doping on thermoelectric properties of $Ca_{0.98}Dy_{0.02}MnO_3$ at high temperature. J. Solid State Chem. **225**, 105–109 (2015)
27. Parker, D., Singh, D.J.: High-temperature thermoelectric performance of heavily doped PbSe. Phys. Rev. B **82**(3), 035204 (2010)

28. Zhang, R.Z., Koumoto, K.: Grain-size-dependent thermoelectric properties of $SrTiO_3$ 3D superlattice ceramics. J. Electron. Mater. **42**(7), 1568–1572 (2013)
29. Tritt, T.M., Subramanian, M.A.: Thermoelectric materials, phenomena, and applications: a bird's eye view. MRS Bull. **31**(3), 188–198 (2006)
30. Romanenko, A.I., Chebanova, G.E., Chen, T., Su, W., Wang, H.: Review of the thermoelectric properties of layered oxides and chalcogenides. J. Phys. D Appl. Phys. **55**(14), 143001 (2021)
31. Choi, W.S., Ohta, H., Moon, S.J., Lee, Y. S., Noh, T.W.: Dimensional crossover of polaron dynamics in Nb: $SrTiO_3/SrTiO_3$ superlattices: possible mechanism of thermopower enhancement. Phys. Rev. B—Condens. Matter Mater. Phys. **82**(2), 024301 (2010)
32. Limelette, P., Hardy, V., Auban-Senzier, P., Jérome, D., Flahaut, D., Hébert, S., Frésard, R., Simon, C., Noudem, J., Maignan, A.: Strongly correlated properties of the thermoelectric cobalt oxide $Ca_3Co_4O_9$. Phys. Rev. B—Condens. Matter Mater. Phys. **71**(23), 233108 (2005)
33. Robert, R., Aguirre, M.H., Hug, P., Reller, A., Weidenkaff, A.: High-temperature thermoelectric properties of Ln (Co, Ni) O_3 (Ln= La, Pr, Nd, Sm, Gd and Dy) compounds. Acta Mater. **55**(15), 4965–4972 (2007)
34. Terasaki, I., Sasago, Y., Uchinokura, K.: Large thermoelectric power in $NaCo_2O_4$ single crystals. Phys. Rev. B **56**(20), R12685–R12687 (1997)
35. Fujita, K., Mochida, T., Nakamura, K.: High-temperature thermoelectric properties of Na_xCoO_2-δ single crystals. Jpn. J. Appl. Phys. **40**(7R), 4644 (2001)
36. Ohtaki, M., Tsubota, T., Eguchi, K., Arai, H.: High-temperature thermoelectric properties of $(Zn_{1-x}\ Al_x)$ O. J. Appl. Phys. **79**(3), 1816–1818 (1996)
37. Lee, K.H., Kim, S.W.: Design and preparation of high-performance bulk thermoelectric materials with defect structures. J. Korean Ceram. Soc. **54**(2), 75–85 (2017)
38. Govardhan, M., Kasiviswanathan, B., Dhanaraj, R., Ibrahim, A.S., Mohanraj, K., Sivakumar, G.: Thermoelectric performance of La- and Ta-co-doped $CaMnO_3$ synthesized using the solid-state reaction method. Inorg. Chem. Commun. **163**, 115218 (2025)
39. Massetti, M., Jiao, F., Ferguson, A. J., Zhao, D., Wijeratne, K., Würger, A., Blackburn, J.L., Crispin, X., Fabiano, S.: Unconventional thermoelectric materials for energy harvesting and sensing applications. Chem. Rev. **121**(20), 12465–12547 (2021)
40. Mallick, M.M., Vitta, S.: Enhancing thermoelectric figure-of-merit of polycrystalline Na_yCoO_2 by a combination of non-stoichiometry and co-substitution. J. Electron. Mater. **47**, 3230–3237 (2018)
41. Shannon, R.D.: Revised effective ionic radii in oxides and fluorides. Acta Crystallogr. A **32**(5), 751–767 (1976)
42. Wu, D., Zhao, D., Zheng, F., Jin, L., Kanatzidis, M.G., He, J.: Understanding nanostructuring processes in thermoelectrics and their effects on lattice thermal conductivity. Adv. Mater. **28**(14), 2737–2743 (2016)
43. Okuda, Y., Nakanishi, K., Miyasaka, S., Tokura, Y.: Large thermoelectric power in $NaCo_2O_4$ single crystals. Phys. Rev. B **63**(11), 113104 (2001)
44. Jiang, F., Peng, Z., Zang, Y., Fu, X.: Progress on rare-earth doped ZnO-based varistor materials. J. Adv. Ceram. **2**(3), 201–212 (2013)
45. Li, Y., Zhang, L., Wang, S.L., Xu, G.Y., Jiang, M.: Na-site doping of $NaCo_2O_4$ thermoelectric materials. Key Eng. Mater. **336–338**, 802–804 (2007)
46. Poudel, B., Hao, Q., Ma, Y., Lan, Y., Minnich, A., Yu, B., Yan, X., Wang, D., Muto, A., Vashaee, D., Chen, X., Ren, Z.: High-thermoelectric performance of nanostructured bismuth antimony telluride bulk alloys. Science **320**(5876), 634–638 (2008)
47. Wood, C.: Materials for thermoelectric energy conversion. Rep. Prog. Phys. **51**(4), 459–539 (1988)
48. Baranovskiy, A., Koresh, I., Amouyal, Y.: Tuning transport properties of nickel-doped zinc oxide for thermoelectric applications. MRS Commun. **8**(3), 858–864 (2018)
49. Constantinescu, G., Galatanu, A., Tobaldi, D., Mikhalev, S., Suarez, D., Paulino, T., Zakharchuk, K., Sergiienko, S., Lopes, D., Kovalevsky, A.: Enhancement of thermoelectric performance of donor-doped ZnO ceramics by involving an in situ aluminothermic reaction during processing. Crystals **12**(11), 1562 (2022)

50. Zhu, B., Chen, C., Yao, Z., Chen, J., Jia, C., Wang, Z., Tian, R., Tao, L., Xue, F., Hng, H.: Multiple doped ZnO with enhanced thermoelectric properties. J. Eur. Ceram. Soc. **41**(7), 4182–4188 (2021)
51. Wu, Q., Zhang, B., Liu, G., Ning, J., Shao, T., Zhang, F., Xue, S.: Study of La doping on the electronic structure and magneto-optical properties of ZnO by GGA+U Method. Crystals **13**(5), 754 (2023)
52. Deng, S., Duan, M., Xu, M., He, L.: Effect of La doping on the electronic structure and optical properties of ZnO. Physica B **406**(11), 2314–2318 (2011)
53. Sulaiman, S., Sudin, I., Basheer, U.M., Omar, M.F.: Review of the nanostructuring and doping strategies for high-performance ZnO thermoelectric materials. Crystals **12**(8), 1076 (2022)
54. Feng, B., Li, G., Kong, D., Xu, C., Kuang, Z., Ma, Y., Chen, Z., Li, Y., Gu, H., Fan, X.: Study on the mechanism of Y doping effect for improving the properties of ZnO thermoelectric ceramics. Scripta Mater. **216**, 114740 (2022)
55. Kumar, R., Singh, R.K., Shukla, S.: Review on grain size effects on thermal conductivity in ZnO thermoelectric materials. RSC Adv. **12**(9), 5428–5438 (2022)
56. Ohtaki, H.: Recent aspects of oxide thermoelectric materials development. J. Ceram. Soc. Jpn. **119**(1393), 770–775 (2011)
57. Pascariu, P., Cojocaru, C., Samoila, P., Romanitan, C.: Nd-doped ZnO nanostructures with enhanced photocatalytic performance for environmental protection. Int. J. Mol. Sci. **24**(7), 6436 (2022)
58. Muta, H., Kurosaki, K., Yamanaka, S.: Thermoelectric properties of rare earth doped $SrTiO_3$. J. Alloy. Compd. **350**(1–2), 292–295 (2003)
59. Zhang, L., Cao, Z., Fu, Q., Li, C., Du, Y., Li, Y., La, T., Zhang, D., Wang, J.: Thermoelectric properties of Yb-La-Nb-doped $SrTiO_3$. J. Eur. Ceram. Soc. **44**(3), 1647–1653 (2024)
60. Liu, J., Wang, C., Peng, H., et al.: Thermoelectric properties of dy-doped $SrTiO_3$ ceramics. J. Electron. Mater. **41**, 3073–3076 (2012)
61. Nag, A.: Perovskite oxide thermoelectric module—a way forward. Catal. Res. **3**(4), 2304024 (2023)
62. Gayner, C., Kar, K.K.: Recent advances in thermoelectric materials. Prog. Mater Sci. **83**, 330–382 (2016)
63. Okuda, T., Nakanishi, K., Miyasaka, S., Tokura, Y.: Large thermoelectric response of metallic perovskites: $Sr_{1-x}La_xTiO_3$ ($0 \leq x \leq 0.1$). Phys. Rev. B **63**(11), 113104 (2001)
64. Gillani, S., Ahmad, R., Rizwan, M., Shakil, M., Rafique, M., Murtaza, G., Jin, H.: First-principles investigation of structural, electronic, optical and thermal properties of Zinc doped $SrTiO_3$. Optik **201**, 163481 (2019)
65. Han, J., Sun, Q., Song, Y.: Enhanced thermoelectric properties of La and Dy co-doped, Sr-deficient $SrTiO_3$ ceramics. J. Alloy. Compd. **705**, 22–27 (2017)
66. Shannon, R.D.: Revised effective ionic radii and systematic studies of interatomic distances in halides and chalcogenides. Acta Crystallogr. A **32**(5), 751–767 (1976)
67. Liu, W., Wang, X., Li, F.: Effects of rare earth element doping on thermoelectric performance of GeTe. Chem. Eng. J. **446**, 137278 (2022)
68. He, J., Qiu, J., Xu, J., Wang, X., Zhao, L.-D.: Enhanced thermoelectric performance of Nb-doped $SrTiO_3$ by nanoinclusion engineering. Sci. Rep. **5**, Article 3449 (2019)
69. Kabir, M.R., Zhang, T., Donelson, R., Li, S.: Thermoelectric properties of Yb- and Nb-co-doped $CaMnO_3$. Phys. Status Solidi A **211**(10), 1200–1205 (2014)
70. Gürsul, M., Çetin, G., Ergin, İ, Korkmaz, L., Özçelik, B., Torres, M., Madre, M., Sotelo, A.: Remarkable variation in microstructural, thermoelectric, and magnetic properties of $CaMnO_3$ through Ce doping. Mater. Sci. Eng., B **299**, 116986 (2023)
71. Suprayoga, E., Putri, W.B.K., Singsoog, K., Paengson, S., Hanna, M.Y., Nugraha, A.R.T., Munazat, D.R., Kurniawan, B., Nurhuda, M., Seetawan, T., Hasdeo, E.H.: Investigation of electron and phonon transport in Bi-doped $CaMnO_3$ for thermoelectric applications. Mater. Res. Bull. **141**, 111359 (2021)
72. Amirkhizi, P., Madre, M.A., Dura, O.J., Pelloquin, D., Mora, M., Torres, M.A., Marinel, S., Sotelo, A.: Effect of Y, La, and Yb simultaneous doping on the thermal conductivity and thermoelectric performances of $CaMnO_3$ ceramics. J. Alloy. Compd. **946**, 170201 (2023)

73. Zhu, Y., Su, W., Liu, J., Zhou, Y., Li, J., Zhang, X., Du, Y., Wang, C.: Effects of Dy and Yb co-doping on thermoelectric properties of $CaMnO_3$ ceramics. Ceram. Int. **41**(1), 1535–1539 (2014)
74. Wang, Y., Lo, S.H., Yu, S., Fan, H., Wang, X.: High-temperature thermoelectric response of electron-doped $Ca_{1-x}R_xMnO_3$ perovskites (R = rare-earth). Chem. Mater. **21**(20), 5005–5011 (2009)
75. Ogawa, R., Fujihara, S., Hagiwara, M.: Effect of dual doping by rare-earth and sodium ions on thermoelectric properties of $CaMnO_3$ ceramics. J. Ceram. Soc. Jpn. **130**(7), 403–409 (2022)
76. Suzuki, T., Sakai, H., Taguchi, Y., Tokura, Y.: Thermoelectric properties of electron-doped $Sr(Mn_{1-x}Mo_x)O_3$ single crystals with perovskite structure. J. Electron. Mater. **41**(6), 1559–1563 (2012)
77. Bocher, L., Funahashi, R., Marchandier, T., Schulze, C.: High-temperature stability, structure and thermoelectric properties of $CaMn_{1-x}Nb_xO_3$ phases. Acta Mater. **57**(21), 5667–5680 (2009)
78. Radaelli, P.G., Cox, D.E., Marezio, M., Cheong, S.-W., Schiffer, P.E., Ramirez, A.P.: Simultaneous structural, magnetic, and electronic transitions in $La_{1-x}Ca_xMnO_3$ with $x = 0.25$ and 0.50. Phys. Rev. Lett. **75**(24), 4488–4491 (1995)
79. Molinari, M., Tompsett, D.A., Parker, S.C., Azough, F., Freer, R.: Structural, electronic and thermoelectric behaviour of $CaMnO_3$ and $CaMnO_3$–δ. J. Mater. Chem. A **2**(4), 14109–14117 (2014)
80. Leonidov, I.A., Konstantinova, E.I., Patrakeev, M.V., Markov, A.A., Kozhevnikov, V.L.: Seebeck coefficient of $Ca_{1-x}Pr_xMnO_3$–δ paramagnetic manganites. Inorg. Mater. **53**(6), 583–588 (2017)
81. Sui, Y., Wang, Y., Su, W.: High-temperature thermoelectric characteristics of $Ca_{0.9}R_{0.1}MnO_3$ (R = La, Pr, Nd, Sm, Eu, Gd, Tb, Dy, Ho, Er, Yb) perovskites. J. Appl. Phys. **104**(9), 093703 (2008)
82. Peleckis, G., Motohashi, T., Karppinen, M., Yamauchi, H.: Enhanced thermoelectric properties of Na_xCoO_2 whisker crystals. Appl. Phys. Lett. **83**(26), 5416–5418 (2003)
83. Motohashi, T., Naujalis, E., Ueda, R., Isawa, K., Karppinen, M., Yamauchi, H.: Simultaneously enhanced thermoelectric power and reduced resistivity of $Na_xCo_2O_4$ by controlling Na nonstoichiometry. Appl. Phys. Lett. **79**(10), 1480–1482 (2001)
84. Tajima, S., Tani, T., Isobe, S., Koumoto, K.: Thermoelectric properties of highly textured $NaCo_2O_4$ ceramics processed by the reactive templated grain growth (RTGG) method. Mater. Sci. Eng., B **86**(1–3), 20–25 (2001)
85. Koumoto, K., Terasaki, I., Funahashi, R.: Complex oxide materials for potential thermoelectric applications. MRS Bull. **31**(3), 206–210 (2006)
86. Palsson, G., Kotliar, G.: Thermoelectric response near the Mott transition. Phys. Rev. Lett. **80**(21), 4775–4778 (1998)
87. Hébert, S., Lambert, S., Pelloquin, D., Maignan, A.: Large thermopower in a metallic cobaltite: the layered Tl-Sr-Co-O misfit. Phys. Rev. B **64**(17), 172101 (2001)
88. Pršić, S., Savić, S., Branković, Z., Vrtnik, S., Dapčević, A., Branković, G.: Mechanochemically assisted solid-state and citric acid complex syntheses of Cu-doped sodium cobaltite ceramics. J. Alloy. Compd. **640**, 480–487 (2015)
89. Jakubczyk, E., Mapp, A., Chung, C., Sansom, C., Jones, J., Dorey, R.: Enhancing thermoelectric properties of $NaCo_2O_4$ ceramics through Na pre-treatment induced nano-decoration. J. Alloy. Compd. **788**, 91–101 (2019)
90. Hong, H.Y., Gwon, S.Y., Won, S.O., Park, K.: Enhanced thermoelectric properties of Ce-doped $Bi_2Sr_2Co_2O_8$ misfit-layer oxides. J. Market. Res. **19**, 1873–1883 (2022)
91. Novitskii, A., Serhiienko, I., Novikov, S., Kuskov, K., Pankratova, D., Sviridova, T., Voronin, A., Bogach, A., Skryleva, E., Parkhomenko, Y., Burkov, A., Mori, T., Khovaylo, V.: Thermoelectric properties of Sm-doped BiCuSeO oxyselenides fabricated by two-step reactive sintering. J. Alloy. Compd. **912**, 165208 (2022)
92. Zhang, X., Lemmens, P., Gnezdilov, V., Choi, K., Keimer, B., Chen, D., Lin, C., Chou, F.: Phonon scattering and stability of $Na_{0.5}CoO_2$. Physica B **359–361**, 424–426 (2005)

93. Perać, S., Savić, S.M., Branković, Z., Bernik, S., Radojković, A., Kojić, S., Vasiljević, D., Branković, G.: Microstructural, thermoelectric, and mechanical properties of Cu-substituted $NaCo_2O_4$. Materials **15**(13), 4470 (2022)
94. Yang, J., Song, W.H., Ma, Y.Q., Zhang, R.L., Zhao, B.C., Sheng, Z.G., Zheng, G.H., Dai, J.M., Sun, Y.P.: Structural, magnetic, and transport properties in the Pr-doped manganites $La_{0.9-x}Pr_xTe_{0.1}MnO_3$ ($0 \le x \le 0.9$). Phys. Rev. B **70**(14), 144421 (2004)
95. Assadi, M.H.N., Li, S., Yu, A.B.: Selecting the suitable dopants: Electronic structures of transition metal and rare earth doped thermoelectric sodium cobaltate. RSC Adv. **3**, 1442–1449 (2013)
96. Krasutskaya, N.S., Chizhova, E.A., Zizika, J.A., Buka, A.V., Wang, H., Klyndyuk, A.I.: Improving of thermoelectric efficiency of layered sodium cobaltite through its doping by different metal oxides. Ceramics **8**(3), 86 (2025)
97. Gutiérrez Seijas, J., Prado-Gonjal, J., Ávila Brande, D., Terry, I., Morán, E., Schmidt, R.: Microwave-assisted synthesis, microstructure, and magnetic properties of rare-earth cobaltites. Inorg. Chem. **56**(1), 627–635 (2017)
98. Funahashi, R., Matsubara, I., Ikuta, H., Takeuchi, T., Mizutani, U., Sodeoka, S.: An oxide single crystal with high thermoelectric performance in air. Jpn. J. Appl. Phys. **39**(11B), L1127 (2000)
99. Minami, T.: Transparent conducting oxide semiconductors for transparent electrodes. Semicond. Sci. Technol. **20**(4), S35 (2005)
100. Hafdallah, A., Djefaflia, F., Saidane, N.: Structural and optical properties of ZnO thin films deposited by pyrolysis spray method: effect of substrate temperature. Optics **7**(2), 68–73 (2018)
101. Bakhshi, H., Sarraf-Mamoory, R., Yourdkhani, A., Song, S., Tseng, Y., Mozharivskyj, Y.: Improvements in the thermoelectric efficiency of $SrTiO_3$ through donor doping. Ceram. Int. **48**(4), 5831–5839 (2022)
102. Löhnert, R., Stelter, M., Töpfer, J.: Evaluation of soft chemistry methods to synthesize Gd-doped $CaMnO_3-\delta$ with improved thermoelectric properties. Mater. Sci. Eng., B **223**, 185–193 (2017)
103. Buscaglia, M.T., Maglia, F., Anselmi-Tamburini, U., Marré, D., Pallecchi, I., Ianculescu, A., Canu, G., Viviani, M., Fabrizio, M., Buscaglia, V.: Effect of nanostructure on the thermal conductivity of La-doped $SrTiO_3$ ceramics. J. Eur. Ceram. Soc. **34**(2), 307–316 (2014)
104. Jana, S.S., Maiti, T.: Enhanced thermoelectric performance in oxide composites of La and Nb codoped $SrTiO_3$ by using graphite as the electron mobility booster. ACS Appl. Mater. Interfaces. **14**(12), 14174–14181 (2022)
105. Xu, G., Funahashi, R., Matsubara, I., Shikano, M., Zhou, Y.: High-temperature thermoelectric properties of the $Ca_{1-x}Bi_xMnO_3$ system. J. Mater. Res. **17**(5), 1092–1095 (2002)
106. Kamarulzaman, N., Kasim, M.F., Rusdi, R.: Band gap narrowing and widening of ZnO nanostructures and doped materials. Nanoscale Res. Lett. **10**, 346 (2015)
107. Wang, N., Chen, H., He, H., Norimatsu, W., Kusunoki, M., Koumoto, K.: Enhanced thermoelectric performance of Nb-doped $SrTiO_3$ by nano-inclusion with low thermal conductivity. Sci. Rep. **3**(1), 1–5 (2013)
108. Terasaki, I.: Layered cobalt oxides as a thermoelecric material. In: Narlikar, A.V. (eds) Frontiers in Magnetic Materials. Springer, Berlin, Heidelberg (2005)
109. Zhan, B., Lan, J., Liu, Y., Lin, Y., Shen, Y., Nan, C.: High temperature thermoelectric properties of Dy-doped $CaMnO_3$ ceramics. J. Mater. Sci. Technol. **30**(8), 821–825 (2014)
110. Ravichandran, J.: Thermoelectric and thermal transport properties of complex oxide thin films, heterostructures and superlattices. J. Mater. Res. **32**, 183–203 (2017)
111. Yazdani, S., Pettes, M.T.: Nanoscale self-assembly of thermoelectric materials: a review of chemistry-based approaches. Nanotechnology **29**(43), 432001 (2018)
112. Singh, K., Kumar, K., Srivastava, S., Chowdhury, A.: Effect of rare-earth doping in CeO_2 matrix: correlations with structure, catalytic and visible light photocatalytic properties. Ceram. Int. **43**(18), 17041–17047 (2017)

113. Ohta, H., Kim, S., Mune, Y., Mizoguchi, T., Nomura, K., Ohta, S., Nomura, T., Nakanishi, Y., Ikuhara, Y., Hirano, M., Hosono, H., Koumoto, K.: Giant thermoelectric Seebeck coefficient of a two-dimensional electron gas in $SrTiO_3$. Nat. Mater. **6**, 129–134 (2007)
114. Liu, K., Liu, Z., Zhang, F., Zhang, J., Yang, X., Zhang, J., Shi, J., Ren, G., He, T., Duan, J.: Improved thermoelectric performance in Pr and Sr Co-doped $CaMnO_3$ materials. J. Alloy. Compd. **808**, 151476 (2019)
115. Lan, J., Lin, H., Fang, H., Mei, A., Nan, W., Liu, Y., Xu, S., Peters, M.: High-temperature thermoelectric behaviors of fine-grained Gd-doped $CaMnO_3$ ceramics. J. Am. Ceram. Soc. **93**(8), 2121–2124 (2010)

Role of RE-Doped Oxides in Piezoelectric and Triboelectric Energy Devices

Poonam Kumari, Ivan Castillo, Jyotsna, Priya, and Aseem Vashisht

Abstract This chapter analyzes rare earth (RE)-doped metal oxides in piezoelectric and triboelectric energy devices. RE elements, with unique 4f electron configurations, significantly affect the structural and functional properties of oxides. Their incorporation in systems like MOFs, perovskites, and nanoparticles enhances interfacial interactions, charge mobility, and polarization for efficient energy harvesting. The chapter examines how RE doping improves electromechanical coupling and raises output voltages by modulating the dielectric and ferroelectric characteristics that are crucial for piezoelectric devices. Through processes like band structure tuning, the formation of oxygen vacancies, and surface energy modification, RE-doped oxides improve surface charge density and charge transfer efficiency in triboelectric nanogenerators (TENGs). Improved carrier dynamics, surface potential, and dielectric constant—parameters that directly affect piezoelectric and triboelectric performance—are demonstrated in case studies of materials such as Ce-doped ZnO, Gd-doped CeO_2, and Eu-doped La_2O_3. Multipurpose systems like Cu-doped WO_3 and Sm-doped $Sc_2Mo_3O_{12}$ show that RE doping enhances energy conversion and adds functionalities such as magnetocaloric effects and negative thermal expansion. This chapter provides valuable insights into the synthesis and design of RE-doped oxides for advanced energy devices. It is especially useful for researchers developing self-powered, sustainable electronics, and sensors.

Keywords RE-doped metal oxides · Piezoelectric · Triboelectric · Energy-harvesting materials

---

P. Kumari (✉) · Jyotsna · Priya
Department of Physics, Akal College of Basic Sciences, Eternal University, Rajgarh, Himachal Pradesh, India
e-mail: p19phy@gmail.com

I. Castillo
Department of Physics, University of Puerto Rico, San Juan, USA

A. Vashisht
University Centre for Research and Development (UCRD), Chandigarh University, Gharuan, Mohali, Punjab, India

© The Author(s), under exclusive license to Springer Nature Singapore Pte Ltd. 2026
N. Kumar et al. (eds.), *Rare Earth-Doped Metal Oxide Nanostructures*, Engineering Materials, https://doi.org/10.1007/978-981-95-6192-6_9

1 Introduction

In the development of sophisticated energy-harvesting systems, RE-doped oxides have become a game-changing class of materials, especially in the domains of piezoelectric and triboelectric devices. The structural, electrical, and functional characteristics of host oxides are greatly impacted by the distinct 4f electron configurations and localized orbitals seen in RE elements [1]. Since most of the eighteenth and nineteenth centuries, 15 lanthanide elements ($Z = 57$ through 71) and yttrium ($Z = 39$) were initially isolated as oxides from uncommon rocks; these are referred to as RE elements [2]. This chapter discusses the function of RE-doped oxides in these superior energy systems. The incorporation of RE-doped oxides has introduced a new method to improve these devices' performance by modifying their structural, dielectric, and surface properties. This chapter delves into the fundamental mechanisms, material strategies, and application prospects of RE-doped oxides within piezoelectric and triboelectric energy devices, underscoring their pivotal role in next-generation smart materials and nanogenerators.

1.1 Piezoelectric Effect

As a new technology for supplying low-power electronic devices, piezoelectric energy harvesting (PEH) generates electrical energy from mechanical energy [3]. The piezoelectric effect is when certain materials create an electrical charge when a mechanical strain (compression or tension) is applied to the material; also, when an electrical field is applied to the material, this causes same chemical deformation called the inverse piezoelectric effect [4]. Jaques and Curie discovered piezoelectricity when looking at a single crystal of quartz around the year 1880 [5]. The term "piezoelectricity" makes perfect sense when we consider that "piezo" is Greek for "to press"; hence, electricity is produced by pressure [6]. Piezoelectricity is being used in many types of devices, including transducers, actuators, surface acoustic wave devices, and frequency control. ZnO, $BaTO_3$, and PZT are some of the most important materials with high piezoelectric coefficients and electromechanical coupling capabilities.

This effect is formed in crystals that have no center of symmetry [4] like quartz (SiO_2), barium titanate ($BaTiO_3$), zinc oxide (ZnO), and polyvinylidene fluoride (PVDF). When such a type of material is subjected to mechanical force, it deforms, causing the relative displacement of charges. The total polarization accumulates as surface charge on the material. The crystal reverts to its uncharged state upon removal of the external force. The piezoelectric effect is the name given to this phenomenon.

1.2 Triboelectric Effect

When two materials come into touch or rub against one another, a contact-induced electrification known as the triboelectric effect occurs, which is the general cause of everyday electrostatics [7]. Mechanisms behind the triboelectric effect are contact electrification and dielectric polarization. The use of materials with different electron affinity increases charge transfer [8]. Energy-harvesting devices based on the triboelectric effect are known as triboelectric nanogenerators (TENGs) since the charge generating processes mostly occur on the nanoscale. The dry leaf/polyvinylidene fluoride-based TENG devices have previously produced electric power of up to 14 mW [9].

TENGs efficiently transform mechanical energy into electric power or messages by fusing the effects of contact electrification and electrostatic induction. From basic scientific knowledge to cutting-edge technologies and applications, TENG development has advanced quickly during the last ten years. The mechanics of contact electrification and electrodynamics, applications, potential future developments, and constraints are all briefly covered in this primer on TENGs. Advances in the theoretical and experimental aspects of TENGs are anticipated because the area is transdisciplinary. For instance, a theoretical grasp of physical principles and concepts can be combined with technologies based on Maxwell's equations for a mechano-driven, slow-moving system. This allows for the creation of general simulation models and related studies to optimize TENGs for various applications (Fig. 9.1).

2 Rare Earth Doping in Oxides

Functional oxides like ZnO, $BaTiO_3$, and TiO_2 can have their structural and electrical properties tailored by RE elements including La, Nd, Sm, Gd, and Dy. This is especially important for improving piezoelectric and triboelectric responses in micro-energy-harvesting (MEH) devices. Their incorporation influences critical parameters such as crystallinity, oxygen vacancy concentration, and bandgap modulation key factors for boosting electroactive behavior. This is largely attributed to the partially filled 4f orbitals of RE ions, which induce localized energy states and modify defect landscapes in host lattices. RE doping in ZnO systems, for instance, has shown negative formation energies and compliance with mechanical stability criteria, demonstrating improved ductility and redshifted conduction bands due to Ce, Nd, Sm, Eu, and Gd additions [10].

Binary RE oxides (RE_2O_3), like La_2O_3, Sm_2O_3, and Nd_2O_3, have been effectively integrated into flexible polymer matrices such as PDMS to develop TENGs [11], offering mechanical flexibility and enhanced surface charge density. Meanwhile, Gd-doped CeO_2 stands out for its ability to fine-tune optical, thermoelectric, and magnetic properties due to its high unpaired 4f electron count, whereas Eu-doped

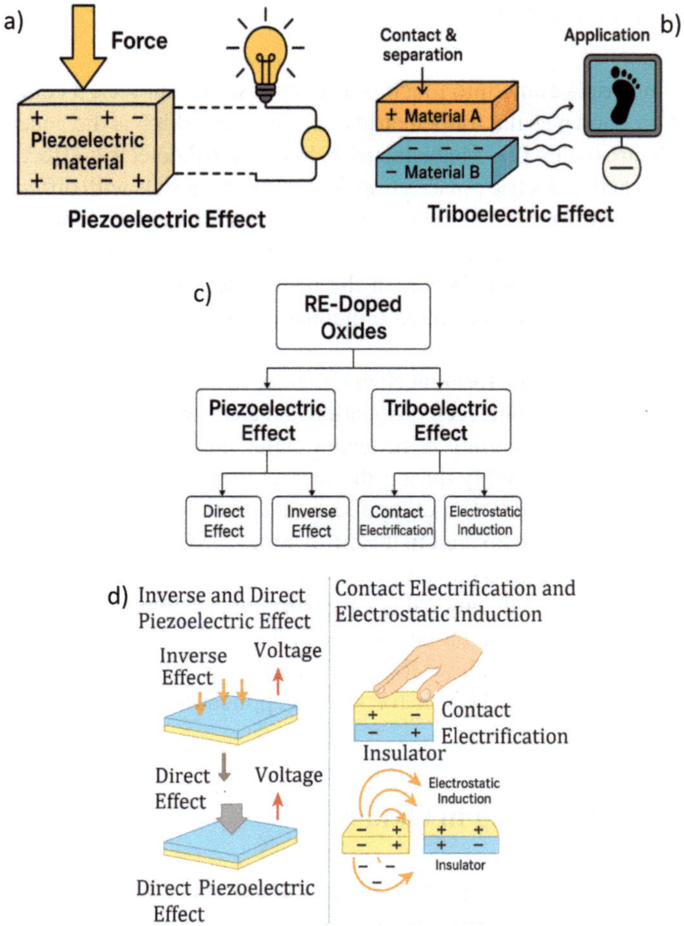

Fig. 9.1 Underlies the (**a**) piezoelectric phenomenon, in which two materials come into contact and separate, causing charge transfer because of differences in their electron affinities, and the (**b**) triboelectric effect, in which mechanical stress causes electric polarization in non-centrosymmetric materials. The mechanism of charge creation and separation in both effects is depicted in subfigure (**c**), emphasizing surface charge exchange (triboelectric) and interior dipole alignment (piezoelectric). The two effects are compared and linked in subfigure (**d**), which shows how the energy conversion pathways are identical while the origin, material needs, and application modalities are different

La_2O_3 provides a model system for probing interstitial doping and its effects on band structures [12–14].

Doping mechanisms typically fall into two categories: **substitution doping**, where RE^{2+} ions replace host cations like Zn^{2+}, generating oxygen vacancies and lattice strain; and **interstitial doping**, where RE ions occupy voids in the lattice, introducing defect dipoles and localized distortions. As shown schematically in Fig. 9.2, both doping pathways contribute to enhanced structural and electronic reconfiguration,

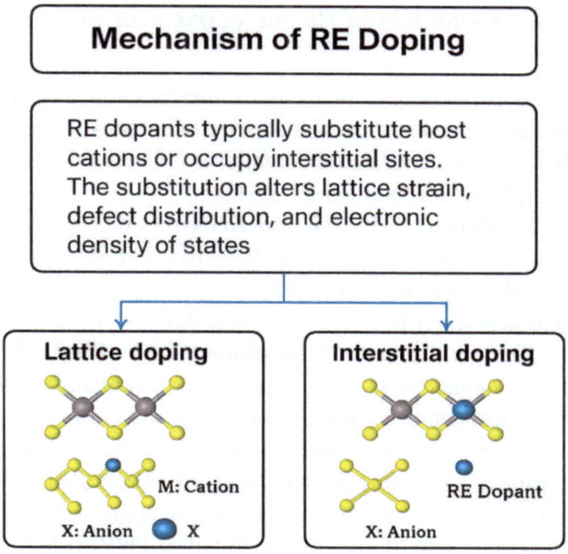

Fig. 9.2 Mechanism of rare earth (RE) doping in oxides

underscoring the versatility of RE-doped oxides in advanced energy and sensing technologies.

3 Classification of RE-Doped Piezoelectric and Triboelectric

Although rare earth ion-doped piezoelectric materials have garnered significant interest due to their exceptional electrical and luminescent properties, comprehensive reviews in this field remain limited. Existing literature primarily focuses on specific types of piezoelectric materials or fabrication techniques. For example, Lv et al. reviewed lead-free perovskite piezoelectric bulk materials, emphasizing the relationships among phase boundaries, domain configurations, and electrical properties [15]. Liu et al. discussed advancements in lead-free textured piezoelectric ceramics, highlighting improvements in piezoelectric performance [16]. Chen et al. examined the development of piezoelectric materials produced through additive manufacturing. On another hand, the current study offers a thorough analysis of how rare earth doping affects the electrical and optical characteristics of thin films, single crystals, and piezoelectric ceramics [17].

3.1 *Lead-Based Piezoelectric and Triboelectric*

RE dopants have drawn a lot of interest lately as a way to enhance and broaden the useful characteristics of lead-based ceramics. An extensive overview of RE-doped lead-based piezoelectric ceramics is given in this section, with particular attention to PZT, PMN-PT, and BS-PT systems.

RE-Doped PMN-PT and KNN-Based Ceramics: These lead-free piezoelectric materials exhibit improved dielectric and electromechanical characteristics when doped with ions such as Sm^{3+}, La^{3+}, or Er^{3+}. These qualities can work in concert with triboelectric effects in hybrid devices [18–21].

The multifunctional **RE-doped PZT ($Pb(Zr,Ti)O_3$)**: La^{3+}, Nd^{3+}, Sm^{3+}, and Eu^{3+} dopants have been demonstrated to enhance piezoelectric constants, dielectric behavior, and even introduce photoluminescence [22–25].

In recognition of their superior electromechanical qualities, **$Pb(Zr_xTi_{1-x})O_3$** ceramics are among the most studied piezoelectric materials. For example, PZT-5A ceramics exhibit a Curie temperature of approximately 365 °C and a strong piezoelectric coefficient ($d_{33} = 375$ pC/N). The $Pb(Zn_{1/3}Nb_{2/3})O_3$- $Pb(Zr,Ti)O_3$ (PZN-PZT) system offers enhanced performance compared to conventional PZT [26], exhibiting improved piezoelectric characteristics. Similarly, $Pb(Mg_{1/3}Nb_{2/3})O_3$-$PbTiO_3$ (PMN-PT) ceramics are well known for their high piezoelectric constant ($d_{33} \approx 663$ pC/N) and relative permittivity ($\varepsilon_r \approx 5260$) [27]. To address the relatively low Curie temperature of PMN-PT, $BiScO_3$-$PbTiO_3$ (BS-PT) ceramics have been developed, providing both a significantly higher Curie temperature (~ 450 °C) and strong piezoelectric response [28]. Figure 9.3 illustrates the classification and functional outcomes of RE-doped piezoelectric materials. The chart outlines three major material types—ceramics, single crystals, and films—further subdivided into lead-based and lead-free systems. RE doping (with ions like Sm, Eu, Nd, Er, Dy) enhances piezoelectric and dielectric properties, introduces luminescence, and enables multifunctionality, with applications in sensors, actuators, energy harvesting, and biomedical imaging. Table 9.1 summarizes the characteristics of a few lead-based piezoelectric ceramics doped with rare earth elements.

3.2 *Lead-Free Piezoelectric and Triboelectric*

Due to the increasing environmental concerns surrounding lead pollution, the development of lead-free piezoelectric ceramics with excellent functional properties has become both important and urgent. Among the various alternatives, three lead-free ceramic systems have gained widespread attention as potential replacements for lead-based counterparts:

- Nd_2O_3-doped $BaTiO_3$-based ceramics (e.g., BCTSZ): Nd^{3+} doping enhances grain growth and domain alignment, achieving high piezoelectric coefficients (e.g., $d_{33} \approx 463$ pC/N) and strong electromechanical coupling [43].

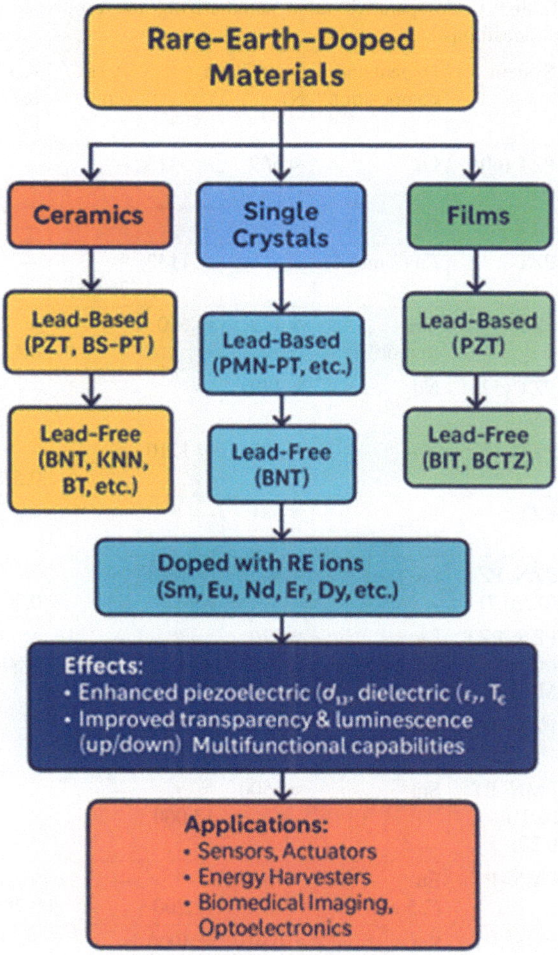

Fig. 9.3 Taxonomy of RE-doped piezoelectric materials

- Sm^{3+}, La^{3+}, Ce^{3+}, and Pr^{3+} doped $BaTiO_3$ or KNN systems: These dopants help form morphotropic phase boundaries (MPBs), which are known to boost piezoelectric performance. Sm^{3+} improves dielectric properties and raises depolarization temperatures [18].
- RE-doped $BiFeO_3$ ceramics: Substituting Bi^{3+} with RE ions like Sm^{3+}, Yb^{3+}, or Ho^{3+} can suppress impurity phases and enhance piezoelectricity in this otherwise challenging material [44].
- Gd-doped CeO_2: Although centrosymmetric, this oxide can exhibit induced piezoelectricity under electric field bias due to oxygen vacancy migration, achieving extraordinarily high apparent d_{33} values at low frequencies [45].

Table 9.1 Comparative table summarizing the properties of various rare earth-doped PZT compositions

System	Dopant and composition	d_{33}(pC/N)	ε_r	P_r (μC/cm^2)	K_p/K_t/K_{33}	Tc (°C)	Conductivity (S/m)	Ref(s)
PZT (60/40)	La	≈ 569	≈ 3413	—	—	~ 156	—	[29]
PZT	Dy	—	—	—	—	≈ 368	—	[29]
PZT	Eu (2 mol%)	—	≈ 1132	≈ 26.05	—	—	—	[30]
PZT	Sm (6 mol%)	≈ 172	≈ 610	—	≈ 0.42	—	—	[31]
PZT (52/48)	Nd	≈ 149	—	—	—	—	~ 10^{-3}	[31]
PZT	Yb	≈ 308	≈ 1210	—	≈ 0.57	—	—	[32]
PZT	La (10 mol%)	≈ 381	—	—	—	—	—	[33, 34]
PZN–PZT (0.3/0.7)	La	≈ 845	≈ 4088	—	≈ 0.70	—	—	[35]
PZN–PZT (near MPB)	La	≈ 570	—	—	≈ 0.60	—	—	[36]
PZN–PZT	Li_2CO_3 + Sm_2O_3	≈ 483	≈ 2524	—	—	—	—	[37]
PMN–PT (0.71/0.29)	Sm	≈ 1500	≈ 13,000	—	—	≈ 89	—	[38]
PMN–PT	Eu (2.5 mol%)	≈ 1420	≈ 12,200	—	≈ 0.78	—	—	[39]
PMN–PT (ternary)	Sm	≈ 0.910	≈ 4090	—	—	≈ 0 184	—	[24]
BS–PT	La	≈ 0.361	—	—	≈ 0.40	≈ 0 467	—	[40]
BS–PT	Nd	≈ 0.350	—	—	≈ 0.68 (K_t)	—	—	[41]
BS–PT	Y	≈ 0.147	—	—	≈ 0.28	≈ 0 490	—	[42]

- Er^{3+}/Eu^{3+} co-doped $Sr_{1.85}Ca_{0.15}NaNb_5O_{15}$ (SCNNO:EE): Used in hybrid nanogenerators, this material exhibits strong dielectric and luminescent properties, making it suitable for energy harvesting and pressure-activated lighting applications.

- RE-doped $BaTiO_3$ (e.g., Er^{3+}, Yb^{3+}, Tm^{3+}): These dopants enhance upconversion luminescence and dielectric behavior, contributing to improved triboelectric output in composite films [46].
- RE-doped In_2O_3 and $Yb_3Ga_5O_{12}$ nanoparticles: Ho^{3+} and Yb^{3+} doping modifies the crystal structure and bandgap, enabling upconversion luminescence and potentially influencing triboelectric charge generation [47].
- $(Na_{0.5}Bi_{0.5})TiO_3$-$BaTiO_3$ (NBT-BT): NBT-BT ceramics are among the most promising lead-free systems, known for high piezoelectric constants ($d_{33} > 100$ pC/N) and Curie temperatures (Tc > 280 °C) [48].
- $(K_{0.5}Na_{0.5})NbO_3$ (KNN): KNN-based ceramics offer high Curie temperatures (~ 400 °C) and good piezoelectric response ($d_{33} \approx 80$ pC/N, $k_p \approx 0.36$), positioning them as promising lead-free candidates [49–51].
- $BaTiO_3$ (BT): $BaTiO_3$ (BT) ceramics are known for high dielectric constants, excellent insulation, and have been widely used in PTCR thermistors, MLCCs, piezoelectric devices, and optoelectronic components [52].
- Among the functional oxides with perovskite, rutile, CaF_2, spinel, and wurtzite structures, ZnO is special because it possesses both piezoelectric and semiconducting characteristics [53]. ZnO, $BaTiO_3$, $SrTiO_3$, and TiO_2 are frequently used due to their strong dielectric and piezoelectric responses, ease of doping, and robust structural properties. Restrictions encompass limited piezoelectric response at the nanoscale level, high leakage currents, and environmental instability in some oxide systems.
- RE-doped inorganic materials, such as $Na_{0.5}Bi_{2.5}Nb2O_9{:}Pr^{3+}$ and $K_{0.5}Na_{0.5}NbO_3{:}Er^{3+}$, exhibit notable photoconductive (PC) properties along with excellent chemical and thermal stability. Enhanced PC performance has been observed in $Na_{0.5}Bi_{0.5}TiO_3$ (NBT)-based ceramics under applied electric fields. NBT is a promising lead-free alternative to $Pb(Zr,Ti)O_3$ due to its high electric-field-induced strain, large coercive field (~ 73 kV/cm), and significant remanent polarization (38 $\mu C/cm^2$). Its low phonon energy (~ 578.6 cm^{-1}) also supports efficient RE ion emission. Additionally, the fluorescence intensity ratio (FIR) technique using thermally coupled RE levels offers high-accuracy temperature sensing. In this study, Ho^{3+}/Yb^{3+} co-doping in NBT ceramics was used to achieve strong upconversion luminescence under 980 nm excitation, with investigations into PC modulation and temperature-sensing performance. The properties of a few lead-free piezoelectric ceramics doped with rare earth elements are compiled in Table 9.2.

Table 9.2 Comparative table summarizing the properties of various rare earth-doped lead-free piezoelectric and triboelectric ceramics

System	Dopant (Content)	d_{33} (pC/N or pm/V)	ε_r	K_p /K_t	tan δ	Tc (°C)	Other Features/ Applications	Ref.(s)
NBT-BT	La (1 at%)	125	—	—	—	> 280	Improved from 117 pC/N baseline	[54]
NBT-BT	CeO_2 (0.4 wt%)	≈120	—	≈ 0.52	—	—	—	[55]
NBT-BT	Sm	202	—	≈ 0.30	—	—	Significant piezoelectric enhancement	[56]
NBT-BT	Nd	175	—	≈ 0.31	—	—	—	—
NBT-BT	La_2O_3 (0.6 wt%)	167	—	≈ 0.30	—	—	—	—
KNN	Dy	95	—	—	—	~ 400	Slight performance increase	[57]
KNN	Er	91	~ 1400	—	≈ 0.03	—	Low dielectric loss	[57]
KNN	Eu	84	—	—	—	—	—	[57]
KNN	Pr	94	—	—	—	—	—	[57]
KNN	Tb	140 (pm/V)	—	—	—	—	High piezoelectric response	[58]
KNN	Tm	127 (pm/V)	—	—	—	—	—	[59]
BT	Sm	—	—	—	—	—	Grain growth suppression, semi-conductivity	[60]
BT	Tb	—	70,000–80,000	—	< 0.04	—	High ε_r, low loss	[61]
BT	Yb/Mn (co-doped)	—	> 2600	—	0.0183	—	High resistivity (4.38×10^{11} Ω cm)	[17]
BT	La_2O_3 (0.3 mol%)	—	≈ 5092	—	—	—	—	[62]

(continued)

Table 9.2 (continued)

System	Dopant (Content)	d_{33} (pC/N or pm/V)	ε_r	K_p /K_t	tan δ	Tc (°C)	Other Features/ Applications	Ref.(s)
BT	Eu	—	≈ 8581	—	—	—	Uniform grains (~ 800 nm)	[63]
NBT (Ho/Yb co-doped)	Yb (x = 0.015), Ho = 0.005	—	—	—	—	230	Up conversion emission, photochromism, temp. sensing	—
SCNNO:Er/Eu	Er^{3+}/Eu^{3+} co-doped	—	High	—	—	—	Hybrid nanogenerator; optical thermometry and lighting	—

4 Applications

4.1 Triboelectric Nanogenerators (TENGs)

This chapter provides a concise introduction to triboelectric nanogenerators (TENGs), which harness the combined effects of contact electrification and electrostatic induction to convert mechanical motion into usable electrical energy or signals. Over the past decade, TENGs have evolved significantly, transitioning from basic scientific concepts to advanced devices with real-world applications. The primer outlines the working principles behind TENGs, particularly focusing on their electrodynamic behavior and the role of Maxwell's equations in modeling slow mechanical systems. It also emphasizes the interdisciplinary nature of TENG research, pointing to future developments in both theoretical modeling and experimental optimization. Through simulation frameworks and tailored experiments, TENG technologies are poised to enhance energy-harvesting and self-powered sensing systems across various fields.

4.2 Piezoelectric-Structured Energy-Harvesting Device (PSEHD)

A piezoelectric-structured energy-harvesting device (PSEHD) and its modified, self-activated variant (SAPSEHD), intended for solid-state lighting, were among the energy-harvesting devices that used these films. The SAPSEHD, upon pressure activation, not only generates electricity but also produces visible light through phosphor activation. With capabilities like pressure-sensitive glowing patterns via custom electrodes, such devices show promise for applications in interactive lighting, smart signage, and scalable energy solutions.

4.3 Multifunctional Hybrid Nanogenerators (HNGs)

This chapter explores a novel approach to multifunctional hybrid nanogenerators (HNGs) by integrating luminescent and triboelectric materials. Researchers developed an Er^{3+}/Eu^{3+} co-doped $Sr_{1.85}Ca_{0.15}NaNb_5O_{15}$ (SCNNO:EE) phosphor that emits green-yellow light and exhibits strong temperature-sensing properties. Due to its high dielectric constant, SCNNO:EE was used as a filler in polydimethylsiloxane to create composite films, which were then optimized for maximum electrical output (75 $\mu C/m^2$, 5.05 μA, and 170 V).

The growing importance of compact, green energy-harvesting technologies, particularly hybrid nanogenerators (HNGs), to meet rising energy demands sustainably. Traditional systems like hydropower and wind turbines, though effective, are limited by their bulkiness and lack of portability. In contrast, TENGs have emerged as promising alternatives by converting mechanical energy—such as vibration or pressure—into electricity, suitable for portable and self-powered devices.

To further improve efficiency and functionality, HNGs integrating both triboelectric and piezoelectric effects have been developed. Among various material strategies, phosphor-based composites stand out due to their photoluminescent, dielectric, and triboelectric properties. However, challenges like environmental degradation of phosphor films are addressed by embedding rare earth-doped phosphors into a stable, optically transparent polymer matrix, such as PDMS.

The study presents a multipurpose HNG device comprising Er^{3+}/Eu^{3+} co-doped $Sr_{1.85}Ca_{0.15}NaNb_5O_{15}$ (SCNNO:EE) phosphors within a PDMS matrix. These lead-free, biocompatible composite films exhibit tunable luminescence and strong dielectric behavior. Devices fabricated from these films, including a piezoelectric-structured HNG (PSEHD) and its self-activating version (SAPSEHD), demonstrate high electrical output and pressure-sensitive lighting capabilities. The SAPSEHD, equipped with UV-LEDs and custom electrode patterns, activates visible luminescence upon mechanical stimulation, making it ideal for human activity monitoring, smart lighting, and scalable self-powered systems.

4.4 Semiconductors and Luminescent Nanodevices

The influence of RE dopants-specifically Ho^{3+} and Yb^{3+} on the structural and optical properties of indium and gallium oxide-based nanoparticles. Using a solvothermal method, researchers synthesized various doped nanoparticles, including Ho^{3+}-doped and Yb^{3+}/Ho^{3+}-co-doped In_2O_3, and Ho^{3+}-doped $Yb_3Ga_5O_{12}$. These RE dopants significantly altered the host crystal structures and enhanced upconversion luminescence (UCL) by enabling complex energy transfer mechanisms, such as from In_2O_3 defects and Yb^{3+} to Ho^{3+}, and via Ho^{3+}-Ho^{3+} cross-relaxation. The study identified optimal doping concentrations for maximizing UCL performance, finding that In_2O_3:2%Ho^{3+} and In_2O_3:5%Yb^{3+},3%Ho^{3+} yielded different luminescence enhancements through distinct mechanisms. Additionally, RE doping broadened the optical bandgaps of both In_2O_3 and $Yb_3Ga_5O_{12}$, expanding their potential as multifunctional materials in semiconductors and luminescent nanodevices. These insights offer valuable strategies for fine-tuning UCL behavior in RE-doped oxide systems.

Table 9.2 highlights advancements in piezoelectric and triboelectric systems doped with rare earth ions to achieve functional enhancements across photonic, optoelectronic, and sensor applications. In the PZT/PLZT family, dopants like Nd, Dy, Tm, Ho, Er, Yb, and combinations thereof enable emission in the NIR to mid-IR range, improved transparency, and enhanced electro-optic behavior for use in laser gain media, EO modulators, and photonic devices. PMN-PT systems doped with Sm, Eu, La, and other ions exhibit outstanding piezoelectric coefficients (up to ~ 1500 pC/N), high dielectric constants, and smart optical properties, making them suitable for actuators, transducers, IR detectors, and EO applications. NBT-BT-based systems use Dy, Er, and Sm dopants for dual-mode luminescence and visible emissions, targeting sensor technologies and fluorescent ceramics. KNN-based materials, known for being lead-free, achieve high transparency and vibrant emissions (red, green, orange) through La, Eu, Pr, Er, and Sm doping-useful for display, biosensing, and optical thermometry. Lastly, doped $BaTiO_3$ with Er, Yb, and Tm demonstrates efficient upconversion from IR to visible light, making it ideal for sensors and laser applications.

Figure 9.4 is structured according to five main systems: $BaTiO_3$ (BT), PMN-PT, NBT-BT, KNN-based oxides, and PZT/PLZT. The effect of RE doping on material characteristics and application performance is investigated for each system. It emphasizes how RE dopants, such as Nd, Sm, Dy, Eu, Er, Yb, and Pr, enhance characteristics including electro-optic performance, transparency, emission wavelengths, and piezoelectric coefficients. These doped materials are used in energy-harvesting technologies, lasers, sensors, electro-optic modulators, smart optics, and upconversion (UC) devices. The chapter focuses on how RE doping customizes the functionality of materials for cutting-edge photonic and electrical applications.

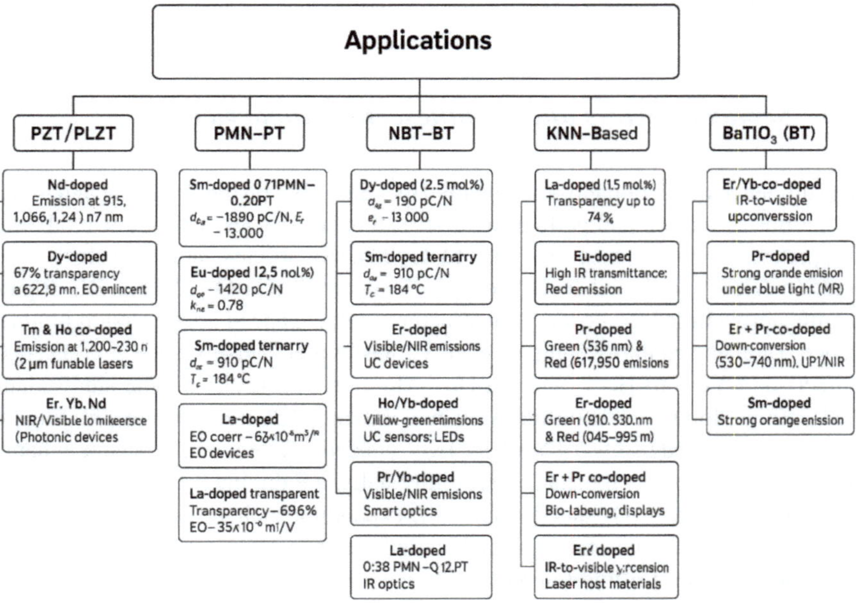

Fig. 9.4 Effect of RE doping on material characteristics and application performance is investigated for PZT/PLZT, PMN-PT, NBT-BT, KNN-based oxides

5 Conclusion

Metal oxides doped with rare earth (RE) have become revolutionary materials for developing triboelectric and piezoelectric energy-harvesting systems. The exact control of electrical, structural, and optical properties in host oxides is made possible by their distinct 4f electron configurations and potent localization effects. RE doping greatly improves energy conversion efficiency and multifunctionality by adjusting crucial parameters including dielectric constant, charge mobility, and defect states. The synthesis methods, material classifications, and performance metrics of RE-doped oxides in lead-based and lead-free systems were methodically examined in this chapter. The various uses of doped ZnO, $BaTiO_3$, PZT, and developing hybrid nanogenerators in self-powered sensors, smart lighting, and next-generation micro-energy systems are demonstrated by these works. RE-doped oxides have a lot of potential for developing compact and sustainable energy sources in the future. Their potential for multipurpose, high-performance energy-harvesting and sensing systems will be further unlocked by ongoing research into doping mechanisms, environmental stability, and device integration.

References

1. Petukhov, A., Lambrecht, W., Segall, B.: Electronic structure of rare-earth pnictides. Phys. Rev. B **53**, 4324 (1996)
2. Castor, S.B., Hedrick, J.B.: Rare earth elements. Ind. Minerals Rocks **7**, 769–792 (2006)
3. Bhatnagar, R., Yadav, V., Kumar, U., Carrasco, M.F.: Piezoelectric energy harvesting: from fundamentals to advanced applications. Energ. Technol. **13**, 2401455 (2025)
4. Bera, B., Sarkar, M.D.: Piezoelectric effect, piezotronics and piezophototronics: a review. Imperial J. Interdiscip. Res. (IJIR) **2**, 1407–1410 (2016)
5. Qiu, W.: High-performance and environmentally-friendly bulk-wave-acoustofluidic devices driven by lead-free piezoelectric materials. Small **21**(10), 2407453 (2025)
6. Newnham, R.E., Bowen, L.J., Klicker, K.A., Cross, L.E.: Composite piezoelectric transducers. Mater. Des. **2**(2), 93–106 (1980)
7. Wu, C., Wang, A.C., Ding, W., Guo, H., Wang, Z.L.: Triboelectric nanogenerator: a foundation of the energy for the new era. Adv. Energy Mater. **9**(1), 1802906 (2019)
8. Logothetis, I., Vassiliadis, S., Siores, E.: Triboelectric effect in energy harvesting. In: IOP Conference Series: Materials Science and Engineering, vol. 254, no. 4, p. 042021 (2017)
9. Slabov, V., Kopyl, S., Soares dos Santos, M.P., Kholkin, A.L.: Natural and eco-friendly materials for triboelectric energy harvesting. Nano-Micro Lett. **12**, 1–18 (2020)
10. Fernandez-Garcia, M., Martinez-Arias, A., Hanson, J., Rodriguez, J.: Nanostructured oxides in chemistry: characterization and properties. Chem. Rev. **104**, 4063–4104 (2004)
11. Hao, Y., Hong, K., Yang, J., Jia, T., Lu, X., Guo, Z., Wang, Z., Qin, Y., Su, W., Yang, D., Li, X.: Evaluation and rehabilitation system for ulnar-innervated muscles facilitated by rare earth oxide-enhanced triboelectric sensor. Adv. Mater. 2505613 (2025)
12. Sriram, S.R., Parne, S.R., Pothukanuri, N., Edla, D.R.: Synthesis and characterization of pure and Cu-doped WO_3 thin films for high performance of toxic gas sensing applications. Appl. Surf. Sci. Adv. **15**, 100411 (2023)
13. Qureshi, M.T., Farooq, U., Al Elaimi, M., Affan, H., Salem, M.S., Khalil, M., Bibi, S., Saleem, M., Saleem, M.: DFT and experimental investigations on Gd-doped CeO_2 for electronic, thermoelectric, and optical properties. Ceram. Int. **51**, 18 937–18 946 (2025)

14. Ahmad, S., Rahman, A.U., Azam, S.: Electronic structure modulation and enhanced optical-thermoelectric performance through relativistic band engineering in Eu-doped La_2O_3 (1.25%, 2.25%): advancing PC-LED technology via GGA+ U+ SOC analysis. J. Phys. Chem. Solids **201**, 112621 (2025)
15. Lv, X., Zheng, T., Zhao, C., Yin, J., Wu, H., Wu, J.: Multiscale structure engineering for high-performance Pb-free piezoceramics. Acc. Mater. Res. **3**, 461–471 (2022)
16. Liu, L.: Progress on the fabrication of lead-free textured piezoelectric ceramics: perspectives over 25 years. J. Mater. Sci.: Mater. Electron. **26**, 4425–4437 (2015)
17. Chen, Y., Zhang, D., Peng, Z., Yuan, M., Ji, X.: Review of research on the rare-earth doped piezoelectric materials. Front. Mater **8**, 679167 (2021)
18. Cao, Z., Wang, J., Zhang, C., Mao, X., Luo, L.: Flexible piezoelectric nanogenerator based on the Er^{3+}doped lead-free $(Na_{0.5}Bi_{0.5})TiO_3$-$BaTiO_3$ piezoelectric nanofibers with strong upconversion luminescence. J. Alloy. Compd. **918**, 165766 (2022)
19. Zhou, S., Lin, D., Su, Y., Zhang, L., Liu, W.: Enhanced dielectric, ferroelectric, and optical properties in rare earth elements doped PMN-PT thin films. J. Adv. Ceram. **10**, 98–107 (2021)
20. Li, C., Xu, B., Lin, D., Zhang, S., Bellaiche, L., Shrout, T.R., Li, F.: Atomic-scale origin of ultrahigh piezoelectricity in samarium-doped PMN-PT ceramics. Phys. Rev. B **101**(14), 140102 (2020)
21. Hu, M., Chang, Z., Nie, N., Wan, Z., Dong, W., Fu, Q.: La-doped PMN-PT transparent ceramics with ultra-high electro-optic effect and its application in optical devices. J. Adv. Ceram. **12**, 1441–1453 (2023)
22. Peláiz-Barranco, A., Guerra, J., García-Zaldívar, O., Calderón-Piñar, F., Mendoza, M., Hall, D., Araújo, E.: Phase transition and dielectric properties of La-doped Pb (Zr, Ti) O_3 antiferroelectric ceramics. Solid State Commun. **149**, 1308–1311 (2009)
23. Zhang, S., Cheng, M., Yan, Y., Li, Z., Wang, P., Murakami, R.-I., Zhang, D.: Improving the piezoelectric properties of Pb (Ni, Nb)O_3-Pb (Hf, Ti) O_3 ceramics through Nd_2O_3 addition. J. Alloy. Compd. **960**, 170778 (2023)
24. Guo, Q., Li, F., Xia, F., Gao, X., Wang, P., Hao, H., Sun, H., Liu, H., Zhang, S.: High-performance Sm-doped $Pb(Mg_{1/3}Nb_{2/3})O_3$-$PbZrO_3$-$PbTiO_3$-based piezoceramics. ACS Appl. Mater. Interfaces. **11**, 43359–43367 (2019)
25. Myint, T., Gunawidjaja, R., Eilers, H.: Light-induced structural changes in Eu-doped (Pb, La)(Zr, Ti) O_3 ceramics. Appl. Phys. Lett. **98**, 171906 (2011)
26. Sahoo, S., Badapanda, T., Rout, S.K., Tripathy, S.N.: Evaluation of composition-induced relaxor behavior in (1–x) $BaTiO_3$-xBi $(Zn_{2/3}Nb_{1/3})O_3$ solid solutions. Appl. Phys. A **131**(5), 410 (2025)
27. Perantie, J., Hagberg, J., Uusimaki, A., Jantunen, H.: Electric-field-induced dielectric and temperature changes in a <011>-oriented $Pb(Mg_{1/3}Nb_{2/3})O_3$-$PbTiO_3$ single crystal. Phys. Rev. B **82**(13), 134119 (2010)
28. Hu, L., Jin, R., Qiu, C., Wang, S., Xu, Z., Yan, Y.: High-performance Mn-doped $BiScO_3$-$PbTiO_3$ piezoelectric ceramics with enhanced thermal stability and electric-field endurance. Sens Actuators A: Phys. **393**, 116765 (2025)
29. Shannigrahi, S., Tay, F., Yao, K., Choudhary, R.: Effect of rare earth (La, Nd, Sm, Eu, Gd, Dy, Er and Yb) ion substitutions on the microstructural and electrical properties of sol-gel grown PZT ceramics. J. Eur. Ceram. Soc. **24**, 163–170 (2004)
30. Khazanchi, R., Sharma, S., Goel, T.: Effect of rare earth Europium substitution on the microstructure, dielectric, ferroelectric and pyroelectric properties of PZT ceramics. J. Electroceram. **14**, 113–118 (2005)
31. Pandey, S., Thakur, O., Bhattacharya, D., Prakash, C., Chatterjee, R.: Structural and electrical properties of Sm^{3+} substituted PZT ceramics. J. Alloy. Compd. **468**, 356–359 (2009)
32. He, J., Guo, X., Yu, J., Qian, S., Hou, X., Cui, M., Yang, Y., Mu, J., Geng, W., Chou, X.: A high-resolution flexible sensor array based on PZT nanofibers. Nanotechnology **31**, 155503 (2020)
33. Kour, P., Kumar, P., Sinha, S., Kar, M.: Study of dielectric and impedance spectroscopy of La substituted nanocrystalline Pb $(Zr_{0.52}Ti_{0.48})O_3$ ceramics. J. Mater. Sci.: Mater. Electron. **26**, 1304–1310 (2015)

34. Kour, P., Pradhan, S., Kumar, P., Sinha, S., Kar, M.: Enhanced ferroelectric and piezoelectric properties in La-modified PZT ceramics. Appl. Phys. A **122**, 1–7 (2016)
35. Deng, Z., Lu, Y., Wei, K.K., Zhang, J.: Understanding customer satisfaction and loyalty: an empirical study of mobile instant messages in China. Int. J. Inf. Manage. **30**, 289–300 (2010)
36. Wang, N., Sun, Q., Ma, W., Zhang, Y., Liu, H.: Investigation of La-doped 0.25 Pb $(Zn_{1/3}Nb_{2/3})O_3$–0.75 Pb $(Zr_x\ Ti_{1-x})\ O_3$ ceramics near morphotropic phase boundary. J. Electroceram. **28**, 15–19 (2012)
37. Fan, G., Shi, M., Lu, W., Wang, Y., Liang, F.: Effects of Li_2CO_3 and Sm_2O_3 additives on low-temperature sintering and piezoelectric properties of PZN-PZT ceramics. J. Eur. Ceram. Soc. **34**, 23–28 (2014)
38. Ming, C., Yang, T., Luan, K., Chen, L., Wang, L., Zeng, J., Li, Y., Zhang, W., Chen, L.-Q.: Microstructural effects on effective piezoelectric responses of textured PMN-PT ceramics. Acta Mater. **145**, 62–70 (2018)
39. Guo, Q., Hou, L., Li, F., Xia, F., Wang, P., Hao, H., Sun, H., Liu, H., Zhang, S.: Investigation of dielectric and piezoelectric properties in aliovalent Eu^{3+}-modified Pb $(Mg_{1/3}Nb_{2/3})O_3$-$PbTiO_3$ ceramics. J. Am. Ceram. Soc. **102**, 7428–7435 (2019)
40. Yao, Z., Liu, H., Hao, H., Cao, M.: Structure, electrical properties, and depoling mechanism of $BiScO_3$-$PbTiO_3$-Pb $(Zn_{1/3}Nb_{2/3})O_3$ high-temperature piezoelectric ceramics. J. Appl. Phys. **109**, 014105 (2011)
41. Kaleva, G., Politova, E., Mosunov, A., Sadovskaya, N., Segalla, A., Zeng, J.: Effect of low-melting additives on the structure, phase transitions, and dielectric properties of 0.36 $BiScO_3\cdot$ 0.64 $PbTiO_3$ ceramics. Inorg. Mater. **48**, 953–959 (2012)
42. Shahzad, K., Li, H., Li, Z., Khan, M.N.: Structural and electrical properties of cation and anion doped $BiScO_3$-$PbTiO_3$ ceramics. J. Alloy. Compd. **762**, 780–785 (2018)
43. He, Q., Ye, W., Han, Y., Chen, K., Nie, J., Zhu, Z., Zhou, S.: Synthesis, characterization, and electrical properties of Nd^{3+}-doped $(Bi_{0.4}Na_{0.2}K_{0.2}Ba_{0.2})TiO_3$ high-entropy ceramics. Int. J. Appl. Ceram. Technol. **22**, e14962 (2025)
44. Adineh, Z., Gholizadeh, A., Hosseini, S.: Rare Earth/Al co-substituted bismuth ferrite for enhanced photocatalytic degradation of methyl orange. J. Sol-Gel. Sci. Technol. **15**, 1–5 (2025)
45. Korobko, R., Patlolla, A., Kossoy, A., Wachtel, E., Tuller, H.L., Frenkel, A.I., Lubomirsky, I.: Giant electrostriction in Gd-doped ceria. Adv. Mater. **24**, 5857–5861 (2012)
46. Ward, R.E., Freeman, C.L., Dean, J.S., Sinclair, D.C., Harding, J.H.: Using metadynamics to obtain the free energy landscape for cation diffusion in functional ceramics: dopant distribution control in rare earth-doped $BaTiO_3$. Adv. Func. Mater. **30**, 1905077 (2020)
47. Li, Y., Xu, J., Hu, T., Shi, G., Peng, Y., Qiao, F., Zhang, L., Zhao, C., Kuang, Y., Shen, L.: Synthesis and upconversion luminescence fine-tuning of Yb^{3+}/Ho^{3+}-doped indium and gallium oxide nanoparticles. Inorg. Chem. **63**, 17032–17042 (2024)
48. Jo, W., Rödel, J.: Electric-field-induced volume change and room temperature phase stability of $(Bi_{1/2}Na_{1/2})\ TiO_3$-x mol.% $BaTiO_3$ piezoceramics. Appl. Phys. Lett. **99**, 042901 (2011)
49. Niranjana, V.S., Ponnan, S., Mukundan, A., Prabu, A.A., Wang, H.C.: Emerging trends in Silane-modified nanomaterial-polymer nanocomposites for energy harvesting applications. Polymers **17**(10), 1416 (2025)
50. Ramajo, L., Parra, R., Ramírez, M., Castro, M.: Electrical and microstructural properties of $CaTiO_3$-doped $K_{1/2}Na_{1/2}\ NbO_3$-lead free ceramics. Bull. Mater. Sci. **34**, 1213–1217 (2011)
51. Liu, Y., Wang, C., Zhu, Y., Hou, L., Wei, T., Wu, T., Huan, Y.: Role of ZnO dopant in enhancing piezoelectric characteristics in KNN-based piezoelectric ceramics. J. Am. Ceram. Soc. **108**(2), e20186 (2025)
52. Borah, M.: Doping in barium titanate ($BaTiO_3$): a historical perspective and future directions. J. Adv. Chem. Sci. **11**(1), 838–841 (2025)
53. Wang, Z.L., Kong, X.Y., Ding, Y., Gao, P., Hughes, W.L., Yang, R., Zhang, Y.: Semiconducting and piezoelectric oxide nanostructures induced by polar surfaces. Adv. Func. Mater. **14**, 943–956 (2004)
54. Li, Y., Chen, W., Xu, Q., Zhou, J., Gu, X., Fang, S.: Electromechanical and dielectric properties of $Na_{0.5}Bi_{0.5}TiO_3$-$K_{0.5}Bi_{0.5}TiO_3$-$BaTiO_3$ lead-free ceramics. Mater. Chem. Phys. **94**, 328–332 (2005)

55. Lv, J., Li, Q., Li, Y., Tang, M., Jin, D., Yan, Y., Fan, B., Jin, L., Liu, G.: Significantly improved energy storage performance of NBT-BT based ceramics through domain control and preparation optimization. Chem. Eng. J. **420**, 129900 (2021)
56. Fadhlina, H., Atiqah, A., Zainuddin, Z.: A review on lithium doped lead-free piezoelectric materials. Mater. Today Commun. **33**, 104835 (2022)
57. Du, J., Xu, Z., Chu, R., Hao, J., Li, W., Jiang, G., Zheng, P.: Rare-earth doped ($K_{0.5}$ $Na_{0.5}$) NbO_3 multifunctional ceramics. J. Mater. Sci.: Mater. Electron. **28**, 5288–5294 (2017)
58. Zhai, Y., Feng, Y., Du, J., Xue, J., Shen, J., Lu, Y., Lu, T., Fu, P., Li, W., Hao, J.: The impedance, dielectric and piezoelectric properties of Tb_4O_7 and Tm_2O_3 doped KNN ceramics. J. Mater. Sci.: Mater. Electron. **30**, 4352–4358 (2019)
59. Zhai, Y., Du, J., Chen, C., Hao, J., Fu, P., Li, W., Xu, Z.: Temperature stability and electrical properties of Tm_2O_3 doped KNN-based ceramics. J. Mater. Sci.: Mater. Electron. **30**, 4716–4725 (2019)
60. Liu, Z.-G., Tang, Z.-H., Hu, S.-C., Yao, D.-J., Sun, F., Chen, D.-Y., Guo, X.-B., Liu, Q.-X., Jiang, Y.-P., Tang, X.-G.: Excellent energy storage density and efficiency in lead-free Sm-doped $BaTiO_3$-Bi $(Mg_{0.5}Ti_{0.5})O_3$ ceramics. J. Mater. Chem. C **8**, 13405–13414 (2020)
61. Lu, H., Liu, L., Lin, J., Yang, W., Zheng, W., Weng, L., Zhang, X.: Diffuse characteristics and piezoelectric properties of Tb-doped BCZT ceramics with $CaCl_2$ as sintering aid. Ceram. Int. **43**, 16348–16355 (2017)
62. Ma, C., Du, H., Liu, J., Du, X., Feng, D.: Charge compensation mechanisms of $BaTiO_3$ ceramics co-doped with La_2O_3 and Bi_2O_3. Ceram. Int. **48**, 5428–5433 (2022)
63. Kumari, P., Kumar, P., Rai, R.: Study the structural and dielectric properties of rare-earth La doped $(Ba_{1-x}La_x)(Ti_{0.815}Mn_{0.0025}Nb_{0.0025}Zr_{0.18})_{0.995}O_3$ ceramics. Integr. Ferroelectr. **167**(1), 24–34 (2015)

Environmental Pollutant Detection of Rare Earth-Doped Metal Oxides

Arushi Sharma, Aseem Vashisht, Alex Ibhadon, and S. K. Mehta

Abstract Environmental pollutants threaten the health of humans and disrupt the balance of ecosystems. Their accumulation in water bodies and deposition into the food chain can result in long-term environmental degradation, chronic health disorders, and hormonal imbalance. Consequently, the early detection and continuous monitoring of these contaminants are essential, not only to safeguard public health but also to ensure the sustainability of our environment. Among the various sensing materials, rare earth-doped metal oxides have become an extremely promoting class because of their prominent physicochemical properties, surface activity, and tunable electronic structures. This chapter explores the role of rare earth-doped metal oxides, including metal oxides ZnO, Fe_3O_4, SnO_2, SiO_2, CuO, TiO_2, and NiO, in the detection of an extensive array of environmental pollutants such as pesticides, heavy metals, antibiotics, toxic dyes, microplastics, and industrial effluents. Doping these metal oxides with rare earth elements like Ce, Eu, La, or Tb significantly improves their sensitivity and selectivity by modulating charge transport, extending optical absorption, and creating active surface sites. The chapter provides insights into the synthesis approaches, structural modifications, and morphological tuning that influence the sensing performance of these materials. Various detection platforms, optical, magnetic, electrical, and catalytic responses, are discussed in a generalized context to emphasize the versatility of these nanostructured oxides. Furthermore, the potential of rare earth-doped metal oxides for use in miniaturized, portable, and field-deployable sensing devices was also highlighted.

Keywords Environmental pollutants · Metal oxide · Rare earth elements · Gaseous pollutants · Doping

A. Sharma · S. K. Mehta (✉)
Department of Chemistry and Centre of Advanced Studies in Chemistry, Panjab University, Chandigarh, India
e-mail: skmehta@pu.ac.in

A. Vashisht
University Centre for Research and Development, Chandigarh University, Mohali, India

A. Ibhadon
Department of Chemical Engineering, Faculty of Science and Engineering, University of Hull, Hull, UK

© The Author(s), under exclusive license to Springer Nature Singapore Pte Ltd. 2026
N. Kumar et al. (eds.), *Rare Earth-Doped Metal Oxide Nanostructures*, Engineering Materials, https://doi.org/10.1007/978-981-95-6192-6_10

1 Introduction

The transboundary transfer of pollutants between rich and developing nations has made environmental degradation a major global concern. Unsustainable resource exploitation, fast urbanization, deforestation, industrialization, rising population density, and rising food demands are some of the causes of this expanding problem. Ecosystems and human health are seriously threatened by the long-lasting harmful contaminants that industrial operations, in particular, discharge into the air, water, and soil.

The extensive use of chemical pesticides and fertilizers, as well as the usage of sewage sludge and untreated wastewater in farming operations, all contribute to agriculture's negative environmental effects. Furthermore, the agricultural industry contributes significantly to climate change by releasing greenhouse gases, particularly nitrous oxide (N_2O) and methane (CH_4). The initial step involves the detection and monitoring of environmental contaminants to evaluate their effects to lower ecological risks [1].

1.1 Overview of Environmental Pollution and Its Global Impact

Sources of Environmental Pollution:

The various types of environmental pollutants are pictorially represented in Fig. 10.1, and their detailed description is provided below:

Industrialization and Automobiles: The explosive growth of the automobile industry leads to the overall growth of the global industry. Rising demand for fossil fuels due to faster urbanization has increased the depletion of energy sources. A high quantity of hazardous metals and greenhouse gases released into the environment causes airborne particles to diffuse in our environment. As a result of all these, the quality of water, air, and soil is affected. The necessity of ongoing environmental monitoring is to lessen their effects on ecosystems and public health [2].

Waste Management and Fossil Fuels: Many harmful chemicals such as oxides of nitrogen, sulfur, and carbon monoxide were liberated in the environment due to the burning of coal, natural gas, and petroleum. These detrimental contaminants are a result of acid rain. Moreover, inadequate techniques of waste disposal including open dumping and inappropriate treatment of industrial and municipal waste, are causes in the sources of water and soil. Hence, there is an urgent need for sustainable environmental friendly techniques of waste management [3].

- **Agricultural Practices**: Nowadays, farmers overuse pesticides, insecticides, fertilizers, and irrigation technologies in order to fulfill the basic needs of an overpopulated society. Due to this excessive dependence toxic agrochemicals

Fig. 10.1 Overview of environmental pollutant classes

mainly that contain sulfates, nitrates, and phosphates have resulted in the nearby water bodies which results in the groundwater contamination and badly affects the health of humans [4].

Hazards of Pollution to the Environment and Public Health

- **Impact on Human Health**: Numerous severe health issues can result from exposure to environmental contaminants. Among the chemicals that have been connected to both immediate and long-term health problems are lead, arsenic, mercury, and nitrates. These include reproductive toxicity, respiratory disorders, nervous system issues, and cancer risks [5]. Contaminants frequently enter the body through the consumption of tainted food or water, and they can particularly harm vulnerable populations such as children, pregnant women, and the elderly

[6]. Additionally, it is commonly known that prolonged exposure to air pollutants, such as methane, fine particulate matter, and nitrous oxide, significantly increases the risk of respiratory and cardiac illnesses [7]. Therefore, protecting human health requires timely detection and proper management of environmental pollutants.
- **Ecological Impacts**: As the pollutants build up in soil, water, and the atmosphere and eventually harm both plant and animal life, persistent pollutants are a major threat to ecosystems [8]. Overuse of fertilizer frequently causes eutrophication, which in turn causes algal blooms that lower oxygen levels and upset aquatic ecosystems. Furthermore, soil contamination lowers fertility and hinders agricultural productivity, while air pollution disrupts climate regulation systems. These environmental stresses eventually reduce the resilience of ecosystems and endanger vital natural services like pollination, carbon storage, and water purification [9, 10].
- **Threat to World Health**: One of the biggest risks to world health in the twenty-first century is pollution. According to the World Health Organization (WHO), nearly one-third of urban areas have dangerously high levels of lead contamination, and around half of the world's cities have carbon monoxide levels over health-safe thresholds. When released into the environment, these contaminants change the biological, chemical, and physical characteristics of natural resources, making them dangerous or useless [11].

2 Role of Different Materials in the Detection of Pollutants

The development of different materials, especially nanoparticles and porous architectures, has greatly increased the detection of environmental toxins. These substances improve the sensitivity and selectivity of the detection of a variety of pollutants, such as organic chemicals, pesticides, and heavy metals. The following sections describe key materials and their applications in the detection of environmental pollutants:

2.1 Diverse Classes of Materials for Sensing Applications

Sensing materials span a wide range of classes, including metal oxides, carbon-based nanomaterials, conducting polymers, quantum dots, MOFs, COFs, and perovskite materials as depicted in Fig. 10.2. Table 10.1 presents a comparative analysis of various classes of sensing materials, highlighting their respective advantages and disadvantages.

- **Metal Oxides**: The remarkable physicochemical features, surface reactivity, and varied sensing mechanisms of metal oxides have made them one of the most interesting substances for detecting pollution in the environment. Metal oxides

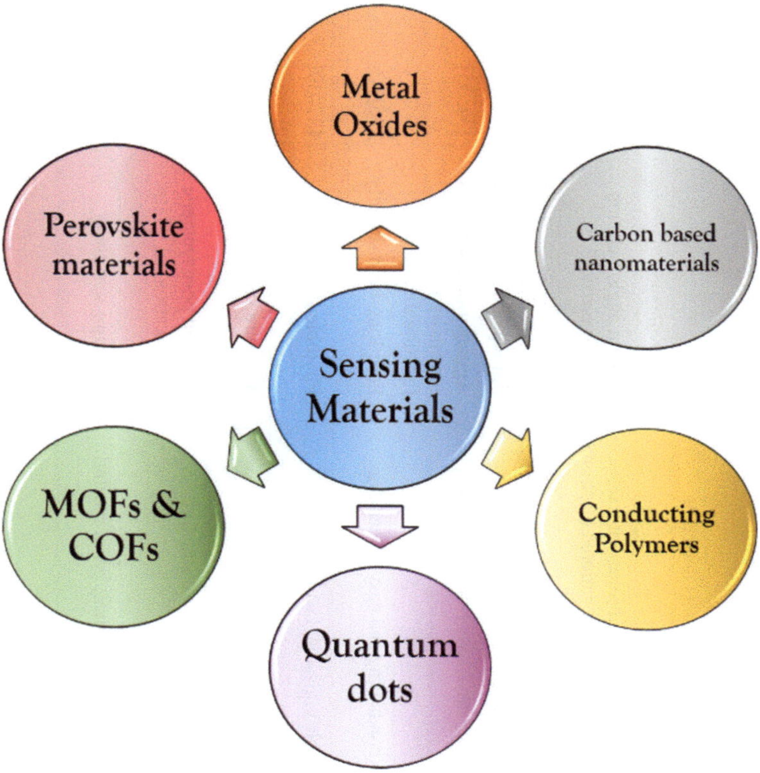

Fig. 10.2 Advanced materials for environmental sensing

such as tin oxide (SnO_2), iron oxide (Fe_2O_3), zinc oxide (ZnO), and titanium oxide (TiO_2) have found various applications due to their high surface reactivity and tunable electronic properties. These materials exhibit notable sensitivity to gases, including nitrogen dioxide (NO_2), ammonia (NH_3), carbon monoxide (CO), and Volatile Organic Compounds (VOCs), where their electrical resistance alters in response to interactions with specific gas molecules on their surface [12]. Building upon this, recent studies have exemplified that these nanostructured metal oxides enhanced selectivity and sensitivity toward heavy metal ions, humidity variations, and pH changes, making them highly useful for advanced sensing applications [13]. However, alone metal oxide-based sensors often face limitations, the requirement for high operating temperatures, and nanoparticle aggregation, which collectively hinder their practical performance. To overcome these challenges, doping with other elements is necessary to enhance their structural stability, improve charge carrier mobility, and introduce defect states that facilitate greater sensitivity and selectivity toward target analytes [14].

- **Carbon-Based Nanomaterials**: Carbon-based nanomaterials such as graphene oxide (GO), carbon nanotubes (CNTs), reduced graphene oxide (rGO), carbon

Table 10.1 Comparative analysis of various material classes used in environmental pollutants detection

Class of materials	Examples	Types	Advantages	Disadvantages	References
Rare earth-doped metal oxides	Ce-doped SnO_2, La-doped TiO_2	Doped semiconducting oxides	Enhanced sensitivity and selectivity; fast electron transfer kinetics; tailored electronic and optical properties	Dopant agglomeration may be possible	[31, 32]
Metal oxides (undoped)	Fe_2O_3, ZnO, TiO_2	p and n-type semiconductors	Excellent surface reactivity, cost-effective and chemically stable	Poor selectivity and limited optical tunability	[33]
Carbon-based materials	CNTs, graphene	1D and 2D	High surface area, easy functionalization	Aggregation issues, low sensitivity without functionalization	[34, 35]
Conducting polymers	PANI, PPy, Polythiophene	Homopolymers, Copolymers	Tunable conducting and easy processing	Poor long-term stability	[36]
Quantum dots	CdSe, ZnS, Graphene QDs	II-VI, IV-VI, Carbon-based	Size-tunable emission, fast response	Photobleaching	[37–39]
MOFs	ZIF-8, UiO-66, MIL-101	Based on metals like Zr, Co, Zn, Fe	Ultrahigh porosity	Moisture sensitive	[40–42]
COFs	COF-1, COF-LZU1	2D/3D porous crystalline organic networks	Lightweight; tailorable pore structure	Difficult post synthetic modification	[43, 44]
Perovskite materials	$LaFeO_3$, $CsPbI_3$	Organic–inorganic hybrids	Strong photoluminescence; broad spectral absorption	Moisture sensitivity	[45, 46]

dots, and fullerene derivatives have attracted a lot of interest in the field of environmental sensing [15, 16]. These materials are suitable for creating sensitive and selective sensors because of their large surface area, exceptional mechanical strength, thermal and electrical conductivity, and extraordinary chemical stability [17]. These materials exhibit diverse surface functionalities that enable strong interactions with various analytes. They are appropriate for the transmission of signals because of their π–π conjugated architectures, which are essential for promoting effective electron transfer [18]. Furthermore, these materials are highly compatible with other nanomaterials, allowing the development of composite structures that can significantly improve sensing performance through synergistic effects. Nevertheless, unaltered carbon nanomaterials might not always show great sensitivity toward certain target molecules, even with all of their benefits. To solve this, it is frequently required to surface functionalized or hybridize with other nanostructures in order to customize their selectivity and raise overall sensing effectiveness [19].

- **Conducting Polymers**: These substances belong to a class of organic polymers that can conduct electricity because of delocalized π-electron systems along their molecular backbones. Polymers such as poly(3,4-ethylenedioxythiophene) (PEDOT), polyaniline (PANI), polypyrrole (PPy), and polythiophene (PT) are frequently used in environmental sensing applications [20]. By changing their optical, electrical, or electrochemical properties, redox reactions or interactions with certain pollutants change their conductivity, which makes them valuable for identifying contaminants.
- **Quantum Dots (QDs)**: Quantum dots (QDs), which are semiconductor materials at the nanoscale with a diameter usually between 2 and 10 nm, have unique size-dependent optical and electrical properties due to the quantum confinement effect [21]. It is feasible to carefully modify their photoluminescence characteristics by changing components like as particle size, core composition, or surface functionalization [22]. QDs are distinguished by their strong fluorescence intensity, excellent photostability, and exceptional quantum efficiency. These qualities make them highly beneficial for the rapid and incredibly sensitive detection of environmental contaminants [23]. However, a detailed evaluation of their possible toxicity and environmental effects is necessary to guarantee their safe use.
- **Covalent Organic Frameworks (COFs) and Metal Organic Frameworks (MOFs)**: Due to their extremely high surface areas, adjustable porosity, and versatile chemical architectures, metal organic frameworks (MOFs) and covalent organic frameworks (COFs) have garnered a lot of interest in the field of environmental sensing. MOFs are composed of well-ordered crystalline networks that are produced by coordinating metal ion clusters with organic linkers. Numerous active sites that are suitable for absorbing pollutants and can be customized to interact with certain target molecules are provided by their porous structure [24]. MOFs are both electrochemical and luminous sensors because they can induce detectable signal changes when they bond with an analyte. Of the several MOFs,

those that contain elements like lanthanides, zirconium, and zinc have demonstrated encouraging outcomes in the detection of heavy metal ions, organic pollutants, and poisonous gases (such as NO_2 and NH_3) [25]. COFs, on the other hand, are made entirely of light elements like carbon, boron, and nitrogen that are joined by strong covalent bonds. They are ideal for environmental applications where chemical flexibility and structural integrity are necessary because of their completely organic, stable structure [26]. They can be used in optical and electrochemical sensors because these frameworks frequently have conjugated π-systems that promote charge transfer and π–π interactions [27]. Despite its potential, creating MOFs with reliable pore architectures and increasing their production for use in industry is still a difficult technical task [28]. Furthermore, unless paired with conductive additives like polymers, many common MOFs are poor conductors, which restricts their direct application in electrochemical sensors.

- **Perovskite Materials**: Perovskites, which have the general formula ABX_3, are showing great promise as environmental sensing materials because of their exceptional optoelectronic characteristics. These include strong light-harvesting capabilities, adjustable energy band gaps, and high charge carrier mobility. These characteristics make it possible to detect a variety of environmental contaminants effectively [29]. According to recent studies, doping perovskites with rare earth elements can improve their chemical endurance and sensitivity even more [30]. However, for their actual implementation, problems including moisture sensitivity, restricted thermal stability, and the difficulties of long-term environmental deployment need to be resolved.

3 Fundamentals of Rare Earth-Doped Metal Oxides

The partially filled 4f orbitals of rare earth elements give them special optical, magnetic, and electrical characteristics. It alters charge carrier dynamics, introduces localized energy levels, and affects crystal structure and defect chemistry when doped into metal oxides.

3.1 Structural and Electronic Properties

Structural Modifications

- The host lattice frequently exhibits structural aberrations as a result of rare earth doping. In TiO_2, for instance, Er^{3+} ions take the place of Ti ions, reducing local symmetry and changing the crystal structure [47].
- Ce and La doping in $LiCoO_2$ causes local structural alterations and cell volume expansion, which may have an impact on lithium ion migration [48].

Electronic Properties

- Doping causes impurity states to be introduced, which may cause the band gap to move. For example, Ce and Pr doping in TiO_2 decreases, increases visible light and closes the band gap absorption, but Yb, Tm, and Ce doping in ZnO increases it [49].
- The n-type conductivity is caused by the presence of 4f electrons from rare earth elements, which are crucial in altering the electronic structure [50].

Thermoelectric Performance

- Doping with rare earth elements increases electrical conductivity and decreases thermal conductivity in thermoelectric substances like $CaMnO_3$, which results in higher thermoelectric values.
- Rare earth-doped materials show enhanced optoelectronic characteristics, such as better absorption coefficients and decreased electron–hole recombination rates.

3.2 Influence of Rare Earth Dopants on Metal Oxides

3.2.1 Defect Engineering and Structural Alterations

According to the charge compatibility and ionic radius, rare earth ions usually either occupy interstitial sites or replace host metal cations when added to metal oxide matrices [51]. In order to make up for the charge imbalance, this substitution frequently causes lattice distortion and produces vacant oxygen spaces or other point defects.

3.2.2 Increase in Electronic and Optical Characteristics

The distinct 4f electron configurations of rare earth ions are protected from the outside world by occupied 5 s and 5p orbitals [52]. They are therefore perfect for luminescent-based sensors since they have stable, crisp emission bands. In photoactive oxides, doping with ions such as Er^{3+}, Sm^{3+}, and Tb^{3+} can improve charge separation and increase absorption of visible light while lowering electron–hole recombination [53]. Increased sensitivity and quick reaction times result from this.

3.2.3 Modified Surface Chemistry

By altering the redox potential, basicity, and acidity of the metal oxides' surfaces, the rare earth elements can also change their surface chemistry. These mitigations enabled the methodical identification of various environmental contaminants, such as industrial dyes, volatile organic compounds, and hazardous metal cations and anions. They also greatly improved the interaction between the sensor's surface and certain analytes [54].

3.2.4 Charge Transport and Electrical Conductivity

The electrical conductivity of materials like a nanocrystalline form titanium dioxide can be greatly increased by adding rare earth metals, frequently by 40–50 times when contrasted to their undoped equivalents. The main cause of this improvement is the higher charged transfer procedures that aid in the removal of deep trap states and improve electron mobility [55].

3.2.5 Magnetic and Multifunctional Effects

The combination of magnetic separation and pollution detection in multipurpose sensor platforms confirms that loading non-magnetic oxides of metal with elements from rare earths can impart magnetic characteristics. For example, paramagnetic behavior is introduced by materials such as gadolinium and neodymium, which is especially helpful in the creation of sophisticated biosensors [56]. This method is useful for real-world applications since it not only improves the ability to sense but also makes it simpler to recover and reuse the sensor material.

4 Fabrication Methods for Producing of Rare Earth-Doped Metal Oxides

4.1 Sol–Gel Method

A system transitions from a liquid "sol" phase to a solid "gel" phase during the sol–gel process. It is a commonly used strategic technique for the synthesis of rare earth-doped metal because it ensures solution consistency and controls the structure of the final material. This process hydrolyzes metal precursors and subsequently undergoes polycondensation reactions to form a three-dimensional gel matrix [57]. The sol–gel process usually entails its co-hydrolysis of transition metals and rare earth precursors in the context of rare element doping, resulting in the creation of homogeneous xerogels. Well-defined crystalline powders are produced by successive gelation and heat treatment processes, frequently maintaining the desired perovskite or other desired crystal structures [58]. By carefully controlling the dopant concentration, this technique has been effectively used to fabricate La-doped ZnO thin films, where the hexagonal wurtzite phase was preserved even at varying lanthanum contents [59]. Furthermore, this method has proven successfully used to create Er^{3+} and Yb^{3+} co-doped Y_2O_3 nanoparticles [60]. Although it offers precise control at the molecular level, allowing customization of phase composition, density, and microstructure, this method is adaptable and efficient for developing novel compounds for the detection of environmental pollutants.

4.2 *Hydrothermal and Microwave-Assisted Methods*

Chemical reactions that occur in solutions of water at high pressures take place in a sealed container known as an autoclave during hydrothermal synthesis. However, in contrast to traditional heating techniques, microwave-aided synthesis uses microwave radiation to quickly and evenly heat reaction mixtures, thus shortening reaction times and increasing yields [61]. The production of titanium dioxide doped with ytterbium (TiO_2-Yb) using this technique is a noteworthy example. In comparison to traditional thermal techniques, this technology not only guarantees controllable particle morphology and dopant distribution, but it also drastically cuts down on synthesis time [62]. Eu^{3+} doped Y_2O_3 nanorods were successfully produced via a microwave-assisted hydrothermal technique, followed by thermal breakdown, in under six minutes. Compared to the agglomerated structures produced at greater calcination temperatures, these nanostructures exhibit superior luminous properties [63].

4.3 *Co-precipitation and Combustion Method*

The co-precipitation method enables the controlled production of homogenous, mixed-component nanoparticles by precipitating numerous solutes from a solution simultaneously. These methods are employed because they are reasonably priced and may produce consistent materials with modifiable properties [64]. The co-precipitation method for producing rare earth-doped metal oxides is exemplified by Er3 + -doped gadolinium oxide nanorods, which also have favorable optical and structural characteristics. These nanorods have interesting light qualities that could be used for environmental pollution monitoring applications because of Gd2O3's low phonon energy [65]. In light of this, the co-precipitation and combustion method of incorporating rare earth elements into metal oxides is scalable and effective, enabling the production of multifunctional materials with remarkable sensitivity and selectivity toward dangerous contaminants. The many methods for producing rare earth-doped metal oxides are shown in Fig. 10.3.

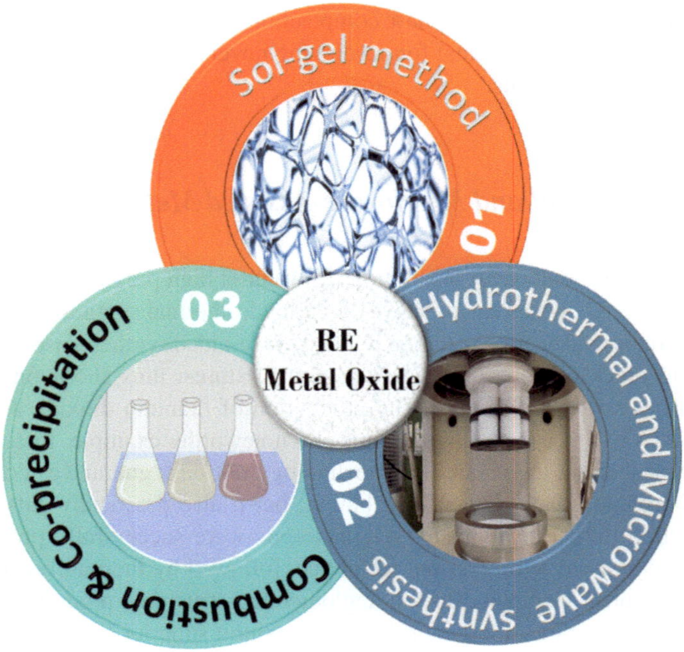

Fig. 10.3 Synthetic approaches for the fabrication of rare earth (RE)-doped metal oxide

5 Detection of Specific Pollutants

For the identification of specific environmental contaminants, such as pigments, pesticides, heavy metals, gases, and volatile organic compounds, sensitive and selective platforms are required. Advanced materials like rare earth-doped metal oxides enable tailored quantification of these contaminants. The representative pollutants and their detection methods are listed in Fig. 10.4.

5.1 Detection of Heavy Metal Ions

The ecosystem and human health are seriously threatened by heavy metal ions found in soil and water. Non-biodegradable ions that tend to accumulate in biological tissues and cause major physiological and neurological diseases include lead (Pb^{2+}), cadmium (Cd^{2+}), chromium (Cr^{6+}), and arsenic (As^{3+}). Rare earth-doped metal oxides have lately been used in the creation of advanced sensors for the detection of heavy metal ions because of their unique physicochemical and electrical properties. Rare earth elements like Gd^{3+}, Eu^{3+}, Sm^{3+}, Tb^{3+}, and Ce^{3+} have a lot of electronic transitions and a lot of coordination flexibility because of their partially filled 4f orbitals [66]. When synthesized in metal oxide matrices like SnO_2, Fe_2O_3, ZnO, and

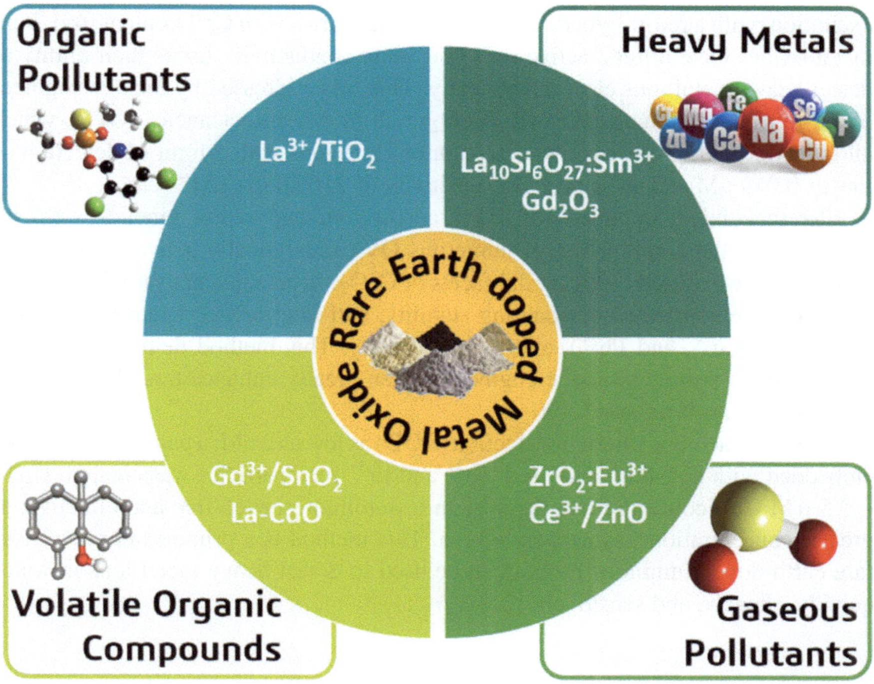

Fig. 10.4 Various rare earth-doped metal oxide classes for pollution detection

TiO_2, rare earth metals dramatically alter the band structure, produce new energy states, and boost charge carrier mobility. $La_{10}Si_6O_{27}$:Sm^{3+} (LSNP) nanoparticles, which are produced via an environmentally safe solution combustion technique, have shown several environmental applications, such as the detection of heavy metal ions. When paired with graphite-modified electrodes, these small particles can identify mercury and ions containing Pb^{2+} at concentrations as low as 1–6 mM. Furthermore, the sensor showed excellent repeatability, proving the cost-effectiveness and efficiency of LSNPs in environmental monitoring systems for the detection of dangerous heavy metal ions [67]. Eu^{3+}-doped Gd_2O_3 nanoparticles were made using different weight percentages of Eu^{3+} in the Gd_2O_3 nanoparticles, ranging from 1 to 5%, in addition to being used as a photoluminescent sensor for the identification of ferric and chromate ions in aqueous environments. Cr^{3+} and Fe^{3+} have respective detection limits of 6.56 ppm and 1.48 ppm. An exceptional recovery rate of over 90% is achieved in tap and distilled water, but only about 87% is achieved in sewage water [68].

One common water contaminant that poses major dangers to human health and the environment is chromium (III) ions (Cr^{3+}). In order to detect Cr^{3+} in aqueous environments, this study emphasizes the use of $NaGdF_4$:Yb^{3+}, Er^{3+} up-converting nanoparticles (UCNPs) as a ratiometric luminous sensor. Selective reabsorption at ~ 580 nm causes these UCNPs to emit visible light upon 980 nm near-infrared (NIR) excitation, with a red-to-green emission ratio that varies with Cr^{3+} content [69]. Pure metal oxides have limited active sites and weak conductivity limits their ability to detect heavy metal ions electrochemically. This was addressed by the development of Ru-loaded CeO_2 nanocubes (Ru/CeO_2) rich in oxygen vacancies (OVs), which allowed for the highly sensitive detection of Hg^{2+} ions with a limit of detection as low as 0.019 μM and an impressive sensitivity of 2169.8 μA μM^{-1} cm^{-2}.

Electrochemical conductivity, Hg^{2+} adsorption, and redox kinetics were all improved by the combination of numerous OVs, catalytically active Ru nanoparticles, and the exposed (100) crystal facets of CeO_2. In actual water sample analysis, the sensor demonstrated outstanding stability, anti-interference against coexisting ions (Cd^{2+}, Cu^{2+}, and Pb^{2+}), and reproducibility. This method demonstrates how OV-rich rare earth metal oxide engineering can greatly enhance trace heavy metal ion detection performance [70].

In order to achieve a limit of detection (LOD) as low as 2 nM, a paper-based sensor embedded with non-functionalized Yb^{3+} and Er^{3+} nanoparticles was created. Up to ~ 0.5 μM Cr^{3+} concentration, the system is autofluorescence-free and effective; at greater concentrations, saturation is seen. This method is a prime example of how rare earth-doped luminous probes can be used to detect heavy metal ions in water quickly, cheaply, and sensitively.

5.2 *Detection of Volatile Organics and Gaseous Pollutants*

Air pollution is mostly caused by gaseous pollutants, including Carbon Monoxide (CO), Sulfur Dioxide (SO_2), Ammonia (NH_3), Nitrogen Oxides (NO_x), and Volatile Organic Compounds (VOCs). By increasing the conductivity of electricity and surface activity, rare earth element doping of metal oxides enhances gas sensing applications and reduces temperature requirements for gas detection. [71]. Ce doping in ZnO enhances adsorption energy and charge transfer, which makes it a promising option for detecting nitrogen oxide (NO) [72]. The ability of metallic oxide sensors, such as SnO_2, to detect gases is significantly impacted by oxygen vacancies, which act as catalyst sites for gas absorption. Gd-doped SnO_2 thin films were prepared by beam electron evaporation. According to the spectroscopic analysis, Gd doping modifies the kind and quantity of oxygen vacancies, particularly encouraging sub-bridging oxygen vacancies. These sub-bridging vacancies were found to significantly improve the selectivity and sensitivity of 3% Gd-doped SnO_2 films toward isopropanol among the several VOCs under investigation [73].

Cadmium oxide (CdO) thin films doped with rare earth metal ions and produced by spray pyrolysis have exceptional gas sensing properties due to their surface shape and adjustable electrical conductivity. Ce^{3+}, Sm^{3+}, La^{3+}, and Pr^{3+} ion doping had a major impact on the CdO films structural orientation, grain size, and carrier concentration. La-doped CdO thin films outperformed other studied volatile organic compounds (VOCs), acetone, ethanol, and toluene, exhibiting the lowest resistivity (3.01×10^{-4} Ω.cm) and the maximum sensitivity (~ 71.42%) toward formaldehyde at 130 °C.

The better surface reactivity and higher electron donor density are responsible for this improvement in results, demonstrating the effectiveness and selectivity of rare earth-doped CdO thin films for formaldehyde (HCHO) gas sensor applications [74].

The synthesis of rare earth-doped indium oxide (In_2O_3) nanospheres using a straightforward solvothermal technique has demonstrated remarkable promise for HCHO gas detection. With a detection limit of 10.9 ppb, a quick response (30 s), and a recovery (160 s) at 170 °C, 6% La-doped In_2O_3 had the best response among the variants, 16 times that of pure In_2O_3. Increased surface area and oxygen vacancies that encourage chemisorbed oxygen species and active locations for HCHO interaction are responsible for the improvement in sensing effectiveness, providing a more calculated approach to the design of effective VOC sensors [75].

Recent advances in multimodal gas sensing applications, which combine resistance and cataluminescence (CTL) responses in a single sensor, have made europium-doped zirconium oxide (ZrO_2:Eu^{3+}) nanomaterials popular. While maintaining a similar resistance-based response, the Eu^{3+}-doped ZrO_2 demonstrated better CTL sensitivity, especially toward acetone and ethanol, as compared to undoped ZrO_2 and WO_3. Although WO_3 is still useful in sensor arrays, its restricted CTL response made it less suitable for multimodal applications, despite its strong resistance responses at low analyte concentrations [76].

Notwithstanding notable progress, issues with rare earth-doped metal oxide sensors continue to exist, including excessive operating temperatures, selectivity for particular gases in mixed settings, and long-term stability in the face of fluctuating humidity levels. Nonetheless, they are very helpful in next-generation sensing technologies due to their controllable electrical characteristics, improved surface reactivity, and capacity to introduce regulated oxygen vacancies. Their potential for selective, low-power, real-time gas monitoring will be further unlocked by ongoing studies into nano-engineering, doping techniques, and device integration. By providing more intelligent and sustainable pollution detection methods, these materials provide useful benefits for improving environmental safety.

5.3 Identification of Organic Pollutants

Sol–gel-synthesized La-Y co-doped TiO_2 nanoparticles are capable of detecting the hazardous organic contaminant 2-chlorophenol. High sensitivity, long-term operational stability, and rapid response-recovery cycles were among the remarkable performance characteristics of a sensor co-doped with 0.2 mol% yttrium (Y) and lanthanum (La). The improved sensing properties are attributed to the combined effects of La-Y co-doping and modified anatase-to-rutile phase composition in TiO_2, which raise surface activity and electron mobility. This suggests that a viable route to the development of next-generation organic pollutant monitoring sensors is co-doping with rare earth metals [77]. One particularly effective technique for recognizing hazardous chemicals, such as pesticides, is to combine rare earth-doped metal oxides with carbon-based nanomaterials. In one technique, La^{2+}-doped TiO_2

nanoparticles were applied to a glassy carbon electrode and secured to functionalized multi-walled carbon nanotubes (f-MWCNTs) to construct an extremely sensitive electrochemical sensor. This hybrid sensor showed exceptional paraoxon-ethyl (PE) detection capability with a surprisingly low detection limit of 0.0019 μM and a large sensitivity of 7.6690 $\mu A\ \mu M^{-1}\ cm^2$. The improved electrocatalytic response of TiO_2 was caused by La^{3+} doping, while the remarkable stability, selectivity, and repeatability were maintained by the conductive framework provided by the f-MWCNT network [78].

6 Challenges and Future Perspectives

The utilization of rare earth-doped metal oxide compounds in environmental pollution detection holds significant promise for the development of sensor technology. However, for these materials to be used successfully in real-world situations, a number of significant obstacles need to be removed. Soil, wastewater, and air are examples of environmental samples that usually contain a complex mixture of inorganic and organic materials. Non-specific interactions, sensor fouling, or interference brought on by these complex matrices may make it more difficult to detect target pollutants precisely. Converting lab-scale synthesis methods into large-scale, cost-effective manufacturing processes suitable for device integration and commercialization presents significant scientific and financial challenges.

Combining sensor arrays with machine learning and artificial intelligence can offer powerful capabilities for recognizing complex patterns, classifying various pollutants, and enabling predictive diagnostics, especially in challenging environments with multiple analytes or background interference. In order to encourage sustainable development and reduce ecological impact, it is equally important to prioritize eco-friendly fabrication methods, the reuse of rare earth minerals, and the employment of green synthesis procedures throughout the sensor's lifespan.

7 Conclusion

The unique structural and functional characteristics of rare earth-doped metal oxides are drawing a lot of interest as novel materials for environmental contamination monitoring. The materials' high specific surface area, exceptional redox characteristics, flexible electronic topologies, and extraordinary luminous and catalytic activity are some of their evident sensing advantages. By boosting detection selectivity, sensitivity to specific analytes, and guaranteeing quick signal response, the deliberate addition of rare earth elements to conventional metal oxide matrices greatly enhances sensor performance. These benefits have been shown for a range of contaminants, including heavy metal ions, industrial dyes, agrochemicals, and airborne pollutants.

This chapter emphasized the significance of rare earth doping in changing the surface and structural properties of metal oxides.

The development of multifunctional and hybrid nanostructures, advanced material design, and the combination of digital and sensing technologies are likely to be the main drivers of future advancements in this subject. Collectively, these advancements will contribute to reducing environmental pollutants and are an important step in maintaining the ecosystem's health and human well-being.

References

1. Rathod, S.V., Saras, P., Gondaliya, S.M.: Environmental pollution. In: Eco-restoration of Polluted Environment, 1st ed., pp. 1–34. CRC Press, Boca Raton (2024). https://doi.org/10.1201/9781003423393-1
2. Antoci, A., Galeotti, M., Sordi, S.: Environmental pollution as engine of industrialization. Commun. Nonlinear Sci. Numer. Simul. **58**, 262–273 (2018). https://doi.org/10.1016/j.cnsns.2017.06.016
3. Gyawali, K., Acharya, P., Poudel, D.: Environmental pollution and its effects on human health, interdisc. Res. Educ. **8**, 84–94 (2023). https://doi.org/10.3126/ire.v8i1.56729
4. Abbasi, A., Sajid, A., Haq, N., Rahman, S., Misbah, Z., Sanober, G., Ashraf, M., Kazi, A.G.: Agricultural pollution: an emerging issue. In: Ahmad, P., Wani, M.R., Azooz, M.M., Tran, L.-S.P. (eds.), Improvement of Crops in the Era of Climatic Changes, pp. 347–387. Springer New York, New York, NY (2014). https://doi.org/10.1007/978-1-4614-8830-9_13
5. Kampa, M., Castanas, E.: Human health effects of air pollution. Environ. Pollut. **151**, 362–367 (2008). https://doi.org/10.1016/j.envpol.2007.06.012
6. Gilliland, F., Avol, E., Kinney, P., Jerrett, M., Dvonch, T., Lurmann, F., Buckley, T., Breysse, P., Keeler, G., De Villiers, T., McConnell, R.: Air pollution exposure assessment for epidemiologic studies of pregnant women and children: lessons learned from the centers for children's environmental health and disease prevention research. Environ. Health Perspect. **113**, 1447–1454 (2005). https://doi.org/10.1289/ehp.7673
7. Filonchyk, M., Peterson, M.P., Zhang, L., Hurynovich, V., He, Y.: Greenhouse gases emissions and global climate change: Examining the influence of CO2, CH4, and N2O. Sci. Total Environ. **935**, 173359 (2024). https://doi.org/10.1016/j.scitotenv.2024.173359
8. Edwards, C.A.: Assessing the effects of environmental pollutants on soil organisms, communities, processes and ecosystems. Eur. J. Soil Biol. **38**, 225–231 (2002). https://doi.org/10.1016/S1164-5563(02)01150-0
9. Chae, Y., An, Y.-J.: Current research trends on plastic pollution and ecological impacts on the soil ecosystem: a review. Environ. Pollut. **240**, 387–395 (2018). https://doi.org/10.1016/j.envpol.2018.05.008
10. Phillips, P.J., Bode, R.W.: Pesticides in surface water runoff in south-eastern New York State, USA: seasonal and stormflow effects on concentrations. Pest Manag. Sci. **60**, 531–543 (2004). https://doi.org/10.1002/ps.879
11. Ahmed, S.: Environmental pollution and global issue (With special regard to air pollution). PIJST **1**, 1–13 (2024). https://doi.org/10.62796/pijst.2024v1i301
12. Bratovcic, A.: Recent developments on metal oxide - based gas sensors for environmental pollution control. In: Karabegović, I. (ed.) New Technologies, Development and Application IV, pp. 952–963. Springer International Publishing, Cham (2021). https://doi.org/10.1007/978-3-030-75275-0_105
13. Raturi, P., Choudhuri, B., Chinnamuthu, P.: Sensing of environmental contaminants using advanced nanomaterial. In: Advanced Materials for a Sustainable Environment, 1st ed., pp. 57–76. CRC Press, Boca Raton (2022). https://doi.org/10.1201/9781003206385-3

14. P. Baraneedharan, J. Manikandan, Metal oxide semiconductor nanostructures surface properties for gas sensing—a review. In: 2022 Sixth International Conference on I-SMAC (IoT in Social, Mobile, Analytics and Cloud) (I-SMAC), pp. 856–859. IEEE, Dharan, Nepal (2022). https://doi.org/10.1109/I-SMAC55078.2022.9986495
15. Jadhav, S., Jadhav, E.B., Sonone, S.S., Sankhla, M.S., Kumar, R.: Carbon-based nanomaterials: nanobiosensors detecting environmental pollution. In: Nanotechnology for environmental pollution decontamination, 1st ed., pp. 263–297. Apple Academic Press, New York (2022). https://doi.org/10.1201/9781003279563-12
16. Mangaiyarkkarasi, J.: Harnessing carbon nanomaterials for next-generation environmental sensing. In: Arun, J., Nirmala, N., Dawn, S.S. (eds.) Advances in Chemical and Materials Engineering, pp. 144–166. IGI Global (2024). https://doi.org/10.4018/979-8-3693-3625-0.ch006
17. Gaber, A., Bilge, S., Bayramoğlu, D., Donar, Y.O., Sınağ, A.: Synthesis and characterization of ordered mesoporous carbon-based materials for electrochemical detection of environmental pollutants. Trends Environ. Anal. Chem. **42**, e00233 (2024). https://doi.org/10.1016/j.teac.2024.e00233
18. Sancar, E., Oktay, B., Ahlatcıoğlu Özerol, E.: Carbon based materials in conducting tissue engineering, Mater. Res. Express **11**, 122001 (2024). https://doi.org/10.1088/2053-1591/ad9549.
19. Bilge, S., Karadurmus, L., Sınağ, A., Ozkan, S.A.: Green synthesis and characterization of carbon-based materials for sensitive detection of heavy metal ions. TrAC, Trends Anal. Chem. **145**, 116473 (2021). https://doi.org/10.1016/j.trac.2021.116473
20. Ensafi, A.A., Mousaabadi, K.Z., Fazel-Zarandi, R.: Application of conductive polymers in electrochemistry. In: Amiri, A., Hussain, C.M. (eds.) ACS Symposium Series, pp. 185–217. American Chemical Society, Washington, DC (2022). https://doi.org/10.1021/bk-2022-1405.ch008.
21. Gogoi, M., Deb, P., Kostka, A.: Differential tunability effect on the optical properties of doped and undoped quantum dots. Physica Status Solidi (a) **209**, 1543–1551 (2012). https://doi.org/10.1002/pssa.201228074
22. Zhu, S., Song, Y., Wang, J., Wan, H., Zhang, Y., Ning, Y., Yang, B.: Photoluminescence mechanism in graphene quantum dots: quantum confinement effect and surface/edge state. Nano Today **13**, 10–14 (2017). https://doi.org/10.1016/j.nantod.2016.12.006
23. Wang, W., Wang, B., Embrechts, H., Damm, C., Cadranel, A., Strauss, V., Distaso, M., Hinterberger, V., Guldi, D.M., Peukert, W.: Shedding light on the effective fluorophore structure of high fluorescence quantum yield carbon nanodots. RSC Adv. **7**, 24771–24780 (2017). https://doi.org/10.1039/C7RA04421F
24. Samanta, P., Desai, A.V., Let, S., Ghosh, S.K.: Advanced porous materials for sensing, capture and detoxification of organic pollutants toward water remediation. ACS Sustainable Chem. Eng. **7**, 7456–7478 (2019). https://doi.org/10.1021/acssuschemeng.9b00155
25. Jia, F., Jiang, W., Zhang, S., Jiang, J., Liu, Y., Jia, F., Yao, H.: Applications of metal-organic frameworks-based sensors for aqueous pollutants: a review. J. Environ. Chem. Eng. **12**, 112925 (2024). https://doi.org/10.1016/j.jece.2024.112925
26. Wang, H., Wang, H., Wang, Z., Tang, L., Zeng, G., Xu, P., Chen, M., Xiong, T., Zhou, C., Li, X., Huang, D., Zhu, Y., Wang, Z., Tang, J.: Covalent organic framework photocatalysts: structures and applications. Chem. Soc. Rev. **49**, 4135–4165 (2020). https://doi.org/10.1039/D0CS00278J
27. Babu, H.V., Bai, M.G.M., Rajeswara Rao, M.: Functional π-conjugated two-dimensional covalent organic frameworks. ACS Appl. Mater. Interfaces **11**, 11029–11060 (2019). https://doi.org/10.1021/acsami.8b19087
28. He, Q., Zhan, F., Wang, H., Xu, W., Wang, H., Chen, L.: Recent progress of industrial preparation of metal–organic frameworks: synthesis strategies and outlook. Materials Today Sustainability **17**, 100104 (2022). https://doi.org/10.1016/j.mtsust.2021.100104
29. Bui, T.H., Shin, J.H.: Perovskite materials for sensing applications: recent advances and challenges. Microchem. J. **191**, 108924 (2023). https://doi.org/10.1016/j.microc.2023.108924

30. Zeng, Z., Xu, Y., Zhang, Z., Gao, Z., Luo, M., Yin, Z., Zhang, C., Xu, J., Huang, B., Luo, F., Du, Y., Yan, C.: Rare-earth-containing perovskite nanomaterials: design, synthesis, properties and applications. Chem. Soc. Rev. **49**, 1109–1143 (2020). https://doi.org/10.1039/C9CS00330D
31. Chen, N., Liu, B., Zhang, P., Wang, C., Du, Y., Chang, W., Hong, W.: Enhanced photocatalytic performance of Ce-doped SnO2 hollow spheres by a one-pot hydrothermal method. Inorg. Chem. Commun. **132**, 108848 (2021). https://doi.org/10.1016/j.inoche.2021.108848
32. Liqiang, J., Xiaojun, S., Baifu, X., Baiqi, W., Weimin, C., Honggang, F.: The preparation and characterization of La doped TiO2 nanoparticles and their photocatalytic activity. J. Solid State Chem. **177**, 3375–3382 (2004). https://doi.org/10.1016/j.jssc.2004.05.064
33. Daou, I., El-Kaddadi, L., Zegaoui, O., Asbik, M., Zari, N.: Structural, morphological and thermal properties of novel hybrid-microencapsulated phase change materials based on Fe2O3, ZnO and TiO2 nanoparticles for latent heat thermal energy storage applications. J. Energy Storage **17**, 84–92 (2018). https://doi.org/10.1016/j.est.2018.02.011
34. Niu, J.J., Wang, J.N., Jiang, Y., Su, L.F., Ma, J.: An approach to carbon nanotubes with high surface area and large pore volume. Microporous Mesoporous Mater. **100**, 1–5 (2007). https://doi.org/10.1016/j.micromeso.2006.10.009
35. Ares, P., Novoselov, K.S.: Recent advances in graphene and other 2D materials. Nano Materials Science **4**, 3–9 (2022). https://doi.org/10.1016/j.nanoms.2021.05.002
36. Hong, X., Liu, Y., Li, Y., Wang, X., Fu, J., Wang, X.: Application progress of polyaniline, Polypyrrole and Polythiophene in Lithium-Sulfur batteries. Polymers **12**, 331 (2020). https://doi.org/10.3390/polym12020331
37. Li, Y., Ding, Y., Zhang, Y., Qian, Y.: Photophysical properties of ZnS quantum dots. J. Phys. Chem. Solids **60**, 13–15 (1999). https://doi.org/10.1016/s0022-3697(98)00247-9
38. Yan, X., Li, B., Li, L.: Colloidal graphene quantum dots with well-defined structures. Acc. Chem. Res. **46**, 2254–2262 (2013). https://doi.org/10.1021/ar300137p
39. Landry, M.L., Morrell, T.E., Karagounis, T.K., Hsia, C.-H., Wang, C.-Y.: Simple syntheses of CdSe quantum dots. J. Chem. Educ. **91**, 274–279 (2014). https://doi.org/10.1021/ed300568e
40. Lai, Z.: Development of ZIF-8 membranes: opportunities and challenges for commercial applications. Curr. Opin. Chem. Eng. **20**, 78–85 (2018). https://doi.org/10.1016/j.coche.2018.03.002
41. Rego, R.M., Kurkuri, M.D., Kigga, M.: A comprehensive review on water remediation using UiO-66 MOFs and their derivatives. Chemosphere **302**, 134845 (2022). https://doi.org/10.1016/j.chemosphere.2022.134845
42. Zorainy, M.Y., Gar Alalm, M., Kaliaguine, S., Boffito, D.C.: Revisiting the MIL-101 metal–organic framework: design, synthesis, modifications, advances, and recent applications. J. Mater. Chem. A **9**, 22159–22217 (2021). https://doi.org/10.1039/d1ta06238g
43. Zhang, Y., Xu, X., Liao, Q., Wang, Q., Han, Q., Chen, P., Xi, K.: New potential of boron-based COFs: the biocompatible COF-1 for reactive oxygen generation and antimicrobial applications. J. Mater. Chem. B **10**, 3285–3292 (2022). https://doi.org/10.1039/d1tb02808a
44. Wu, G., Lu, X., Li, Y., Jia, Z., Cao, X., Wang, B., Zhang, P.: Two-dimensional covalent organic frameworks (COF-LZU1) based mixed matrix membranes for pervaporation. Sep. Purif. Technol. **241**, 116406 (2020). https://doi.org/10.1016/j.seppur.2019.116406
45. Boulahouache, A., Benlembarek, M., Salhi, N., Djaballah, A.M., Rabia, C., Trari, M.: Preparation, characterization and electronic properties of LaFeO3 perovskite as photocatalyst for hydrogen production. Int. J. Hydrogen Energy **48**, 14650–14658 (2023). https://doi.org/10.1016/j.ijhydene.2022.12.327
46. Yao, Z., Zhao, W., (Frank) Liu, S.: Stability of the $CsPbI_3$perovskite: from fundamentals to improvements. J. Mater. Chem. A **9**, 11124–11144 (2021). https://doi.org/10.1039/d1ta01252e
47. Chen, G.-P., Yang, J.-N., Qiao, C.-B., Huang, L.-J., Yu, J.: First-principles calculations of local structure and electronic properties of Er^{3+}-doped TiO_2. Acta Phys. Sin. **71**, 246102 (2022). https://doi.org/10.7498/aps.71.20221847
48. Ning, F., Xu, B., Shi, J., Wu, M., Hu, Y., Ouyang, C.: Structural, electronic, and li migration properties of re-doped (RE = Ce, La) $LiCoO_2$ for Li-ion batteries: a first-principles investigation. J. Phys. Chem. C **120**, 18428–18434 (2016). https://doi.org/10.1021/acs.jpcc.6b05091

49. Khuili, M., Fazouan, N., Makarim, H.A.E., Atmani, E.H., Rai, D.P., Houmad, M.: DFT study of rare earth (Tm, Yb, Ce) doped ZnO: structural, optoelectronic and electrical properties (2020). https://doi.org/10.48550/arXiv.2004.10728
50. Xie, K., Jia, Q., Wang, Y., Zhang, W., Xu, J.: The electronic structure and optical properties of Anatase TiO2 with rare earth metal dopants from first-principles calculations. Materials **11**, 179 (2018). https://doi.org/10.3390/ma11020179
51. Zheng, B., Fan, J., Chen, B., Qin, X., Wang, J., Wang, F., Deng, R., Liu, X.: Rare-earth doping in nanostructured inorganic materials. Chem. Rev. **122**, 5519–5603 (2022). https://doi.org/10.1021/acs.chemrev.1c00644
52. Wang, H., Yang, S., Fan, W., Cui, Y., Gong, G., Jiao, L., Chen, S., Qi, J.: Sight into a rare-earth-based catalyst with spatial confinement effect from the perspective of electronic structure. ACS Appl. Mater. Interfaces **17**, 14749–14772 (2025). https://doi.org/10.1021/acsami.4c17065
53. Reszczyńska, J., Grzyb, T., Sobczak, J.W., Lisowski, W., Gazda, M., Ohtani, B., Zaleska, A.: Visible light activity of rare earth metal doped (Er3+, Yb3+ or Er3+/Yb3+) titania photocatalysts. Appl. Catal. B **163**, 40–49 (2015). https://doi.org/10.1016/j.apcatb.2014.07.010
54. Jiang, Y., Fu, H., Liang, Z., Zhang, Q., Du, Y.: Rare earth oxide based electrocatalysts: synthesis, properties and applications. Chem. Soc. Rev. **53**, 714–763 (2024). https://doi.org/10.1039/D3CS00708A
55. Hossain, M.K., Raihan, G.A., Akbar, M.A., Kabir Rubel, M.H., Ahmed, M.H., Khan, Md.I., Hossain, S., Sen, S.K., Jalal, M.I.E., El-Denglawey, A.: Current applications and future potential of rare earth oxides in sustainable nuclear, radiation, and energy devices: a review. ACS Appl. Electron. Mater. **4**, 3327–3353 (2022). https://doi.org/10.1021/acsaelm.2c00069
56. Lee, Y.-Y., Sriram, B., Wang, S.-F., Kogularasu, S., Chang-Chien, G.-P.: A comprehensive review on emerging role of rare earth oxides in electrochemical biosensors. Microchem. J. **193**, 109140 (2023). https://doi.org/10.1016/j.microc.2023.109140
57. Bokov, D., Turki Jalil, A., Chupradit, S., Suksatan, W., Javed Ansari, M., Shewael, I.H., Valiev, G.H., Kianfar, E.: Nanomaterial by sol-gel method: synthesis and application. Adv. Mater. Sci. Eng. **2021,** 5102014 (2021). https://doi.org/10.1155/2021/5102014
58. Rocca, A., Licciulli, A., Politi, M., Diso, D.: Rare earth-doped $SrTio_3$ perovskite formation from Xerogels. ISRN Ceram. **2012**, 1–6 (2012). https://doi.org/10.5402/2012/926537
59. Shabbir, A., Khan, Z.S., Pervaiz, H., Haseeb, H.M.: Rare-earth ion-doped zinc oxide as a viable photoanode material for dye-sensitized solar cells: optoelectrical and wettability aspects. In: IEEC 2023, p. 19. MDPI (2023). https://doi.org/10.3390/engproc2023046019
60. Benammar, I., Salhi, R., Deschanvres, J.-L., Maalej, R.: The effect of rare earth element (Er, Yb) doping and heat treatment on suspension stability of Y2O3 nanoparticles elaborated by sol-gel method. J. Market. Res. **9**, 12634–12642 (2020). https://doi.org/10.1016/j.jmrt.2020.08.105
61. Yang, G., Park, S.-J.: Conventional and microwave hydrothermal synthesis and application of functional materials: a review. Materials **12**, 1177 (2019). https://doi.org/10.3390/ma12071177
62. Wang, X., Chen, C., Bi, X.: Preparation of rare earth Yb doped TiO2 photo-catalyst by microwave hydrothermal method and its photo-catalytic activity. In: Proceedings of the 7th International Conference on Management, Education, Information and Control (MEICI 2017). Atlantis Press, Shenyang, China (2017). https://doi.org/10.2991/meici-17.2017.170
63. Vadivel Murugan, A., Viswanath, A.K., Ravi, V., Kakade, B.A., Saaminathan, V.: Photoluminescence studies of Eu3+ doped Y2O3 nanophosphor prepared by microwave hydrothermal method. Appl. Phys. Lett. **89**, 123120 (2006). https://doi.org/10.1063/1.2356694
64. Li, Z., Sun, Y., Ge, S., Zhu, F., Yin, F., Gu, L., Yang, F., Hu, P., Chen, G., Wang, K., Volinsky, A.A.: An overview of synthesis and structural regulation of magnetic nanomaterials prepared by chemical Coprecipitation. Metals **13**, 152 (2023). https://doi.org/10.3390/met13010152
65. Boopathi, G., Raj, S.G., Kumar, G.R., Mohan, R.: Synthesis of Er doped Gd2O3 nanostructured materials by co-precipitation technique. In: 2014 2nd International Conference on Devices, Circuits and Systems (ICDCS), pp. 1–5. IEEE, Coimbatore, India (2014). https://doi.org/10.1109/ICDCSyst.2014.6926179

66. Gupta, S.K., Grover, V., Shukla, R., Srinivasu, K., Natarajan, V., Tyagi, A.K.: Exploring pure and RE co-doped (Eu3+, Tb3+ and Dy3+) gadolinium scandate: luminescence behaviour and dynamics of energy transfer. Chem. Eng. J. **283**, 114–126 (2016). https://doi.org/10.1016/j.cej.2015.07.043
67. Alam, M.W., Nivetha, A., BaQais, A., Ansari, S.A., Yewale, M.A., Sadaf, S.: Development and analysis of novel Sm doped LaSiO for photocatalytic degradation and electrochemical sensing of heavy metals. Ceram. Int. **50**, 48628–48639 (2024). https://doi.org/10.1016/j.ceramint.2024.09.212
68. Chaudhary, S., Kumar, S., Umar, A., Singh, J., Rawat, M., Mehta, S.K.: Europium-doped gadolinium oxide nanoparticles: a potential photoluminescencent probe for highly selective and sensitive detection of Fe3+ and Cr3+ ions. Sens. Actuators, B Chem. **243**, 579–588 (2017). https://doi.org/10.1016/j.snb.2016.12.002
69. López-Peña, G., Rodríguez García, A., García Solé, J., Martín Rodríguez, E.: Luminescent rare-earth-doped up-converting nanoparticles for Cr3+ sensing in water. J. Lumin. **274**, 120701 (2024). https://doi.org/10.1016/j.jlumin.2024.120701
70. Sun, Y.-F., Li, J.-J., Xie, F., Wei, Y., Yang, M.: Ruthenium-loaded cerium dioxide nanocomposites with rich oxygen vacancies promoted the highly sensitive electrochemical detection of Hg(II). Sens. Actuators, B Chem. **320**, 128355 (2020). https://doi.org/10.1016/j.snb.2020.128355
71. Sharma, N., Choudhury, S.P.: Gas sensing using metal oxide semiconductor doped with rare earth elements: a review. Mater. Sci. Eng., B **307**, 117505 (2024). https://doi.org/10.1016/j.mseb.2024.117505
72. Sharma, N., Dwivedi, U.K., Nakate, U.T., Singh, M., Choudhury, S.P.: Gas-sensing properties of NO on Ce-doped zinc oxide: a DFT study. Part & Part Syst Charact **42**, 2400126 (2025). https://doi.org/10.1002/ppsc.202400126
73. Sohal, M.K., Mahajan, A., Gasso, S., Nahirniak, S.V., Dontsova, T.A., Singh, R.C.: Rare earth-tuned oxygen vacancies in gadolinium-doped tin oxide for selective detection of volatile organic compounds. J. Mater. Sci. Mater. Electron. **31**, 8446–8455 (2020). https://doi.org/10.1007/s10854-020-03379-7
74. Periyasamy, V., Babu, R.R., Ahmad, A., Albaqami, M.D., Alotabi, R.G., Elamurugu, E.: Spray-deposited rare earth metal ion (Sm3+, Ce3+, Pr3+, La3+)-Doped CdO thin films for enhanced formaldehyde gas sensing characteristics. ACS Omega **7**, 35191–35203 (2022). https://doi.org/10.1021/acsomega.2c04303
75. Ma, X., Zhu, H., Yu, L., Li, X., Ye, E., Li, Z., Loh, X.J., Wang, S.: Rare-earth-doped indium oxide nanosphere-based gas sensor for highly sensitive formaldehyde detection at a low temperature. Nanoscale **15**, 1609–1618 (2023). https://doi.org/10.1039/D2NR04972D
76. Fois, M., Cox, T., Ratcliffe, N., De Lacy Costello, B.: Rare earth doped metal oxide sensor for the multimodal detection of volatile organic compounds (VOCs). Sens. Actuators B: Chem. **330**, 129264 (2021). https://doi.org/10.1016/j.snb.2020.129264
77. Jing, Z., Ling, B., Yu, Y., Qi, W., Zhang, S.: Preparation and gas sensing activity of La and Y co-doped titania nanoparticles. J. Sol-Gel Sci. Technol. **73**, 112–117 (2015). https://doi.org/10.1007/s10971-014-3501-7
78. Nehru, R., Chen, S.-M.: A La^{3+} -doped TiO_2 nanoparticle decorated functionalized-MWCNT catalyst: novel electrochemical non-enzymatic sensing of paraoxon-ethyl. Nanoscale Adv. **2**, 3033–3049 (2020). https://doi.org/10.1039/D0NA00260G

Smart Hybrid Systems for Dual Functionality: Energy and Sensing

Anup Singh, Partha Sarathi Padhi, Navjot Kaur, and Amit Vats

Abstract Materials that can simultaneously offer energy-related performance and sensing capabilities hold immense promise for real-time system monitoring, efficient operation, and intelligent integration, especially in domains such as biomedical devices, aerospace, wearable electronics, and next-generation energy platforms. These multifunctional materials acting as a technological bridge between human/machine or plant/machine interfaces provide a transformative pathway for modern smart systems. Recent advances of rare-earth-doped metal-oxide nanostructures have effectively demonstrated their dual functionality. Doping with rare-earth ions (e.g., La^{3+}, Ce^{3+}, Eu^{3+}, Gd^{3+}, Nd^{3+}) significantly alters the structural, electronic, and surface properties of host oxides such as TiO_2, SnO_2, CeO_2, ZnO, and complex compounds like the silicon oxynitride phosphors ($SrSi_2O_2N_2$:Eu^{2+}, Dy^{3+}; $BaSi_2O_2N_2$:Eu^{2+}, Dy^{3+}; and (Sr,Ba)$Si_2O_2N_2$:Eu^{2+}, Dy^{3+}; plus $SrSi_2O_2N_2$:Yb^{2+}, Dy^{3+}). These modifications concurrently enhance energy storage/harvesting capabilities and sensing performance in terms of selectivity and detection limits. Thanks to their tunable physicochemical traits, these advanced materials have been employed as active electrodes in supercapacitors, lithium-ion batteries, and photoelectrochemical cells, while also serving as chemiresistive and optoelectronic sensors. However, substantial opportunities remain to uncover further functionality. This chapter offers a comprehensive overview of the current research landscape on smart hybrid materials based on rare-earth-doped oxides and related systems, emphasizing design strategies and their integration into self-powered, dual-function devices.

A. Singh (✉)
University Centre for Research and Development, Chandigarh University, Gharuan, Punjab, India
e-mail: anup.e18325@cumail.in

P. S. Padhi
CoCooN Research Group, Department of Solid State Sciences, Ghent University, Ghent, Belgium

N. Kaur
Department of Physics, Sri Guru Granth Sahib World University, Fatehgarh Sahib, Punjab, India

A. Vats
School of Material Science and Technology, National Institute of Technology Kurukshetra, Haryana, India

© The Author(s), under exclusive license to Springer Nature Singapore Pte Ltd. 2026

N. Kumar et al. (eds.), *Rare Earth-Doped Metal Oxide Nanostructures*, Engineering Materials, https://doi.org/10.1007/978-981-95-6192-6_11

Keywords Energy storage · Sensing · Energy harvesting · Rare-earth doping · Dual application

1 Introduction

For a system to be smart, it needs to be multifunctional in terms of materials and applications. Smart hybrid systems embody this innovation by enabling a single material or composite to perform multiple tasks [1]. Sensors, and control components have been treated as separate entities within electronic systems as compared to traditional energy storage units and also it increases the complexity, size, and energy consumption of devices making them very expensive [2]. The new approach to smart hybrid systems offers a streamlined approach by integrating these functions into a unified architecture [3]. There is a growing need for devices that are multifunctional; for example, an energy harvester system needs to have an efficient energy storage unit integrated within a compact single system, this can drastically change the course of energy sustainability [4]. In the healthcare sector, electronics can be enhanced by incorporating dual monitoring and sensing capabilities, enabling early diagnosis and timely treatment [5]. Dual functionality can be of great significance if the energy and sensing materials are incorporated in the single system, they can offer assistance in unlimited domains form electronics for early assessment of problem in electronics, portable electronics, electric vehicles, wearable technologies, and automated diagnostics, in biomedical sensing, environmental pollution monitoring, where both efficient energy management and real-time responsiveness are crucial [6, 7]. Sensing demands materials with frequent and efficient response to monitor real-time problem, while energy storage materials require materials with efficient ion mobility and specific surface area in order to store charge efficiently, rare-earth (La, Eu, Gd, Nd, Dy, Er)-doped materials have remarkable application in metallurgy, magnetism, catalysis, luminescence to future electrochemical pseudo capacitance energy storage [8, 9]. These materials can upon doping alter the nanoparticle size, shape, phase porosity, and strength, thereby resulting in the modification of the properties of pristine material, and they can be synthesized with methods like vapor deposition, sol–gel, hydrothermal, etc. [10–13]. But, to fabricate a hybrid multifunctional system, material selection plays a key role and a wise combination of which can lead to a smart hybrid dual-functional material. Consideration of factors like interfacial resistance, signal transduction, ionic mobility, mechanical/thermal, and structural stability facilitates miniaturization of key requirements for next-generation multifunctional electronics [14]. Smart hybrid systems are typically developed by combining electroactive materials such as transition metal oxides, perovskites, especially rare-earth-based perovskites, and forming composites with graphene, reduced graphene oxide, carbon nanotubes, Mxene, etc., are currently dominant materials with application in the field of energy storage, sensing, catalysis, and biomedical [15, 16].

This chapter explores recent progress and applications of smart hybrid systems that unite energy storage and sensing functionalities. It highlights design strategies involving rare-earth dopants and composite materials. Ultimately, the fusion of energy and sensing in multifunctional materials is ushering in a new era of adaptive, self-aware systems for energy-efficient, health-responsive applications. Figure 1 shows the schematic diagram of different types of energy storage systems, like electrochemical, thermal, mechanical, and wireless, etc., along with various sensing, showing that hybrid systems can dually store energy and perform sensing operation in future electronics for sustainability.

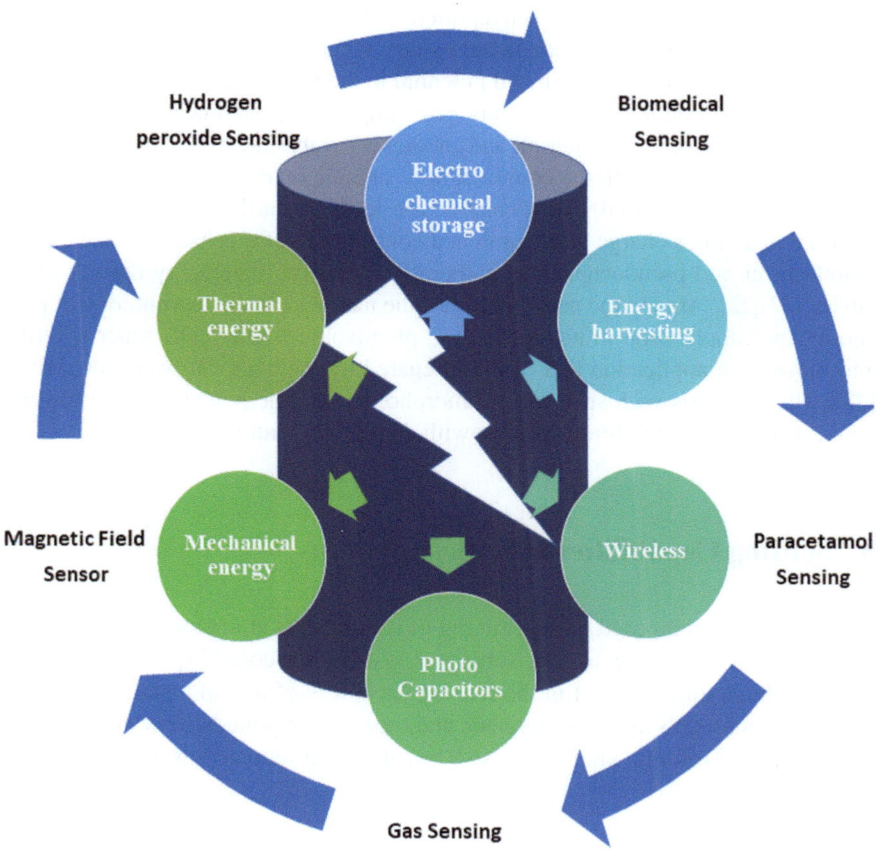

Fig. 1 Schematic diagram of different hybrid energy storage and sensing application of materials

2 Hybrid Storage System

Storage is one of the primary components of the hybrid system as it holds the potential to make a sustainable hybrid system. The components of the storage system have been divided into electrode, electrolyte, and substrate material, among which the electrode material is the key component [17]. Material science has led to the origin of material with different morphologies such as nanorods, nanowires, spheres, hollow spheres, and porous spheres, different types of morphology have their significance in different applications such as storage, harvesting, sensing, and biomedical application, and there are ways to fine-tune morphology as per the requirement [18–20]. Many electrode materials have been reported for their efficient charge storage characteristics. Materials like transition metal oxides, metal–organic framework, quantum dots, and chalcogenides have shown potential in the storage application. One of the aspects of electrode material is to fabricate simple electrode material (either using only organic or inorganic material) and another is to fabricate hybrid electrode material which comprises with the conjunction of organic (carbon-based) and inorganic material (metal ion-based); such conjunction leads to the formation of composite material that stores charge with combined contribution of mechanisms like electric double layer and pseudocapacitance, thereby enhancing the energy density of the device [21, 22]. Among the nanostructures, the nanowire is the morphology of great significance application in nanogenerator, photovoltaic, solar cell, water splitting, and biosensing application and can be prepared by methods like vapor deposition [23]. The materials with such refined morphology are the potential materials with application for smart hybrid materials with dual functionality.

3 Sensing Capabilities of Hybrid Systems

Hybrid systems composed of intermixing of components at microscopic scale [24]. Sensors consist of interconnects, substrates, and sensor core [25]. In recent years, there is a huge improvement in the sensing capability of hybrid materials [26, 27]. The hybrid materials made of inorganic and organic compounds show good sensing behavior [28]. Different kinds of sensors based on hybrid materials are discussed as (Fig. 2).

3.1 Physical Sensors: Pressure, Temperature, Strain

The most prominent physical sensors are: pressure, temperature, and strain. Strain sensors are used for the conversion of mechanical input to electrical signals [25]. The conducting materials used for wearable strain and pressure sensors are conductive polymers, CNTs, and graphene [29, 30].

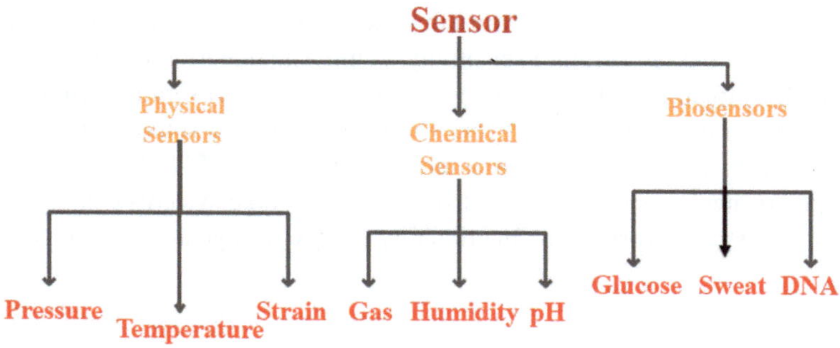

Fig. 2 Sensor and its types

Strain sensors can be wearable or stretchable. Graphene woven fabrics (GWFs) are another example of strain sensors that are utilized as a sensing material [31]. Strain sensors are also beneficial for paralyzed patients to detect face expressions [32, 33]. Another example of wearable strain sensor is the hybrid mixture of Ag-NW (silver-based nanowires) that is utilized for the detection of hand gestures and is attached on fingers [34]. There are different types of stretchable strain sensors, such as capacitive, resistive, and optical [35]. Resistive strain sensors convert mechanical input to electrical signal by varying the resistance while deformation; for example, TPU electrospun fibrous mats that are enhanced with reduced graphene oxide [35]. Capacitive strain sensors work on the principle of varying capacitance due to spatial modifications; a sensor consisting of polydimethylsiloxane (PDMS) dielectric layer formed by depositing gold film through thermal evaporation on the surface of parylene electrodes acting as a hybrid device [35]. Optical strain sensors are based upon the changes in the transmission of light during deformation; optical fibers, such as thermoplastic elastomers are good examples of optical strain sensors. Pressure sensors are also used to monitor physiological pressure that can recognize the health status of the cardiovascular system by evaluating blood pressure and pulse rate [36, 37]. Polyvinylidene fluoride is a very usual material that is used in pressure sensors [38]. Other examples of pressure sensor are, EGaln that is a conductive liquid utilized for determining pressure while gripping [39], microfiber sensors are best for detecting high sensitivity in which nanosuspension of graphene oxide are used as an conductive element [40], air vibrations and radial pulse by utilizing piezoelectric sensor [41], triboelectric pressure sensors in which deposited kapton sheets are used for identifying pulses at fingertips, wrist, etc. [42]. The other uses of pressure sensors are analyzing pressure sores, examining vocal fold disorders, etc. [43–45]. Pressure sensors are of two types: capacitive sensor and resistance sensor [24]. Capacitive sensors that depend on mineral hydrogels are useful for sensing minor changes in pressure [46].

Temperature sensors that depend on nanoparticles can detect the changes in temperature occurring due to change in electrical conductivity or resistance [47]. Yttrium aluminum garnets doped with rare earth such as Erbium (Er), Thulium (Tm), Holmium (Ho), and Ytterbium (Yb) can be used as temperature sensors

[48]. Fiber Bragg grating sensors are fabricated to sense and measure temperature [49]. To detect high temperature sensitivity, composite material, i.e., flake graphite/ polydimethylsiloxane (PDMS)/carbon nanotubes based sensors are used [24]. To measure temperature accurately, hybrid-based temperature sensors are used [50]. Fiber-optic cables are also used as a temperature sensor [51]. The integration of such sophisticated sensing materials inside energy storage systems can result in the fabrication of smart hybrid dual-functionality energy storage and sensing systems.

4 Dual Functionality

4.1 Electrochemical Energy Storage and Sensing

Electrochemical energy storage and sensing have been a matter of huge interest these days. For this, the synthesized electrode material must have good response to both energy storage and sensing. A prominent example is the hydrothermally synthesized graphene oxide–gadolinium oxide europium nanocomposite ($GO–Gd_2O_3:Eu^{3+}$). Here, Gd_2O_3 enhances energy storage via its high dielectric constant and chemical stability, while Eu^{3+} imparts luminescent properties enabling optical sensing. This material demonstrates excellent supercapacitive performance, with a specific capacitance of 512.14 Fg^{-1} in 1 M H_2SO_4 and a Coulombic efficiency of 94.54% after 5000 cycles. Simultaneously, its photoluminescence enables real-time monitoring, allowing it to function as both an energy storage unit and a biosensor [51]. An equally compelling system is the $La@NiCoFe_2O_4@Ti_3C_2$ hybrid, which synergistically merges the high redox activity of lanthanum-doped spinel ferrites with the superior conductivity and two-dimensional architecture of MXenes. This composite achieves an exceptional specific capacitance of 2900 Fg^{-1} and an energy density of 75 Wh/kg. Beyond energy storage, it exhibits precise electrochemical detection of creatinine, a critical biomarker for renal function within a range of 0–1000 μM, with a correlation coefficient (R^2) of 0.99. This duality highlights its potential in non-invasive biomedical diagnostics, where both energy and analytical performance are required in a compact format [52]. At the molecular level, the outstanding multifunctionality of these materials arises from synergistic effects such as heterojunction formation, improved ion and electron transport, and increased electrochemically active surface area [53]. Rare-earth dopants like europium and lanthanum not only enhance charge transfer kinetics through multivalent redox states but also introduce distinct luminescent signatures via upconversion or photoluminescence mechanisms crucial for optical sensing. For instance, Eu^{3+} is known for its sharp red emission and environmental sensitivity, enabling the detection of parameters like pH and specific biomolecules through changes in fluorescence. In parallel, luminescent materials such as quantum dots, metal–organic frameworks (MOFs), and aggregation-induced emission (AIE) systems have emerged as highly responsive platforms for sensing temperature, pressure, or chemical analytes. When integrated with supercapacitor systems, these materials further expand the functional scope of hybrid devices, supporting both energy storage and high-resolution sensing in real time. Importantly, the combination of energy and sensing capabilities is not merely additive, it represents a conceptual leap from passive components to active, intelligent systems [51–54].

4.2 Biosensors to Storage Duality

In wearable or implantable electronics, for instance, materials must not only store and supply energy but also detect physiological changes, provide feedback, and ensure biocompatibility. Biosensors are used to detect biochemical signals at microscopic level and its sensing element consists of biological materials, i.e., antigens, biofilms, proteins, antibodies, and enzymes. For the determination of glucose changes, sensors depend on electrodes of carbon cloth that are transformed by layered double hydroxide that consists of both aluminum and nickel [55]. Finger-prick test is a very general method to detect glucose level in the body [56]. Sweat, saliva, tears, and interstitial fluid contain glucose levels that are similar to blood glucose level [57]. For the determination of blood sugar level, sweat-depending glucose sensors are used [58]. Tear-dependent flexible sensors are also used to determine glucose that are formed using soft contact lenses [59]. Rare-earth-based nonenzymatic glucose sensors are reported in literature named the material [59]. DNA biosensors are also called as genosensors that are used to transform biochemical reactions to recognizable analytical signals [60]. Medisense commercialized Exac Tech in 1987 is the first glucose tracking system for diabetic patients [61]. DNA biosensors are used to sense DNA damage using nucleic acid detection mechanisms and are also useful to detect infectious diseases [62, 63]. Graphene-DNA depending hybrids are used as biosensors in different fields like nanobiotechnology, nanomedicine, and immunoassay [64].

There are different methods available to enhance the selectivity of biosensors. One of the techniques is permselective membranes which involves coating of sensor interface that is enzyme membrane using polymers either in layer or multilayer [65, 66]. Selectivity can also be enhanced by introducing mediators in biosensors [67]. Sensitivity of biosensors can be enhanced by improving the transfer of electrons between electrode and target material and this can be done by introducing materials like carbon nanotubes, nanofibers, conducting polymers, and nanorods in the biosensing membrane [68, 69]. There are various types of materials that are being used to make biomarkers for the early diagnosis of fluctuations in vitals and in-time treatment of serious virus or diseases, as the cost-effective rare-earth-doped paper-based sensor of $NaGdF_4$: 5% Er^{3+}, 3% Nd^{3+} was employed for detection of D-glucose [70]. Figure 3 shows the different types of electrode materials that can be employed as a biomarker for various virus or bacteria. If such biomarker materials have charge storage capability then they can be used in the fabrication of smart hybrid dual-functional wearable devices for modern healthcare systems.

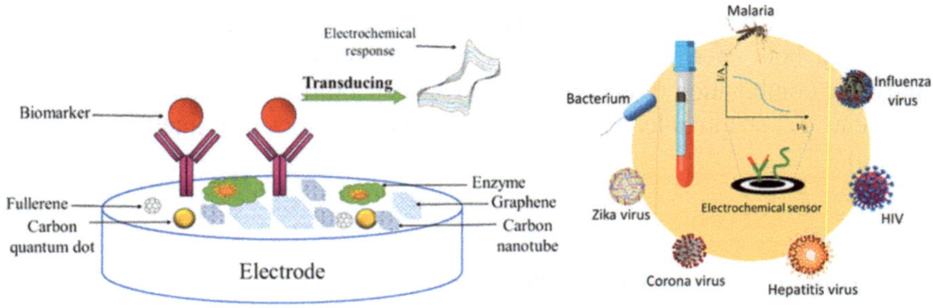

Fig. 3 Different electrode materials that can be used as biomarkers for sensing of different viruses [71]

Dual-functional hybrids thus offer a comprehensive, systems-level solution for sustainable energy and healthcare technologies. Perovskites also mark significance in the energy storage and sensing application as they comprise the combination of rare earth and transition metals in their making. A lanthanum manganese-based perovskite like $LaMnO_3$, $La_{0.95}Co_{0.05}MnO_3$, $La_{0.8}Co_{0.2}MnO_3$, and $La_{0.6}Co_{0.4}MnO_3$ has been reported for dual functionality and it delivered good specific capacitance, along with remarkable paracetamol sensing potential [72].

Electrochemically, it is possible to store hydrogen energy and it is the cleanest form as the byproduct is only water therefore no greenhouse gas is obtained during the process [73]. A novel $CoLa_2O_4$/V-Ag-MOF hybrid nanocomposite is an excellent option for smart hybrid systems as it exhibits dual and even **trifunctional capabilities**. Engineered via a hydrothermal route, the nanocomposite combines the high redox activity and structural advantages of a spinel-type oxide ($CoLa_2O_4$) with the tunable porosity and multi-valence behavior of a vanadium–silver metal–organic framework (V-Ag-MOF). As an energy storage electrode, it achieves an outstanding specific capacity of 1313 C g^{-1} (three-electrode setup) and an energy density of 83.1 Wh kg^{-1} with a maximum power density of 4160 W kg^{-1} in a hybrid supercapacitor configuration. Additionally, the material functions as a selective electrochemical sensor for hydrogen peroxide (H_2O_2), detecting concentrations as low as 1 mM, and shows strong catalytic performance for the hydrogen evolution reaction (HER) with a low overpotential of 43 mV and Tafel slope of 39 mV dec^{-1}. This multifunctionality realized through synergistic interactions and a highly porous, redox-active structure demonstrates the material's promise for integrated smart systems that can store energy, monitor biochemical signals, and participate in electrocatalysis, fulfilling core requirements of next-generation dual- or multifunctional device [74]. A hybrid system comprised of CeO_2–MnO_2 on carbon nanofiber showed dual functionality by sensing hydrogen peroxide and storing charge efficiently [75].

4.3 Gas Sensing and Energy Storage

Air pollution is a crucial aspect for the safety of all the present species and environment, and it needs to be properly regulated. There are different types of harmful gases that are continuously polluting the air quality, which raises the need for highly sensitive sensors [76]. There are various hazardous chemicals that can cause damage to the respiratory tract. For example, gas sensors based upon wearable fiber are used to detect NO_2 gas [77]. For other harmful gases like NH_3, C_2H_5OH, HCHO, and face masks that include single-walled carbon nanotubes decorated with ZnO quantum dots [78]. Lanthanide coordination polymers are also used for the detection of ammonia gas [79]. Gas sensors that depend on inorganic semiconductor metal oxides, i.e., TiO_2, ZnO, WO_3, Fe_2O_3, etc., are used to recognize reduced gases [80]. Some organic conducting polymer (i.e., polypyrrole, polythiophene, and polyaniline)-based gas sensors are useful for chemical sensors at room temperature [81–83]. The other example of a gas sensor is based on Polyaniline/TiO_2 nanohybrid combination [82, 83]. Samarium oxide (Sm_2O_3)-based nanorods are used for the detection of carbon monoxide gas (CO) [84]. Flexible gas sensors based on graphene are used to detect dangerous gases, i.e., CO_2, SO_2, H_2S, NH_3, NO_2, etc. [85, 86]. In case of gas sensors, the sensing mechanism depends upon the gas molecules' adsorption and desorption on the top of material. After the adsorption of gas molecules, material resistance gets modified [87]. The sensing performance can be increased by improving the nanomaterials' surface through doping of different nanoparticles onto the surface [88]. Humidity sensors are used to detect changes in humidity by sensing different physical properties, i.e., electrical conductivity [88, 89]. Capacitive sensors are better sensors than resistive sensors as it has good durability for water vapor condensation at excessive humidity [89]. It has been reported that Yttrium oxide (Y_2O_3) is useful for humidity sensing [90]. There are different flexible humidity sensors, such as sensors that depend on reduced graphene oxide are useful to detect humidity in different objects and human skin, carbon nanotubes based sensors are good to determine electrical properties and multiwalled carbon nanotubes based sensors are efficient to sense environmental humidity [91, 92]. Quartz crystal microbalances are very useful to detect humidity [93]. Sensors based on manganese oxide, i.e., MnO_2 and Mn_2O_3 are beneficial for sensing relative humidity [91, 94].

pH describes the activity of hydronium (H_3O^+) ions, and it is applicable to different areas like pharmaceutical research, marine research, soil analysis, food, and sciences [94]. To determine the activity of hydronium ions, electrochemical pH sensors are used and to determine the acidic and basic concentration of pH-sensitive dye. Fluorescence-dependent pH sensors, i.e., bromothymol blue, phenol red, cresol red, etc., and absorption-dependent sensors, such as a microfluidic device that is integrated with polyaniline sensor, are essential for pH imaging [95, 96]. Eu(III) ofloxacin complex can be used as a luminous pH indicator [97].

The rare-earth-doped materials tend to have the potential as sensing materials Gd, Nd, and Sm are doped with zinc oxide and thin films have been fabricated and they have good sensitivity to ammonia [98]. Furthermore, rare earth appearing as a good candidate due to their dual application is sensing and energy storage, as conventional gas sensors based on metal oxides often suffer from poor sensitivity and environmental interference, while supercapacitor electrodes face limitations in terms of specific capacitance and charge retention. Gd, Nd, and Sm overcome these challenges, this study focuses on the synthesis of lutetium (Lu^{3+})-doped zinc ferrite ($ZnFe_2O_4$) nanoparticles, aiming to enhance their structural, sensing, and electrochemical properties. The Lu^{3+} ion, with its favorable ionic radius and unique electronic configuration, was selected to modulate the microstructure and surface characteristics of zinc ferrite. Structural and morphological investigations confirmed the formation of single-phase spinel nanoparticles with reduced crystallite size and increased porosity, both of which are essential for improved gas adsorption and charge storage behavior [99].

The doped nanoparticles exhibited excellent sensitivity and selectivity toward carbon monoxide, achieving a maximum gas response of 80% at 300 °C for 20,000 PPMv concentration, with stable performance under varying humidity conditions. Furthermore, electrodes fabricated from Lu-doped $ZnFe_2O_4$ demonstrated outstanding supercapacitor behavior, with the $Lu^{0.7}$ composition reaching a specific capacitance of 280 F/g at a current density of 0.25 A/g. This performance was attributed to enhanced charge transfer and ion diffusion capabilities introduced by Lu^{3+} doping. These findings establish Lu-doped zinc ferrite as a dual-functional material with significant potential in both gas sensing and energy storage applications, paving the way for its integration into advanced environmental and electrochemical devices [99].

4.4 Thermal, Mechanical Energy and Sensing

An innovative bifunctional microcapsule that combines thermal energy storage with photoluminescence was developed by zirconia core–shell doping with rare earth (Er^{3+}, Sm^{3+}, and Yb^{3+}) and fabricated with core–shell morphology, the series showed photoluminescence emission enhancement and thermal energy storage. By encapsulating n-dodecane in a rare-earth-doped zirconia shell, the material offers improved thermal stability, efficient heat storage, and enhanced light emission. This dual functionality makes it suitable for smart applications like sensors, optical devices, and energy-efficient systems [100]. Another Ce^{3+}-doped calcium carbonate ($CaCO_3$:Ce^{3+}) was developed with similar morphology as discussed above, and this material also delivered high thermal sensing and energy storage ability [101].

A remarkable study introduces a rare-earth-doped smart hybrid material system that demonstrates dual functionality in mechanical sensing and energy storage through a newly developed phenomenon called Force-Induced Charge Carrier Storage (FICS). Unlike traditional mechanoluminescent (ML) materials that emit light instantaneously upon mechanical stimulation, the FICS-enabled system captures and stores mechanically excited charge carriers in deep traps at room temperature. These stored charges can later be released as thermally stimulated luminescence, enabling non-real-time stress recording without the need for continuous power supply. The FICS effect was successfully validated in five piezoelectric ML phosphors $SrSi_2O_2N_2/Eu^{2+}$, Dy^{3+}, $BaSi_2O_2N_2/Eu^{2+}$, Dy^{3+}, $(Sr,Ba)Si_2O_2N_2/Eu^{2+}$, Dy^{3+}, $SrSi_2O_2N_2/Yb^{2+}$, Dy^{3+}, and ZnS/Cu, all of which exhibited triboelectricity-driven excitation and deep-trap-mediated charge retention. The emission intensity was directly correlated with mechanical input energy, enabling precise stress mapping. Practical demonstrations such as electronic signature recording, impact detection, and vehicle collision monitoring underscore the system's versatility. Overall, this work establishes a foundational platform for next-generation multifunctional materials, where rare-earth doping provides tunable luminescent and trapping properties, and the hybrid architecture supports robust mechanical and thermal performance opening new avenues in self-powered sensing, mechanical-to-optical energy conversion, and smart environmental interaction systems [102].

4.5 Wireless Energy and Sensing

This work presents a novel smart hybrid material having a magnetic/mechanical/electrical (MME) fiber fabricated using coaxial 3D printing, which integrates magnetic actuation, mechanical flexibility, electrical conductivity, in situ sensing, and wireless energy transmission into a single soft structure. The outer layer of the fiber contains neodymium-based (NdFeB) magnetic particles, a rare-earth element crucial for strong and programmable magnetic responsiveness, while the inner core holds a conductive liquid metal (EGaIn). This unique combination enables multifunctionality: The fiber can bend under magnetic fields, sense deformation through inductance changes, and generate or transmit electrical energy without wires. Demonstrated applications include a magnetically controlled surgical tool for minimally invasive procedures, a somatosensory gripper for object recognition, and a soft robot capable of movement and energy harvesting. This research is significant for the theme of smart hybrid systems or dual-functionality energy and sensing, as it offers a compact, flexible platform that unites multiple advanced capabilities, paving the way for next-generation biomedical tools, soft robotics, and wearable electronics [103].

4.6 Energy Harvesting and Sensing

To form a sustainable energy supply, there is need for a system that can effectively harvest energy. Although the sunlight is in abundance, the conversion efficiency of the energy harvesting system matters the most. To improve this energy harvesting efficiency, currently there is a lot of research ongoing, related to smart and hybrid material with dual functionally that can address the energy harvesting and the magnetic sensing. Recently, energy harvesting and sensing performance enhancement have been reported through rare-earth material based nanogenerators, wherein the nanogenerator is made of $(1 - x)$ $BaTiO_3 - x$ $ErFeO_3$ (where $x = 0.01$, 0.02, 0.03, and 0.04) material composition. The doping of rare-earth-based ferrite into the $BaTiO_3$ made the material magnetically active along with elevating the piezoelectric response; such material was proposed as an efficient material for stray magnetic field sensors in household electronics and energy harvester [104]. A recent study shows that the self-powered sensor which can perform dual sensing and energy harvesting task effectively, this system is called potentiometric sensor in which the concentration of diluted analytes is measured along with harvesting energy from the voltage generated by the signal itself marking the technological advancement, as no external power source is required for the functioning of the device [105].

5 Other Hybrid Dual-Function Materials' Devices and Their Architecture

A multifunctional strontium manganese-based $SrMnO_3$/g-C_3N_4nanocomposite, synthesized via a simple hydrothermal method, demonstrates dual electrochemical functionality and making it highly suitable for smart hybrid systems. This material serves as a high-performance electrode for a hybrid supercapacitor, demonstrating an energy density of 84.8 Wh/kg, power density of 1190.3 W/kg, and excellent cycling stability over 10,000 cycles. Furthermore, this nanocomposite is also known as a sensitive, enzyme-free electrochemical sensor for hydroquinone (HQ), with a low detection limit of 6.32 μM and a wide linear detection range of 1–600 μM. The synergistic interaction between $SrMnO_3$ and g-C_3N_4 enhances redox activity, surface area, and electron transport, enabling efficient energy storage and real-time environmental monitoring on a single-material platform, highlighting its potential in integrated energy-sensing applications for next-generation smart devices [106].

Photo-supercapacitor (PS) acts as a dual-function device as it combines the energy storage capacity of a supercapacitor with the working of a solar cell [66]. Previously, solar cells were joined with supercapacitors by an external circuit but it reduces the supercapacitor's energy storage capability but photo-capacitor-based dual-function devices solve this problem [107, 108]. In PSs, both functions, i.e., energy storage and photo-conversion, are integrated into one device by using common electrodes [66]. There are different types of PSs available, e.g. Dye-sensitized and polymer solar cell type [109]. PSs of type dye-sensitized solar cell contain double-electrode and three-electrode structures [110]. Thermally chargeable supercapacitors (TCSCs) are also an example of dual-function devices in which thermal energy harvesting and energy storage capabilities of supercapacitors are combined [111]. The device consists of an energy storage and energy harvesting unit that are electrically connected and consists of a thermoelectric generator, DC load, temperature sources, battery, or supercapacitor unit [111]. There are two main architecture of TCSCs, i.e., one is sandwich-type in which electrolyte is sandwiched between two electrodes and planar-type in which electrodes (i.e., Cu- or Au-based electrodes) are lined in plane by depositing a layer of electrolyte (for example, ionic gelatin) over it [112]. Second is planar-type TCSCs in which electrodes that are next to each other are coated by the layer of an electrolyte [111]. $NaLuF_4$: Yb, Tm@$NaGdF_4$-(153Sm) nanoparticle was reported that shows application in four modal imaging [113, 114]. In this, Gd^{3+} is situated in the outermost shell that is useful for the identification of signals of MRI contrast. However, a signal of CT contrast is detected by Lutetium (Lu) that is situated in the core. Fluorescence signals are detected by the doping of Tm^{3+} and Yb^{3+}. All these imaging systems are integrated onto a single physical platform, where various Lanthanide (Ln) are combined in one nanoparticle and can be used in theranostics and multimodal imaging [115]. Such kinds of nanoparticles include core–shell architecture. Nanoparticles that are based on Ln-doping are coated with silica-based amorphous material to minimize surface quenching effects [116].

Emerging research underscores the potent synergy arising from rare-earth doping strategies coupled with ultrathin binary metal-oxide nanolaminate architectures, offering multifunctional capabilities in pseudocapacitive energy storage, photocatalytic energy harvesting, and environmental sensing. For instance, Er^{3+}, La^{3+}, and Sm^{3+}-doped TiO_2 nanofibers synthesized by electrospinning exhibit significantly enhanced humidity sensitivity compared to undoped TiO_2, owing to increased permittivity and reduced resistivity [117]. In pseudocapacitive systems, rare-earth-doped TiO_2 electrodes have been shown to improve electronic conductivity and cycling performance by tuning surface chemistry and crystal structure, ultimately boosting device performance in energy storage applications [118]. The La-doped TiO_2/reduced graphene oxide electrodes achieve high specific capacitance (~ 117.6 F/g at 10 mV/s) and outstanding cycling stability (~ 91% retention over 10,000 cycles) [119]. Whereas, Sm-doped TiO_2 electrodes deliver about 194 mAh/g with nearly 100% capacity retention over 50 cycles, attributed to enhanced Ti^{3+} stabilization and improved conductivity [120]. These modifications empower TiO_2 for superior energy storage through pseudo capacitance enhanced by ionic transport across engineered interfaces. Figure 4a depicts a schematic representation of rare-earth doping in TiO_2, showing substitutional incorporation of RE^{3+} (e.g., Sm^{3+}, La^{3+}) into Ti^{4+} lattice sites. This rae earth doping results in defect-state (e.g., Ti^{3+}, oxygen vacancies) and intra-band energy level formation, which enhance visible-light absorption, charge separation, and conductivity, as this mechanisms central to energy storage and sensing applications.

Recent interface-engineered TiO_2/Al_2O_3 bilayer coatings applied to Li-rich cathodes delivered ~ 146 mAh g^{-1} specific capacity with ~ 90.4% capacity retention over 100 cycles, along with reduced voltage decay and interfacial impedance [123]. Meanwhile, Al_2O_3/Tm_2O_3 nanolaminates demonstrated stable electroluminescence (blue emission around 456 nm) and integrated rare-earth activation within ultrathin Al_2O_3 layers, exemplifying the tight layering and achievable defect control in rare-earth layers [122]. Figure 4b demonstrates a schematic representation of an Al_2O_3/Tm_2O_3 nanolaminate architecture on a MOS structure, illustrating a layered

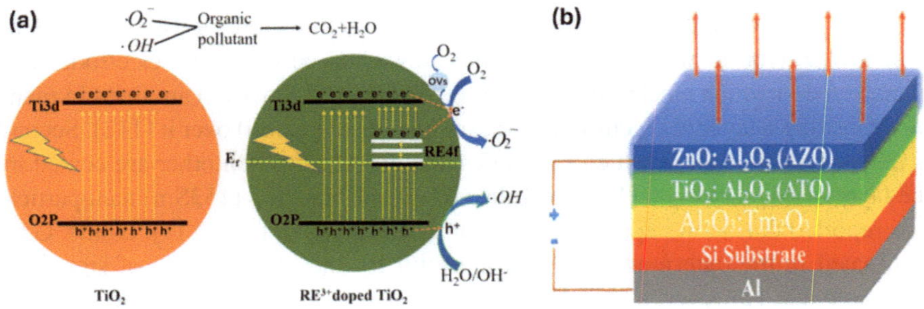

Fig. 4 **a** Schematic representation of rare-earth doping in TiO_2 through substitution of Ti^{4+} by RE^{3+} ions (e.g., Sm^{3+}, La^{3+}), creating defect states (Ti^{3+}, oxygen vacancies) and intra-band energy levels [121]. **b** Schematic representation of an Al_2O_3/Tm_2O_3 nanolaminate illustrating a layered rare-earth doping strategy [122]

rare-earth doping strategy within ultrathin oxide films (e.g., ~ 3 nm Al_2O_3 interlayers) for designing dual-function materials combining energy storage and sensing. For instance, $Al_2O_3TiO_2$ nanolaminates manifest strong interface-confined carrier dynamics, tunable nanobattery phenomena, enhanced conductivity, and negative differential resistance features that support both high-density energy storage and environmental sensing [123–126]. Complementing this, oxygen-vacancy engineered TiO_2-Δ_x devices architectures, closely related to TiO_2/Al_2O_3 composites, exhibit tunable quantum conductance states, highlighting their utility for energy-related applications (e.g., memristive energy storage) as well as resistive sensing under environmental stimuli [127]. These integrated designs pave the way for next-generation multifunction integrated energy–environment devices.

6 Conclusion

To qualify as truly smart, a system must be inherently hybrid, capable of delivering multiple functionalities within a single unit. Among these, energy storage and sensing are the most critical and experimentally explored integrations, yet much remains to be achieved. The fusion of these functions into a unified smart hybrid system represents a transformative advance in multifunctional materials research. As shown throughout this chapter, rare-earth–doped metal-oxide nanostructures offer a compelling route to dual-function platforms delivering robust energy storage performance alongside high-fidelity sensing.

These materials exploit tunable electronic structures, enhanced redox activity, and tailored optical properties, endowed by rare-earth dopants to meet the rising demand for compact, energy-efficient smart systems. By integrating pseudocapacitive and electric double-layer energy storage mechanisms with sensing modalities such as gas detection, biosensing, temperature monitoring, and mechanical deformation recognition, hybrid materials offer a consolidated response to challenges in biomedical diagnostics, environmental monitoring, soft robotics, and wearable electronics.

The diverse examples discussed from lanthanide-based perovskites and MXene hybrids to force-responsive charge-storing systems demonstrate how material design strategies can be precisely tuned to combine specific functionalities. Despite these advances, challenges remain in ensuring long-term stability, biocompatibility, scalable synthesis, and holistic system integration. Future work should emphasize optimizing material interfaces, exploring new dopant systems, and employing AI-driven design and diagnostics to fully unlock the potential of smart hybrid materials.

In sum, the convergence of energy storage and sensing into single-material platforms marks a pivotal milestone toward adaptive, sustainable, autonomous devices. These dual-functional systems not only enhance device efficiency but also catalyze a new era of self-powered, intelligent technologies capable of real-time monitoring and response.

References

1. Ferreira, A.D.B.L., Nóvoa, P.R.O., Marques, A.T.: Multifunctional material systems: a state-of-the-art review. Composite Structures **151**, 3–35 (2016). https://doi.org/10.1016/j.compstruct.2016.01.028
2. Elalfy, D.A., Gouda, E., Kotb, M.F., Bureš, V., Sedhom, B.E.: Comprehensive review of energy storage systems technologies, objectives, challenges, and future trends. Energ. Strat. Rev. **54**, 101482 (2024). https://doi.org/10.1016/j.esr.2024.101482
3. Khalil, M.I., Jhanjhi, N.Z., Humayun, M., Sivanesan, SivaKumar, Masud, M., Hossain, M.S.: Hybrid smart grid with sustainable energy efficient resources for smart cities. Sustain. Energy Technol. Assess. **46**, 101211 (2021). https://doi.org/10.1016/j.seta.2021.101211
4. Shen, G., Djenizian, T., Fan, Z., Ko, H., Yu, C.: Multifunctional energy-integrated devices. Adv. Mater. Technol. **9**, 2401273 (2024). https://doi.org/10.1002/admt.202401273
5. Gao, Z., Zhou, Y., Zhang, J., Foroughi, J., Peng, S., Baughman, R.H., Wang, Z.L., Wang, C.H.: Advanced energy harvesters and energy storage for powering wearable and implantable medical devices. Adv. Mater. **36**, 2404492 (2024). https://doi.org/10.1002/adma.202404492
6. Wang, X., Ding, N., Cai, Y., Li, J., Fu, H., Deng, F.: A unique harvesting-sensing integrated methodology using dual capacitor matching and event-triggered energy management for self-powered sensing. Energy Convers. Manage. **341**, 119964 (2025). https://doi.org/10.1016/j.enconman.2025.119964
7. He, J., Liu, H., Cao, X., Zhu, Z., Liang, Y., Xiao, B., Chen, J., Xiao, M., Zhang, Z.: Dual-mode sensing chips for wide-range and high-accuracy biomarker detection. Adv. Funct. Mater. **34**, 2404909 (2024). https://doi.org/10.1002/adfm.202404909
8. Yuan, M., Zhang, X., Wang, J., Zhao, Y.: Recent progress of energy-storage-device-integrated sensing systems. Nanomaterials **13**, 645 (2023). https://doi.org/10.3390/nano13040645
9. Chai, S., Zhang, W., Yang, J., Zhang, L., Theint, M.M., Zhang, X., Guo, S., Zhou, X., Ma, X.: Sustainability applications of rare earths from metallurgy, magnetism, catalysis, luminescence to future electrochemical pseudocapacitance energy storage. RSC Sustain. **1**, 38–71 (2022). https://doi.org/10.1039/d2su00054g
10. Pandhurnekar, C.P., Pandhurnekar, H.C., Yadao, B.G., Mungole, A.J., Mohabe, P.: Recent advances in the rare earth metal doped nanomaterials and their applications in biomedical

imaging techniques. AIP Conf. Proc. **2974**, 020044 (2024). https://doi.org/10.1063/5.0181801

11. Lorenzi, R., Paleari, A., Golubev, N.V., Ignateva, E.S., Sigaev, V.N., Niederberger, M., Lauria, A.: Non-aqueous sol–gel synthesis of hybrid rare-earth-doped γ-Ga_2O_3nanoparticles with multiple organic–inorganic-ionic light-emission features. J. Mater. Chem. C **3**, 41–45 (2014). https://doi.org/10.1039/c4tc02118e
12. Nagpal, D., Singh, A., Vasishth, A., Singh, R., Kumar, A.: Electrochemical evaluation of hybrid La_2CoCrO_6/Co_3O_4/rGO composite for enhanced supercapacitor performance. Carbon Trends **15**, 100358 (2024). https://doi.org/10.1016/j.cartre.2024.100358
13. Vats, A.K., Kumar, A., Sangwan, N., Kumar, A.: Symmetric/asymmetric energy storage device of reduced graphene oxide assisted $LaNi_{0.9}Co_{0.1}O_3$ perovskite nanomaterials. Appl. Phys. A **127** (2021). https://doi.org/10.1007/s00339-021-05113-4
14. Rehman, A., Zeng, X.: Interfacial composition, structure, and properties of ionic liquids and conductive polymers for the construction of chemical sensors and biosensors: a perspective. Curr. Opin. Electrochem. **23**, 47–56 (2020). https://doi.org/10.1016/j.coelec.2020.03.010
15. Chen, R., Jia, X., Zhou, H., Ren, S., Han, D., Li, S., Gao, Z.: Applications of MXenes in wearable sensing: advances, challenges, and prospects. Mater. Today **75**, 359–385 (2024). https://doi.org/10.1016/j.mattod.2024.04.013
16. Zeng, Z., Xu, Y., Zhang, Z., Gao, Z., Luo, M., Yin, Z., Zhang, C., Xu, J., Huang, B., Luo, F., Du, Y., Yan, C.: Rare-earth-containing perovskite nanomaterials: design, synthesis, properties and applications. Chem. Soc. Rev. **49**, 1109–1143 (2020). https://doi.org/10.1039/c9cs00330d
17. Reenu, N., Sonia, N., Phor, L., Kumar, A., Chahal, S.: Electrode materials for supercapacitors: a comprehensive review of advancements and performance. J. Energy Storage **84**, 110698 (2024). https://doi.org/10.1016/j.est.2024.110698
18. Singh, J., Kumar, A.: Investigation of structural, morphological and electrochemical properties of mesoporous La_2CuCoO_6 rods fabricated by facile hydrothermal route. Int. J. Miner. Metall. Mater. **27**, 987–995 (2020). https://doi.org/10.1007/s12613-020-2011-6
19. Singh, A., Vasishth, A., Kumar, A.: Hydrothermal synthesis and electrochemical performance of mesoporous La_2CrMnO_6 double perovskite for energy storage applications. Phys. Status Solidi (A) **220**, 2300198 (2023). https://doi.org/10.1002/pssa.202300198
20. Singh, A., Nagpal, D., Dinesh, N., Vasishth, A., Ul-Hamid, A., Rajput, P., Gupta, M., Kumar, A.: Synergistic effect of redox electrolyte and reduced graphene oxide on the electrochemical performance of $LaCaCrMnO_6$. J. Energy Storage **87**, 111489 (2024). https://doi.org/10.1016/j.est.2024.111489
21. Kothandam, G., Singh, G., Guan, X., Lee, J.M., Ramadass, K., Joseph, S., Benzigar, M., Karakoti, A., Yi, J., Kumar, P., Vinu, A.: Recent advances in carbon-based electrodes for energy storage and conversion. Adv. Sci. **10**, 230104 (2023). https://doi.org/10.1002/advs.202301045
22. Tie, D., Huang, S., Wang, J., Ma, J., Zhang, J., Zhao, Y.: Hybrid energy storage devices: advanced electrode materials and matching principles. Energy Storage Mater. **21**, 22–40 (2018). https://doi.org/10.1016/j.ensm.2018.12.018
23. Raman, S., Ravi Sankar, A., Sindhuja, M.: Advances in silicon nanowire applications in energy generation, storage, sensing, and electronics: a review. Nanotechnology **34**, 182001 (2023). https://doi.org/10.1088/1361-6528/acb320
24. Ren, Z., Yang, J., Qi, D., Sonar, P., Liu, L., Lou, Z., Shen, G., Wei, Z.: Flexible sensors based on organic–inorganic hybrid materials. Adv. Mater. Technol. **6**, 200088 (2021). https://doi.org/10.1002/admt.202000889
25. Gao, Y., Yu, L., Yeo, J.C., Lim, C.T.: Flexible hybrid sensors for health monitoring: materials and mechanisms to render wearability. Adv. Mater. **32**, 1902133 (2019). https://doi.org/10.1002/adma.201902133
26. Jian, Y., Hu, W., Zhao, Z., Cheng, P., Haick, H., Yao, M., Wu, W.: Gas sensors based on chemi-resistive hybrid functional nanomaterials. Nano-Micro Lett. **12**, 71 (2020). https://doi.org/10.1007/s40820-020-0407-5

27. Lim, H., Kim, H.S., Qazi, R., Kwon, Y., Jeong, J., Yeo, W.: Advanced soft materials, sensor integrations, and applications of wearable flexible hybrid electronics in healthcare, energy, and environment. Adv. Mater. **32**, 1901924 (2019). https://doi.org/10.1002/adma.201901924
28. Wang, S., Kang, Y., Wang, L., Zhang, H., Wang, Y., Wang, Y.: Organic/inorganic hybrid sensors: a review. Sens. Actuators, B Chem. **182**, 467–481 (2013). https://doi.org/10.1016/j.snb.2013.03.042
29. Suzuki, K., Yataka, K., Okumiya, Y., Sakakibara, S., Sako, K., Mimura, H., Inoue, Y.: Rapid-response, widely stretchable sensor of aligned MWCNT/Elastomer composites for human motion detection. ACS Sens. **1**, 817–825 (2016). https://doi.org/10.1021/acssensors.6b00145
30. Wang, S., Xiao, P., Liang, Y., Zhang, J., Huang, Y., Wu, S., Kuo, S., Chen, T.: Network cracks-based wearable strain sensors for subtle and large strain detection of human motions. J. Mater. Chem. C **6**, 5140–5147 (2018). https://doi.org/10.1039/c8tc00433a
31. Wang, Y., Wang, L., Yang, T., Li, X., Zang, X., Zhu, M., Wang, K., Wu, D., Zhu, H.: Wearable and highly sensitive graphene strain sensors for human motion monitoring. Adv. Func. Mater. **24**, 4666–4670 (2014). https://doi.org/10.1002/adfm.201400379
32. Su, M., Li, F., Chen, S., Huang, Z., Qin, M., Li, W., Zhang, X., Song, Y.: Nanoparticle based curve arrays for multirecognition flexible electronics. Adv. Mater. **28**, 1369–1374 (2015). https://doi.org/10.1002/adma.201504759
33. Yin, B., Wen, Y., Hong, T., Xie, Z., Yuan, G., Ji, Q., Jia, H.: Highly stretchable, ultrasensitive, and wearable strain sensors based on facilely prepared reduced graphene oxide woven fabrics in an ethanol flame. ACS Appl. Mater. Interfaces **9**, 32054–32064 (2017). https://doi.org/10.1021/acsami.7b09652
34. Amjadi, M., Pichitpajongkit, A., Lee, S., Ryu, S., Park, I.: Highly stretchable and sensitive strain sensor based on silver nanowire-elastomer nanocomposite. ACS Nano **8**, 5154–5163 (2014). https://doi.org/10.1021/nn501204t
35. Souri, H., Banerjee, H., Jusufi, A., Radacsi, N., Stokes, A.A., Park, I., Sitti, M., Amjadi, M.: Wearable and stretchable strain sensors: materials, sensing mechanisms, and applications. Adv. Intell. Syst. **2**(8), 000039 (2020). https://doi.org/10.1002/aisy.202000039
36. Choong, C., Shim, M., Lee, B., Jeon, S., Ko, D., Kang, T., Bae, J., Lee, S.H., Byun, K., Im, J., Jeong, Y.J., Park, C.E., Park, J., Chung, U.: Highly stretchable resistive pressure sensors using a conductive elastomeric composite on a micropyramid array. Adv. Mater. **26**, 3451–3458 (2014). https://doi.org/10.1002/adma.201305182
37. Wang, X., Gu, Y., Xiong, Z., Cui, Z., Zhang, T.: Silk-molded flexible, ultrasensitive, and highly stable electronic skin for monitoring human physiological signals. Adv. Mater. **26**, 1336–1342 (2013). https://doi.org/10.1002/adma.201304248
38. Kweon, O.Y., Lee, S.J., Oh, J.H.: Wearable high-performance pressure sensors based on three-dimensional electrospun conductive nanofibers. NPG Asia Mater. **10**, 540–551 (2018). https://doi.org/10.1038/s41427-018-0041-6
39. Yeo, J.C., Yu, J., Koh, Z.M., Wang, Z., Lim, C.T.: Wearable tactile sensor based on flexible microfluidics. Lab Chip **16**, 3244–3250 (2016). https://doi.org/10.1039/c6lc00579a
40. Kenry, N., Yeo, J.C., Yu, J., Shang, M., Loh, K.P., Lim, C.T.: Highly flexible graphene oxide nanosuspension liquid-based microfluidic tactile sensor. Small **12**, 1593–1604 (2016). https://doi.org/10.1002/smll.201502911
41. Wang, Z.L.: Progress in piezotronics and piezo-phototronics. Adv. Mater. **24**, 4632–4646 (2012). https://doi.org/10.1002/adma.201104365
42. Ouyang, H., Tian, J., Sun, G., Zou, Y., Liu, Z., Li, H., Zhao, L., Shi, B., Fan, Y., Fan, Y., Wang, Z.L., Li, Z.: Self-powered pulse sensor for antidiastole of cardiovascular disease. Adv. Mater. **29**, 1703456 (2017). https://doi.org/10.1002/adma.201703456
43. Liu, M., Pu, X., Jiang, C., Liu, T., Huang, X., Chen, L., Du, C., Sun, J., Hu, W., Wang, Z.L.: Large-area all-textile pressure sensors for monitoring human motion and physiological signals. Adv. Mater. **29**, 1703700 (2017). https://doi.org/10.1002/adma.201703700
44. Ho, M.D., Ling, Y., Yap, L.W., Wang, Y., Dong, D., Zhao, Y., Cheng, W.: Percolating network of ultrathin gold nanowires and silver nanowires toward "invisible" wearable sensors for detecting emotional expression and apexcardiogram. Adv. Func. Mater. **27**, 1700845 (2017). https://doi.org/10.1002/adfm.201700845

45. Yang, J., Chen, J., Su, Y., Jing, Q., Li, Z., Yi, F., Wen, X., Wang, Z., Wang, Z.L.: Eardrum-inspired active sensors for self-powered cardiovascular system characterization and throat-attached anti-interference voice recognition. Adv. Mater. **27**, 1316–1326 (2015). https://doi.org/10.1002/adma.201404794
46. Karimi, M., Badiei, A., Lashgari, N., Ziarani, G.M.: A chromotropic acid modified SBA-15 as a highly sensitive fluorescent probe for determination of Fe^{3+} and I^{-} ions in water. J. Porous Mater. **25**, 137–146 (2017). https://doi.org/10.1007/s10934-017-0427-9
47. Rahman, M.T., Cheng, C., Karagoz, B., Renn, M., Schrandt, M., Gellman, A., Panat, R.: High performance flexible temperature sensors via nanoparticle printing. ACS Appl. Nano Mater. **2**, 3280–3291 (2019). https://doi.org/10.1021/acsanm.9b00628
48. Chubb, D.L., Wolford, D.S.: Rare earth optical temperature sensor. Rev. Sci. Instrum. **71**, 2233–2240 (2000). https://doi.org/10.1063/1.1150616
49. Ramakrishnan, M., Rajan, G., Semenova, Y., Farrell, G.: Hybrid fiber optic sensor system for measuring the strain, temperature, and thermal strain of composite materials. IEEE Sens. J. **14**, 2571–2578 (2014). https://doi.org/10.1109/jsen.2014.2306892
50. Iuchi, T., Hiraka, K.: Hybrid-type temperature sensor for in situ measurement. Rev. Sci. Instrum. **77**, 114902 (2006). https://doi.org/10.1063/1.2393160
51. Miah, K., Potter, D.: A review of hybrid fiber-optic distributed simultaneous vibration and temperature sensing technology and its geophysical applications. Sensors **17**, 2511 (2017). https://doi.org/10.3390/s17112511
52. Waris, M.H., Muzaffar, N., Mumtaz, M.A., Afzal, A.M., Iqbal, M.W., Mumtaz, S., Ali, M., Alaraidh, I.A., Okla, M.K., Munnaf, S.A.: High performance lanthanum-doped nickel cobalt ferrites on titanium carbide MXene electrode material for superior hybrid device and precision creatinine sensing. Appl. Phys. A **131**, 220 (2025). https://doi.org/10.1007/s00339-025-08306-3
53. Rubel Alam, Md., Gharami, M., Dev, B., Ashikur Rahman, Md., Islam, T.: Next-generation materials for multifunctional applications: design progress and prospects of 2D MXene-enabled hybrid architectures. Adv. Mater. Interfaces **12**, 2400681 (2025). https://doi.org/10.1002/admi.202400681
54. Nadar, N.R., Deepak, J., Sharma, S.C., Radha Krushna, B.R., Puneeth, Sowjanya, R., Sureka Varalakshmi, V., Sahu, S., Sargunam, B., Nagabhushana, H., Kumar Swamy, B.E., Ruthwik, S.S.: Graphene oxide based Gd_2O_3:Eu^{3+} nanocomposites: a multifaceted approach to advanced energy storage and bio sensing applications. Inorganic Chem. Commun. **165**, 112515 (2024). https://doi.org/10.1016/j.inoche.2024.112515
55. Hai, B., Zou, Y.: Carbon cloth supported NiAl-layered double hydroxides for flexible application and highly sensitive electrochemical sensors. Sens. Actuators, B Chem. **208**, 143–150 (2014). https://doi.org/10.1016/j.snb.2014.11.022
56. Bratlie, K.M., York, R.L., Invernale, M.A., Langer, R., Anderson, D.G.: Materials for diabetes therapeutics. Adv. Healthcare Mater. **1**, 267–284 (2012). https://doi.org/10.1002/adhm.201200037
57. Zafar, H., Channa, A., Jeoti, V., Stojanović, G.M.: Comprehensive review on wearable sweat-glucose sensors for continuous glucose monitoring. Sensors **22**, 638 (2022). https://doi.org/10.3390/s22020638
58. Heikenfeld, J.: Let them see you sweat. IEEE Spectr. **51**, 46–63 (2014). https://doi.org/10.1109/mspec.2014.6934933
59. Chitare, Y.M., Magdum, V.V., Jadhav, S.B., Kulkarni, S.P., Lokhande, C.D., Gunj1akar, J.L.: Rare earth element-based nonenzymatic glucose sensor. In: Springer eBooks, pp. 393–410 (2023). https://doi.org/10.1007/978-3-031-23401-9_14
60. Teles, F., Fonseca, L.: Trends in DNA biosensors. Talanta **77**, 606–623 (2008). https://doi.org/10.1016/j.talanta.2008.07.024
61. Mandpe, P., Prabhakar, B., Gupta, H., Shende, P.: Glucose oxidase-based biosensor for glucose detection from biological fluids. Sens. Rev. **40**, 497–511 (2020). https://doi.org/10.1108/sr-01-2019-0017

62. Christopoulos, T.K.: Nucleic acid analysis. Anal. Chem. **71**, 425–438 (1999). https://doi.org/10.1021/a19900161
63. Paleček, E., Fojta, M., Tomschik, M., Wang, J.: Electrochemical biosensors for DNA hybridization and DNA damage. Biosens. Bioelectron. **13**, 621–628 (1998). https://doi.org/10.1016/s0956-5663(98)00017-7
64. Lu, C., Zhu, C., Li, J., Liu, J., Chen, X., Yang, H.: Using graphene to protect DNA from cleavage during cellular delivery. Chem. Commun. **46**, 3116 (2010). https://doi.org/10.1039/b926893f
65. Njagi, J.I., Kagwanja, S.M.: The interface in biosensing: improving selectivity and sensitivity. In: ACS Symposium Series, pp. 225–247 (2011). https://doi.org/10.1021/bk-2011-1062.ch011
66. Sun, Y., Yan, X.: Recent advances in dual-functional devices integrating solar cells and supercapacitors. Solar RRL, **1**, 3–4 (2017). https://doi.org/10.1002/solr.201700002
67. Ivanova, E.V., Schuhmann, W., Ryabov, A.D.: Reagentless enzymatic sensors based on carbon-paste electrodes containing ruthenium mediators for the on-line determination of glycerol. J. Anal. Chem. **64**, 404–409 (2009). https://doi.org/10.1134/s1061934809040157
68. Vamvakaki, V., Tsagaraki, K., Chaniotakis, N.: Carbon nanofiber-based glucose biosensor. Anal. Chem. **78**, 5538–5542 (2006). https://doi.org/10.1021/ac060551t
69. Sangodkar, H., Sukeerthi, S., Srinivasa, R.S., Lal, R., Contractor, A.Q.: A biosensor array based on polyaniline. Anal. Chem. **68**, 779–783 (1996). https://doi.org/10.1021/ac950655w
70. López-Peña, G., Ortiz-Mansilla, E., Arranz, A., Bogdan, N., Manso-Silván, M., Rodríguez, E.M.: Non-invasive paper-based sensors containing rare-earth-doped nanoparticles for the detection of D-glucose. Colloids Surf. B **239**, 113934 (2024). https://doi.org/10.1016/j.colsurfb.2024.113934
71. Baranwal, J., Barse, B., Gatto, G., Broncova, G., Kumar, A.: Electrochemical sensors and their applications: a review. Chemosensors **10**, 363 (2022). https://doi.org/10.3390/chemosensors10090363
72. Deshmukh, V., Harini, H., Nagaswarupa, H., Naik, R., Ravikumar, C.: Development of novel Co^{3+} doped $LaMnO_3$ perovskite electrodes for supercapacitors and sensors: mechanism of electrochemical energy storage and oxygen intercalation. J. Energy Storage **68**, 107805 (2023). https://doi.org/10.1016/j.est.2023.107805
73. Abdelhamid, H.N.: An introductory review on advanced multifunctional materials. Heliyon **9**, e18060 (2023). https://doi.org/10.1016/j.heliyon.2023.e18060
74. Rehman, A.U., Muzaffar, N., Barsoum, I., Afzal, A.M., Ali, M., Iqbal, M.W., Ahmad, Z., Mumtaz, S., Bahajjaj, A.A.A., Munnaf, S.A.: Designing of efficient $CoLa_2O_4$/V-Ag-MOF hybrid electrode for energy storage, hydrogen evolution reaction, and chemical sensors. Fuel **384**, 133991 (2024). https://doi.org/10.1016/j.fuel.2024.133991
75. Kaushal, I., Kumar, V., Saharan, P., Mittal, A., Bhagat, R., Kumar, S., Sharma, A.K.: Electrochemical energy storage and hydrogen peroxide sensing using hybrid framework of CeO_2–MnO_2 on carbon nano fiber composite. J. Alloy. Compd. **934**, 167740 (2022). https://doi.org/10.1016/j.jallcom.2022.167740
76. Gaffney, J.S., Marley, N.A.: Atmospheric chemistry and air pollution. Sci. World J. **3**, 199–234 (2003). https://doi.org/10.1100/tsw.2003.18
77. Yun, Y.J., Hong, W.G., Choi, N., Kim, B.H., Jun, Y., Lee, H.: Ultrasensitive and highly selective graphene-based single yarn for use in wearable gas sensor. Sci. Rep. **5**, 10904 (2015). https://doi.org/10.1038/srep10904
78. Gao, Z., Lou, Z., Chen, S., Li, L., Jiang, K., Fu, Z., Han, W., Shen, G.: Fiber gas sensor-integrated smart face mask for room-temperature distinguishing of target gases. Nano Res. **11**, 511–519 (2017). https://doi.org/10.1007/s12274-017-1661-9
79. Thammakan, S., Sinchow, M., Chiangraeng, N., Yoshinari, N., Konno, T., Ngamjarurojana, A., Nimmanpipug, P., Rujiwatra, A.: Photoluminescent lanthanide coordination polymers for ammonia gas detection at ambient temperature and humidity. Cryst. Growth Des. **23**, 6840–6850 (2023). https://doi.org/10.1021/acs.cgd.3c00680

80. Mousty, C.: Sensors and biosensors based on clay-modified electrodes: new trends. Appl. Clay Sci. **27**, 159–177 (2004). https://doi.org/10.1016/j.clay.2004.06.005
81. Smela, E.: Microfabrication of PPy microactuators and other conjugated polymer devices. J. Micromech. Microeng. **9**, 1–18 (1999). https://doi.org/10.1088/0960-1317/9/1/001
82. Negi, Y.S., Adhyapak, P.V.: Development in polyaniline conducting polymers. J. Macromol. Sci. Part C—Polym. Rev. **42**, 35–53 (2002). https://doi.org/10.1081/mc-120003094
83. Kaushik, A., Kumar, R., Arya, S.K., Nair, M., Malhotra, B.D., Bhansali, S.: Organic–inorganic hybrid nanocomposite-based gas sensors for environmental monitoring. Chem. Rev. **115**, 4571–4606 (2015). https://doi.org/10.1021/cr400659h
84. Ferdiana, N.A., Bahti, H.H., Kurnia, D., Wyantuti, S.: Synthesis, characterization, and electrochemical properties of rare earth element nanoparticles and its application in electrochemical nanosensor for the detection of various biomolecules and hazardous compounds: a review. Sens. Bio-Sens. Res. **41**, 100573 (2023). https://doi.org/10.1016/j.sbsr.2023.100573
85. Singh, E., Meyyappan, M., Nalwa, H.S.: Flexible graphene-based wearable gas and chemical sensors. ACS Appl. Mater. Interfaces **9**, 34544–34586 (2017). https://doi.org/10.1021/acsami.7b07063
86. Molina, A., Escobar-Barrios, V., Oliva, J.: A review on hybrid and flexible CO_2 gas sensors. Synth. Met. **270**, 116602 (2020). https://doi.org/10.1016/j.synthmet.2020.116602
87. Panigrahi, P.K., Chandu, B., Puvvada, N.: Recent advances in nanostructured materials for application as gas sensors. ACS Omega **3**, 3092–3122 (2024). https://doi.org/10.1021/acsomega.3c06533
88. Cai, Z., Park, S.: Synthesis of Pd nanoparticle-decorated SnO_2 nanowires and determination of the optimum quantity of Pd nanoparticles for highly sensitive and selective hydrogen gas sensor. Sens. Actuators, B Chem. **322**, 128651 (2020). https://doi.org/10.1016/j.snb.2020.128651
89. Zafar, Q., Azmer, M.I., Al-Sehemi, A.G., Al-Assiri, M.S., Kalam, A., Sulaiman, K.: Evaluation of humidity sensing properties of TMBHPET thin film embedded with spinel cobalt ferrite nanoparticles. J. Nanopart. Res. **18**, 186 (2016). https://doi.org/10.1007/s11051-016-3488-9
90. Patil, A.S., Patil, A.V., Dighavkar, C.G., Adole, V.A., Tupe, U.J.: Synthesis techniques and applications of rare earth metal oxides semiconductors: a review. Chem. Phys. Lett. **796**, 139555 (2022). https://doi.org/10.1016/j.cplett.2022.139555
91. Çigil, A.B., Cankurtaran, H., Kahraman, M.V.: Photo-crosslinked thiol-ene based hybrid polymeric sensor for humidity detection. React. Funct. Polym. **114**, 75–85 (2017). https://doi.org/10.1016/j.reactfunctpolym.2017.03.002
92. Rosace, G., Trovato, V., Colleoni, C., Caldara, M., Re, V., Brucale, M., Piperopoulos, E., Mastronardo, E., Milone, C., De Luca, G., Plutino, M.R.: Structural and morphological characterizations of MWCNTs hybrid coating onto cotton fabric as potential humidity and temperature wearable sensor. Sens. Actuators, B Chem. **252**, 428–439 (2017). https://doi.org/10.1016/j.snb.2017.05.175
93. Janata, J.: Chemical sensors. Anal. Chem. **64**, 196–219 (1992). https://doi.org/10.1021/ac00036a012
94. Xu, C., Miyazaki, K., Watanabe, T.: Humidity sensors using manganese oxides. Sens. Actuators, B Chem. **46**, 87–96 (1998). https://doi.org/10.1016/s0925-4005(97)00330-4
95. Wencel, D., Abel, T., McDonagh, C.: Optical chemical pH sensors. Anal. Chem. **86**, 15–29 (2013). https://doi.org/10.1021/ac4035168
96. Moradi, V., Akbari, M., Wild, P.: A fluorescence-based pH sensor with microfluidic mixing and fiber optic detection for wide range pH measurements. Sens. Actuators, A **297**, 111507 (2019). https://doi.org/10.1016/j.sna.2019.07.031
97. Yan, B.: Photophysical applications of photofunctional rare-earth hybrid materials. In: Springer Series in Materials Science, pp. 199–255 (2017). https://doi.org/10.1007/978-981-10-2957-8_8
98. Kasirajan, K., Chandrasekar, L.B., Maheswari, S., Karunakaran, M., Sundaram, P.S.: A comparative study of different rare-earth (Gd, Nd, and Sm) metals doped ZnO thin films

and its room temperature ammonia gas sensor activity: synthesis, characterization, and investigation on the impact of dopant. Opt. Mater. **121**, 111554 (2021). https://doi.org/10.1016/j.optmat.2021.111554

99. Nadaf, S., Chethan, B., Swathi, K., Laxmeshwar, S.S., Jagadeesha Angadi, V., Manjunatha, K., Vandana, M., Chattham, N., Hegde, G., Sharma, K., Sathish, T., Gupta, M., Ubaidullah, M., Pattar, V., Basavegowda, N., Wu, S.Y., Manjunatha, S., Bajorek, A.: Synthesis, characterization and application of rare earth (Lu^{3+}) doped zinc ferrites in carbon monoxide gas sensing and supercapacitors. Ceram. Int. **50**, 23208–23221 (2024). https://doi.org/10.1016/j.ceramint.2024.04.045
100. Zhang, Y., Wang, X., Wu, D.: Microencapsulation of n-dodecane into zirconia shell doped with rare earth: design and synthesis of bifunctional microcapsules for photoluminescence enhancement and thermal energy storage. Energy **97**, 113–126 (2016). https://doi.org/10.1016/j.energy.2015.12.114
101. Wei, H., He, F., Li, Y., Zhang, Q., Zhou, Y., Yan, H., He, R., Fan, J., Yang, W.: Bifunctional Paraffin@$CaCO_3$:Ce^{3+} phase change microcapsules for thermal energy storage and photoluminescence. ACS Sustain. Chem. Eng. **7**, 18854–18862 (2019). https://doi.org/10.1021/acssuschemeng.9b03935
102. Zhuang, Y., Tu, D., Chen, C., Wang, L., Zhang, H., Xue, H., Yuan, C., Chen, G., Pan, C., Dai, L., Xie, R.: Force-induced charge carrier storage: a new route for stress recording. Light Sci. Appl. **9**, 182 (2020). https://doi.org/10.1038/s41377-020-00422-4
103. Zhang, Y., Pan, C., Liu, P., Peng, L., Liu, Z., Li, Y., Wang, Q., Wu, T., Li, Z., Majidi, C., Jiang, L.: Coaxially printed magnetic mechanical electrical hybrid structures with actuation and sensing functionalities. Nat. Commun. **14**, 4428 (2023). https://doi.org/10.1038/s41467-023-40109-z
104. Hajra, S., Vivekananthan, V., Sahu, M., Khandelwal, G., Raj, N.P.M.J., Kim, S.: Triboelectric nanogenerator using multiferroic materials: an approach for energy harvesting and self-powered magnetic field detection. Nano Energy **85**, 105964 (2021). https://doi.org/10.1016/j.nanoen.2021.105964
105. Saha, A., Alam, M.A.: Signal as an energy source: dual-functional potentiometric sensor serves as an energy harvester. ACS Appl. Electron. Mater. **6**, 4855–4863 (2024). https://doi.org/10.1021/acsaelm.4c00064
106. Umar, E., Iqbal, M.W., Shaheen, F., Ullah, H., Wahab, R., Hassan, R.U.: Synergistic effect of redox-active $SrMnO_3$/g-C_3N_4 electrode materials for supercapattery hybrid energy storage devices and electrochemical sensing of hydroquinone. J. Power. Sources **640**, 236761 (2025). https://doi.org/10.1016/j.jpowsour.2025.236761
107. Meng, Q., Wu, H., Meng, Y., Xie, K., Wei, Z., Guo, Z.: High-performance all-carbon yarn micro-supercapacitor for an integrated energy system. Adv. Mater. **26**, 4100–4106 (2014). https://doi.org/10.1002/adma.201400399
108. Wee, G., Salim, T., Lam, Y.M., Mhaisalkar, S.G., Srinivasan, M.: Printable photo-supercapacitor using single-walled carbon nanotubes. Energy Environ. Sci. **4**, 413–416 (2010). https://doi.org/10.1039/c0ee00296h
109. O'Regan, B., Grätzel, M.: A low-cost, high-efficiency solar cell based on dye-sensitized colloidal TiO_2 films. Nature **353**, 737–740 (1991). https://doi.org/10.1038/353737a0
110. An, Q., Zhang, F., Zhang, J., Tang, W., Deng, Z., Hu, B.: Versatile ternary organic solar cells: a critical review. Energy Environ. Sci. **9**, 281–322 (2015). https://doi.org/10.1039/c5ee02641e
111. Teixeira, J.S., Costa, R.S., Pires, A.L., Pereira, A.M., Pereira, C.: Hybrid dual-function thermal energy harvesting and storage technologies: towards self-chargeable flexible/wearable devices. Dalton Trans. **50**, 9983–10013 (2021). https://doi.org/10.1039/d1dt01568k
112. Han, C., Qian, X., Li, Q., Deng, B., Zhu, Y., Han, Z., Zhang, W., Wang, W., Feng, S., Chen, G., Liu, W.: Giant thermopower of ionic gelatin near room temperature. Science **368**, 1091–1098 (2020). https://doi.org/10.1126/science.aaz5045
113. Sun, Y., Zhu, X., Peng, J., Li, F.: Core–shell lanthanide upconversion nanophosphors as four-modal probes for tumor angiogenesis imaging. ACS Nano **7**, 11290–11300 (2013). https://doi.org/10.1021/nn405082y

114. Zhao, X., He, S., Tan, M.C.: Design of infrared-emitting rare earth doped nanoparticles and nanostructured composites. J. Mater. Chem. C **4**, 8349–8372 (2016). https://doi.org/10.1039/c6tc02373h
115. Park, Y.I., Lee, K.T., Suh, Y.D., Hyeon, T.: Upconverting nanoparticles: a versatile platform for wide-field two-photon microscopy and multi-modal in vivo imaging. Chem. Soc. Rev. **44**, 1302–1317 (2014). https://doi.org/10.1039/c4cs00173g
116. Liu, J., Bu, W., Shi, J.: Silica coated upconversion nanoparticles: a versatile platform for the development of efficient theranostics. Acc. Chem. Res. **48**, 1797–1805 (2015). https://doi.org/10.1021/acs.accounts.5b00078
117. Pascariu, P., Homocianu, M., Tudorache, F., Bele, A., Rusu, D.: Enhancing humidity sensitivity properties through RE (Er, La, and Sm) dopants in TiO_2 nanofiber composites. J. Mater. Sci. **59**, 2712–2727 (2024). https://doi.org/10.1007/s10853-024-09362-5
118. Pérez, J.C.A., Torres, J.G.T., de la Cruz Romero, D., Perez-Vidal, H., Rocha, M.A.L., López, I.C., Uribe, A.C., Que, Z.G.: Photocatalytic treatment of pesticides using TiO_2 Doped with rare earth. In: Water and Wastewater Treatment. IntechOpen (2019). https://doi.org/10.5772/intechopen.84677
119. Rasheed, S., Ali, G., Kousar, R., Raza, M.A., Kubra, K.T., Iftikhar, F.J.: Synthesis and electrochemical performance evaluation of La doped TiO_2/reduced graphene oxide nanocomposites for supercapacitor application. J. Electroanal. Chem. **928**, 117015 (2023). https://doi.org/10.1016/j.jelechem.2022.117015
120. Prakash, J., Kumar, A., Dai, H., Janegitz, B.C., Krishnan, V., Swart, H.C., Sun, S.: Novel rare earth metal–doped one-dimensional TiO_2 nanostructures: fundamentals and multifunctional applications. Mater. Today Sustain. **13**, 100066 (2021). https://doi.org/10.1016/j.mtsust.2021.100066
121. Li, J., Chu, B., Xie, Z., Deng, Y., Zhou, Y., Dong, L., Li, B., Chen, Z.: Mechanism and DFT study of degradation of organic pollutants on rare earth ions doped TiO_2 photocatalysts prepared by sol-hydrothermal synthesis. Catal. Lett. **152**, 489–502 (2022). https://doi.org/10.1007/s10562-021-03634-4
122. Liu, Y., Ouyang, Z., Yang, L., Yang, Y., Sun, J.: Blue electroluminescent Al_2O_3/Tm_2O_3 nanolaminate films fabricated by atomic layer deposition on silicon. Nanomaterials **9**, 413 (2019). https://doi.org/10.3390/nano9030413
123. Chen, W.M., Hsieh, H.Y., Wu, D.Z., Tang, H.Y., Chang-Liao, K.S., Chi, P.W., Wu, P.M., Wu, M.K.: Advanced TiO_2/Al_2O_3 bilayer ALD coatings for improved lithium-rich layered oxide electrodes. ACS Appl. Mater. Interfaces **16**, 13029–13040 (2024). https://doi.org/10.1021/acsami.3c16948
124. Padhi, P.S., Rai, S.K., Srivastava, H., Ajimsha, R.S., Banik, S., Srivastava, A.K., Misra, P.: Tunable nanobattery effect and negative differential resistance characteristics in interfacial polarization-dominated Al_2O_3/TiO_2 nanolaminates. ACS Appl. Nano Mater **7**, 13928 (2024). https://doi.org/10.1021/acsanm.4c02024
125. Padhi, P.S., Rai, S.K., Ajimsha, R.S., Misra, P.: Engineering Maxwell–Wagner relaxation and interface carrier confinement in Al2O3/TiO2 subnanometric laminates for high-density energy storage applications. RSC Appl. Interfaces **1**(6), 1348–1359 (2024). https://doi.org/10.1039/D4LF00125G
126. Padhi, P.S., Rai, S.K., Vijay, K., Srivastava, H., Banik, S., Ajimsha, R.S., Srivastava, A.K., Misra, P.: Small polaron hopping and tunneling transport in Maxwell–Wagner relaxation dominated Al2O3/TiO2 subnanometric laminates. Appl. Phys. Lett. **124**, 233504 (2024). https://doi.org/10.1063/5.0202776
127. Sahu, M.C., Padhi, P.S., Mallik, S.K., Roy, S., Sahoo, S., Banik, S., Sharma, R.K.: Pulse tunable quantum conductance states in oxygen vacancy engineered TiO_2-Δ_x memristor for artificial neural network applications. ACS Appl. Electron. Mater. **7**, 5271 (2025). https://doi.org/10.1021/acsaelm.5c00694

Electrochemical Sensing Based on Rare Earth-Doped Metal Oxides

Deeksha Nagpal and Astakala Anil Kumar

Abstract Electrochemical sensors have acquired considerable prominence for applications in environmental monitoring, healthcare diagnostics, and industrial safety owing to their rapid response, high sensitivity, and cost-effectiveness. Metal oxides as electrochemical sensors perform better as of their superior redox activity, electrical behavior, and chemical stability. Their electronic structure and defect chemistry, influencing the overall performance have been explained in this chapter. Further, basic concepts of electrochemical transduction are described for more clarification. This chapter elaborates that sensing capability improves with the addition of rare-earth dopants in metal oxides from the available literature, as it can alter lattice defects, improve electrical conductivity, and offer additional active sites for better catalytic efficacy and analyte absorption. Their applications in environmental monitoring, pollution detection, gas, and biosensing have been highlighted, along with their experimental data demonstrating improved sensitivity, selectivity, and stability. The overall chapter provides an insight for developing the next generation electrochemical sensor with enhanced performance from a technology perspective.

Keywords Electrochemical sensors · Rare earth element dopants · Sensing mechanism · Metal oxides · Improved sensitivity

1 Introduction

The growth in sensor technologies is vigorously increasing for detecting chemical and biological contaminants with swift and accurate response, along with a cost-effective approach. The electrochemical sensors have gathered attention among other types of sensors because they respond quickly, have superior sensitivity, are easy to operate,

D. Nagpal (✉)
University Centre for Research and Development, Chandigarh University, Mohali, Punjab, India
e-mail: cu.deekshaphd@gmail.com

A. A. Kumar
School of Advanced Material Engineering, Kookmin University, Seoul, Republic of Korea

© The Author(s), under exclusive license to Springer Nature Singapore Pte Ltd. 2026
N. Kumar et al. (eds.), *Rare Earth-Doped Metal Oxide Nanostructures*, Engineering Materials, https://doi.org/10.1007/978-981-95-6192-6_12

and are employed to fabricate compact and portable devices [1, 2]. These sensors are utilized in environmental monitoring, industrial diagnostics, healthcare diagnosis, food safety, etcetera [3–5].

The performance of electrochemical sensors is dependent upon the electrode material utilized, as it determines how effectively a chemical signal is transformed into an electrical signal. Recently, rare earth (RE)-doped metal oxides have been explored more for enhanced sensing performance as they exhibit exceptional catalytic, electrical, and surface characteristics that improves the interactions between target analyte and electrode [6].

The incorporation of RE dopants in metal oxides sensors along with the basic principle of electrochemical sensing has been thoroughly explained in this chapter. The insight behind the improvement of signal transduction, selectivity, and sensitivity of these sensors have been discussed in more detail.

1.1 Overview of Electrochemical Sensors

A device designed to sense specific materials by converting chemical response involving redox reactions (oxidation and reduction) into an electrical response, is known as an electrochemical sensor [7]. Generally, these redox reactions take place at the surface of the electrode and depend upon the concentration of the target material, allowing electrochemical sensors to deliver quantitative and real-time measurements. In comparison to other sensing methods, such as optical or spectroscopic, electrochemical sensing is more advantageous because it is easy to operate, cost-effective, time-saving, and appropriate for portable and practical applications [8]. Electrochemical sensors offer a quicker and more straightforward method of measuring analytes than traditional chemical sensors, which frequently require intricate processing or indirect detection techniques.

The surface of the electrode is tailored with the materials that can improve electron transfer capability and boost sensitivity in order to get superior efficacy. Catalytically active materials enhance the rate of chemical reactions, whereas coatings such as polymers, enzymes, or selective membranes contribute to improved selectivity by interacting specifically with target molecules [9].

The progress in nanotechnology has led to the widespread usage of nanoscale materials in the development of electrochemical sensor applications [10, 11]. The material, especially in its nanostructured form, possesses an enhanced surface area, improved conductivity, and additional active sites that promote superior interactions with the utilized analyte [11]. Rare earth-doped metal oxides have demonstrated potential in enhancing the electrical conductivity and catalytic efficiency of sensor electrodes, thereby contributing to the development of more precise and dependable detection systems.

1.2 Role of Metal Oxides in Electrochemical Sensing

Metal oxides have gained widespread attention in the field of electrochemical sensing applications to attain superior performance [12–15]. Because of their various physical and chemical characteristics, like adjustable electronic structure, variable oxidation state, and superior thermal stability, they are preferred for the occurrence of redox reactions at the interface of electrolyte and electrode. Instead of only serving as passive substrates, they also serve as active participants in charge transfer interactions with analytes. Also, the semiconductor nature of metal oxides facilitates fine-tuning of charge carrier dynamics, and improved sensitivity in electrochemical signal detection [16].

Metal oxides exhibit valuable properties while synthesizing at the nanoscale level in the form of nanowires, nanoparticles, or thin films, leading to a considerable increase in their surface area [17]. This enhanced surface-to-volume ratio offers more active participation of active sites for analyte interaction, resulting in reduced response time and improved sensitivity [18]. Furthermore, certain metal oxides demonstrate their nature of selective adsorption, this nature allows their selective bonding with specific chemical species, thus proposed as an essential feature for achieving high sensor selectivity [19, 20].

Instead of having numerous benefits, metal oxides have some limitations, like low electrical conductivity and degraded long-term stability. Therefore, to overcome these limitations and to tune the physicochemical characteristics of metal oxides, they are incorporated with conductive materials or suitable dopants. It is evident from the literature that RE dopants have proven to be efficacious as they have increased conductivity, surface activity, and catalytic performance. The discussion on the improved performance and their underlying mechanisms has been done in the following sections.

1.3 Advantages of Rare Earth Element Doping

Incorporating rare earth (RE) elements having unique 4f electrons into metal oxides is advantageous for boosting their electrochemical sensing capabilities [21]. Incorporation of elements such as lanthanum (La), cerium (Ce), neodymium (Nd), and samarium (Sm) to the base metal oxides exhibits distinctive shifts in the electronic structures and chemical properties that can substantially alter the structural, electronic, and catalytic behavior. A key advantage of RE doping lies in its ability to generate well-controlled lattice defects, such as oxygen vacancies and interstitial defects, which act as active centers for redox reactions. These defect sites are critical in enhancing the electrocatalytic efficiency, as the oxygen vacancies improve analyte diffusion and accelerates electron transfer processes, thereby leading to greater sensitivity and faster response times at lower concentrations of analyte, thereby delivering to a highly sensitive sensor material.

Rare-earth doping also alters the electronic properties of metal oxides. Bandgap tuning and increased charge carrier density can be achieved by such doping, leading to an elevated electrical conductivity. These characteristics are particularly advantageous for detecting trace levels of analytes, where signal amplification is essential. Additionally, the larger ionic radii of RE elements induce lattice distortions that enhance surface reactivity and stimulate stronger interactions with target molecules. In addition to functional modifications, rare earth elements contribute to the chemical and thermal stability of the metal oxide framework, which is an important factor for sensor practical implementation and maintaining steady performance over longer duration of time, particularly in challenging environmental conditions.

The intrinsic properties of rare earth-doped metal oxides can be precisely tailored by varying the type and concentration of the dopant, providing an effective approach for optimizing sensor performance for specific analyte detection. Nanostructured integration with conductive materials, such as carbon-based materials also has their significance in sensing. These doped materials form highly efficient electrode materials for the development of electrochemical sensors that exhibit enhanced selectivity, sensitivity, and long-term durability for application in medical and modern optoelectronics.

2 Theoretical Background and Sensing Mechanism

To design advanced material that has superior sensing capabilities, a detailed understanding of the electrochemical sensor mechanism is required. The chapter further explains the basic principle of the working of metal oxide sensors, which is further enhanced with the RE dopants. Moreover, the factors influencing sensor performance, comprising catalytic activity, conductivity, density, and surface area, are discussed along with the discussion on electrochemical redox reactions occurring near the interface of electrolyte and electrode. The sensing performance of metal oxides can be greatly improved with the RE doping, which influences active defect sites, electronic structure, and charge carrier mobility. Both experimental results and theoretical concepts of RE dopants improved the sensitivity and selectivity with better signal transduction and analyte interactions.

2.1 Fundamentals of Electrochemical Transduction

Electrochemical transduction is a process in which the interface of electrode is involved, wherein chemical interactions are converted into a detectable electrical signal [22]. In an electrochemical sensor, this transduction typically involves oxidation or reduction reactions of analyte molecules, which result in the generation or consumption of electrons. These electron flow changes are recorded as current or potential, depending on the measurement technique employed.

There are three main types of electrochemical transducers:

- Amperometric sensors, which measure the current generated by redox reactions at a constant applied potential [23]. These are widely used due to their simplicity, sensitivity, and suitability for continuous monitoring.
- Potentiometric sensors, which detect changes in potential (voltage) between a working and a reference electrode, with no significant current flow [24]. These are particularly useful for ion-selective measurements.
- Conductometric or impedimetric sensors, which monitor changes in electrical conductivity or impedance in response to analyte interaction with the sensing surface [25]. These are useful in label-free detection and real-time sensing.

The efficiency of this transfer depends on several factors, including the electronic conductivity of the electrode material, surface area, availability of catalytic sites, and nature of surface defects. Metal oxides, when used as electrode materials, provide excellent redox activity and chemical stability. However, their performance can be limited by factors such as low intrinsic conductivity or poor active site accessibility. This is where rare earth doping becomes significant. By introducing RE elements, the electronic structure and surface properties of metal oxides can be tailored to enhance electron mobility, increase defect concentrations (such as oxygen vacancies), and improve overall electrocatalytic performance.

Thus, understanding the principles of electrochemical transduction is essential for the rational design of RE-doped metal oxide sensors that are not only more sensitive but also more selective and reliable.

2.2 *Redox Behavior and Charge Transfer in Metal Oxide Interfaces*

The effectiveness of an electrochemical sensor largely depends on the redox activity and charge transfer processes occurring at the electrode–electrolyte interface [26]. In metal oxide-based sensors, these processes are governed by the intrinsic redox properties of the metal cations and the surface chemistry of the oxide material [27].

Metal oxides often contain transition metal ions that can easily change their oxidation states (e.g., Fe^{3+}/Fe^{2+}, Mn^{4+}/Mn^{3+}), making them capable of participating in electron exchange with analyte species [28]. When a target molecule undergoes oxidation or reduction at the sensor surface, electrons are transferred either to or from the electrode material. This electron exchange generates a measurable signal, such as current (in amperometric sensing) or potential change (in potentiometric sensing). However, the rate and efficiency of this charge transfer process depend heavily on the electronic structure and defect density of the metal oxide. While many metal oxides possess favorable redox properties, they often suffer from poor electrical conductivity due to their semiconducting nature. This limitation can hinder the flow of electrons, reducing the sensor's sensitivity and response time [29].

The electronic configuration of metal oxides has been modified through the doping of RE elements as they possess distinct valence states and large ionic radii in comparison to the metal oxides, resulting in distortion of the lattice structure, formation of oxygen vacancies, and confined energy states. All of these characteristics enhance the charge carrier mobility and density and improve redox activity at the electrode surface. Additionally, the band positioning and work function of metal oxides can be modified by the incorporation of RE dopants, which results in an enhanced sensor response, quick charge transfer, and lesser activation energy for redox reactions to occur.

2.3 Electronic Effects of Rare-Earth Dopants

The efficacy of electrochemical sensing has been substantially enhanced by incorporating rare earth elements into metal oxide frameworks, resulting in modification of their electronic structure [30]. The lanthanide series is chosen as dopant materials because it comprises partially occupied 4f orbitals, which are not involved in chemical bonding; however, they highly influence the local electronic environment and charge allocation within the host material.

The major electronic effect is the band structure of the host material, which has been modified by introducing RE ions with different valencies compared to the host element, resulting in a shift of the Fermi level and altered positions of the valence and conduction bands [31]. The semiconductor characteristics of the materials are improved via these bandgap modifications, offering increased electrical conductivity and providing superior electron transfer at the interface of the electrolyte and electrode. Also, in the bandgap of metal oxides, RE dopants introduce localized energy states, resulting in elevated temporary trap charge carriers or electron hopping. The overall sensor response has been improved by introducing all the discussed effects that contribute to enhanced redox reaction kinetics. The increased electronic conductivity is of greater importance for amperometric and impedimetric sensing methods, as it depends on the effective charge transfer required to generate a signal output.

In addition to that, there is an occurrence of oxygen vacancies with adding RE dopants in metal oxides, which serve as electron donors, resulting in enhanced conductivity. Also, these oxygen vacancies act like an active site for absorbing analyte and occurrence of redox reactions, leading to increased sensitivity and selectivity for sensing application [32]. The synergy of host oxide matrix and RE dopant in improved electrochemical efficacy, makes RE-doped metal oxides as favorable material for developing electrochemical sensor applications.

2.4 Enhancement of Defect Chemistry and Catalytic Sites

The electrochemical performance of sensing materials has also been influenced by defect engineering. The defects, including lattice imperfections, interstitial sites, and oxygen vacancies, can improvise the analyte interactions, catalytic, and charge transfer properties of metal oxide-based sensors. The strategic introduction of defects through the RE dopants can be done to significantly improvise the efficacy of the sensors. Incorporation of RE ions having a larger ionic radius and differing valency with the metal oxide disturbs the local atomic structure of the lattice, resulting in lattice strain. This ionic radius and valency mismatch creates additional oxygen vacancies and other structural defects, which are beneficial in improving the electrochemical efficacy of materials utilized in sensor fabrication. Oxygen vacancies serve as catalytic sites for the occurrence of redox reactions and also act as electron donors, thereby increasing electrical conductivity and enhancing charge transfer between the sensing electrode and the analyte [33]. Moreover, the induced lattice strain and alteration of coordination chemistry around the RE dopant site can enhance surface reactivity, allowing better adsorption of target molecules and stronger catalytic interactions. This leads to increased sensitivity and lower detection limits, even in the presence of interfering species.

The chemical stability of the metal oxides is enhanced by introducing RE dopants, so that the fabricated sensor is working under variable environmental circumstances like pH, humidity, and temperature [34]. All these characteristics are required for practical applications, including industrial protection, medical diagnostics, and environmental monitoring. Researchers can regulate the defect chemistry through the dopant material and its concentration to adjust the electrocatalytic characteristics of the sensing material utilized for different applications. Therefore, the doping strategy of metal oxides with RE dopants opens the research in a broad spectrum to improve the overall efficacy of electrochemical sensors via defect engineering.

3 Applications of Rare-Earth-Doped Metal Oxides in Electrochemical Sensors

Metal oxides are efficacious in anchoring electrocatalysts that enhance the selectivity and sensitivity of sensors employed for environmental pollutant detection [35]. For example, electrodes made of Fe_3O_4 have benefits, including nanoscale dimensions, biocompatibility, superparamagnetic behavior, excellent oxygen ion mobility, less toxicity, and strong catalytic activity in oxidation reactions that enhance sensitivity and stability in pesticide detection [36, 37]. Semiconducting metal oxides like TiO_2, ZnO, and SnO_2 demonstrate notable responsiveness to a wide range of pesticides, making them valuable materials in sensor development [38–40]. Rare-earth-based electrochemical sensors have emerged as highly promising analytical tools due to their improved sensitivity and operational stability. These sensors offer multiple benefits, including rapid response, portability, user-friendly operation, high selectivity, cost-effectiveness, minimal reagent use, low detection limits, and enhanced accuracy [41–43]. The lanthanide series, which involves the filling of the 4f electron subshell, enables a wide range of electron transitions [44]. As a result, lanthanide elements exhibit diverse valence states, with the +3 oxidation state being the most stable and commonly observed.

3.1 Gas Sensor

The detection of trace levels of benzene derivatives is crucial for environmental monitoring and protection. In this context, a high-performance gas sensor was developed using gadolinium (Gd)-doped mesoporous tungsten trioxide (WO_3), engineered through heteroatomic doping [45]. The resulting material featured uniform mesopores (15.7–18.1 nm), a tunable surface area (52–55 m^2/g), and well-defined crystalline walls. Gd doping introduced a high density of oxygen vacancies and W^{5+} defect states within the WO_3 lattice, significantly enhancing the sensor's responsiveness to ethylbenzene. The Gd-doped WO_3 sensor demonstrated a strong signal at low concentrations (response of 237 at 50 ppm), fast response and recovery times (13 s/25 s), and a remarkably low theoretical detection limit of 24 ppb.

Praseodymium (Pr) was effectively incorporated into In_2O_3 nanomaterials using a simple co-precipitation method followed by hydrothermal treatment. This doping process led to a reduction in particle size and an increase in the concentration of active surface oxygen species. As a result, the Pr-doped In_2O_3 sensors exhibited significantly improved sensitivity toward ethanol gas within the concentration range of 112.40–50 ppm [46].

Hierarchical microstructures of indium oxide (In_2O_3) co-doped with lanthanum (La) and cerium (Ce) were synthesized via a hydrothermal method. The concentrations of La and Ce were systematically varied to optimize gas sensing performance. Results demonstrated that dual doping notably enhanced the sensitivity of In_2O_3 toward n-butanol detection. Among the samples, $La_1Ce_{4.5}$-In_2O_3 exhibited the highest response (143) to 100 ppm n-butanol at 280 °C—representing an 8.0-, 4.3-, and 1.4-fold increase compared to pure In_2O_3, La_1-In_2O_3, and $Ce_{4.5}$-In_2O_3, respectively [47]. This enhanced performance is primarily attributed to improved band structure, elevated charge carrier concentration, reduced charge transfer resistance, and intensified surface redox activity.

A praseodymium-doped SnO_2 sensor, developed via a solvothermal synthesis route, demonstrated excellent SO_2 sensing performance [48]. It showed a strong response (19.5–50 ppm), high selectivity, rapid response and recovery times, and remarkable long-term stability, making it highly effective for sulfur dioxide detection.

A highly efficient electrochemical sensor was developed using a (Dy, W) co-doped Bi_2O_3 system integrated with Ln_2NiMnO_6 double perovskite oxides as electrode materials for the selective detection of nitric oxide (NO) [49]. This sensing platform operates effectively over a broad temperature range (325–500 °C), offering high sensitivity (50 mV/decade), rapid response time (< 60 s), and a low detection limit of 200 ppb. The enhanced electrocatalytic activity is attributed to the presence of rare-earth cations, particularly the Ni^{3+}–O–Mn^{3+} interactions and the favorable eg^1 electron configuration in La_2NiMnO_6, which contribute significantly to the sensor's redox behavior and overall performance.

Cerium-doped cobalt oxide (Ce–Co_3O_4) microflowers were synthesized via a straightforward hydrothermal method. Among the tested compositions (1, 3, and 5 at.% Ce), the sensor containing 1 at.% Ce showed the highest sensitivity toward NO_2 gas, with a notable 35% response at a low operating temperature of 50 °C and a detection limit as low as 4500 ppb [50]. When cobalt oxide (Co_3O_4), a p-type semiconductor, is employed as a gas-sensing material, surface redox reactions with target gases result in alterations to the carrier concentration, thereby affecting the width of the depletion layer and ultimately changing the material's resistance. Under operational conditions, atmospheric oxygen adsorbs onto the Co_3O_4 surface, extracting electrons from its conduction band, which leads to the formation of ionized oxygen species and the establishment of a surface depletion region, as depicted in the corresponding band diagram (Fig. 12.1a). Upon exposure to nitrogen dioxide (NO_2), a strong electron-withdrawing gas, further electron extraction occurs due to NO_2's high electron affinity, enhancing the sensor's response. Doping Co_3O_4 with cerium (Ce^{3+}) increases carrier concentration and oxygen vacancy density, which contributes to improved sensing performance. This enhancement is likely associated with the Ce ion's ability to alternate between Ce^{3+} and Ce^{4+} oxidation states, promoting charge transfer and active surface reactions. Morphologically, the 1 at.% Ce-doped Co_3O_4 displays a distinct crumpled sheet structure forming microflower-like assemblies (Fig. 12.1b inset), which provides a high density of reactive edge sites conducive to gas adsorption and sensor performance enhancement. The sensor demonstrated excellent repeatability across five cycles and long-term stability (99% over one month). Its low power requirement and effective performance at near-room temperature make it suitable for integration with IoT-enabled gas sensing platforms.

Defects in nanomaterials greatly affect their electrical, optical, and mechanical properties, making them valuable for applications in healthcare, energy, and environmental monitoring [51, 52]. In aquaculture, ammonia—produced by fish metabolism

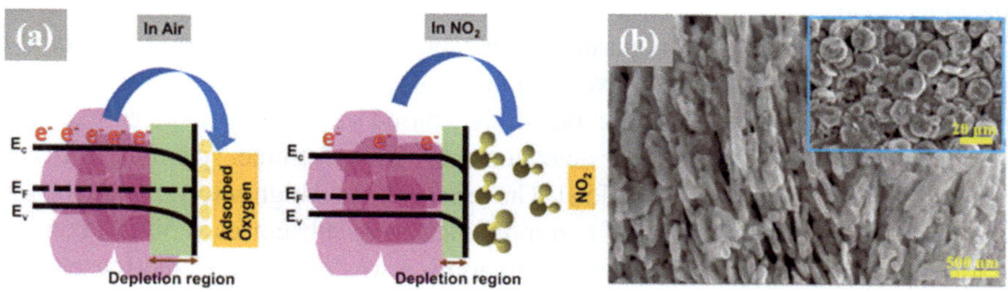

Fig. 12.1 **a** Illustrative band structure diagram depicting the NO_2 sensing mechanism of the 1 at.% cerium-doped sensor, **b** field emission scanning electron microscopic image of Co_3O_4 doped with 1 at.% cerium [50]

and organic waste—can become toxic, threatening aquatic ecosystems and the livelihoods of millions who rely on fishing. To detect such pollutants, researchers are developing sensitive electrochemical sensors using nanostructured materials like SnO_2. This material is widely studied due to its thermal stability, biocompatibility, and conductivity [48]. Doping SnO_2 with samarium, a rare earth element, further enhances its sensing capabilities by increasing oxygen vacancies and ionic mobility in the lattice [53, 54]. Sm-doped SnO_2 nanostructures were synthesized by N. Rani and her co-workers using a co-precipitation method to detect aqueous ammonia [55]. Undoped and samarium-doped SnO_2 nanostructures were synthesized using the co-precipitation method, as illustrated in Fig. 12.2a. The experimental configuration employed for recording the I–V characteristics is depicted in Fig. 12.2b, these measurements were conducted to compare the performance of undoped and Sm-doped samples in ammonia-containing water. Both materials exhibited an increase in current upon exposure to ammonia, with Sm doping significantly enhancing the sensor response. The sensing performance improved with higher Sm concentrations (from 3 to 9%) and increasing ammonia levels (100–500 ppm).

The study of K. Gayathri and her co-workers depicted the improved ammonia (NH_3) gas sensing performance of samarium-doped tin oxide (SnO_2) thin films, synthesized for the first time using a cost-effective automated nebulizer spray pyrolysis technique [56]. Films with different Sm doping levels (0, 1, 3, and 5 wt%) were evaluated, with the 3 wt% doped film showing the best sensitivity, selectivity, and fastest response (14 s) and recovery (28 s) times. The performance enhancement is linked to increased oxygen vacancies, altered electronic properties, and optimized surface morphology due to Sm incorporation.

3.2 Bio Sensors

A modified screen-printed electrode (SPE) incorporating a neodymium-doped Co_3O_4 nanoflower-based nanocomposite was investigated for its dual sensing capability of theophylline and linezolid [57]. The electrode facilitated the oxidation of linezolid at pH 3.0, showing a 155 mV shift toward a less positive potential compared to the bare electrode.

La-doped $NiCoFe_2O_4$ nanoparticles were successfully synthesized by a straightforward hydrothermal method involving controlled adjustment of the metal atomic ratios. The resulting $La@NiCoFe_2O_4@Ti_3C_2$-based sensor demonstrated excellent reproducibility and a broad linear detection range for creatinine, spanning from 0 to 1000 μM as depicted in Fig. 12.3 [58]. The sensor showed high selectivity at 200 μM creatinine even in the presence of common interfering substances at 50 μM concentrations. Notably, only a slight response variation was observed in the presence of urea, confirming minimal interference and highlighting the sensor's reliability for selective creatinine detection.

Ronidazole, a commonly used veterinary antibiotic, is considered potentially carcinogenic and poses environmental and health risks due to its persistence and

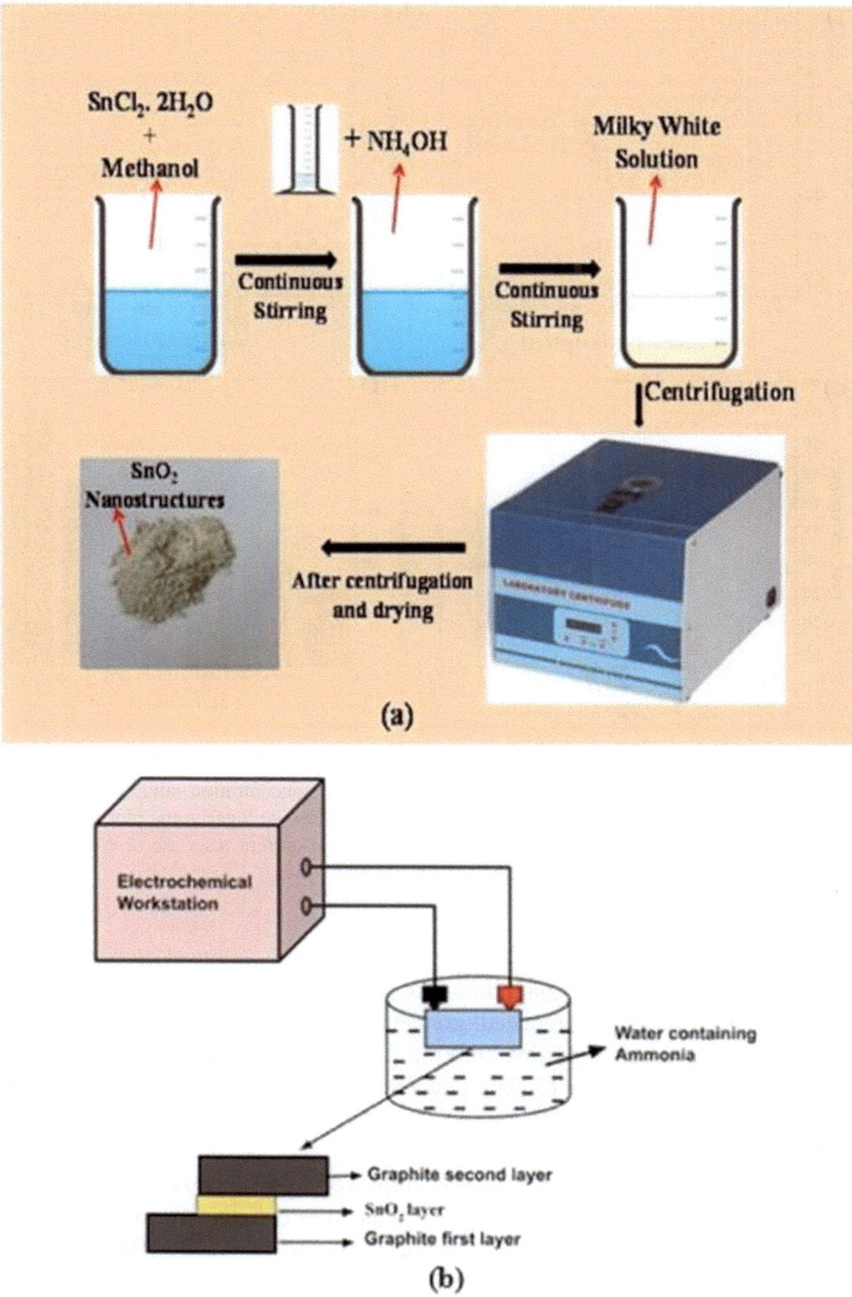

Fig. 12.2 **a** Schematic illustration of the synthesis process of SnO_2 nanostructures via the co-precipitation technique, **b** experimental arrangement utilized for measuring the I–V sensing characteristics [55]

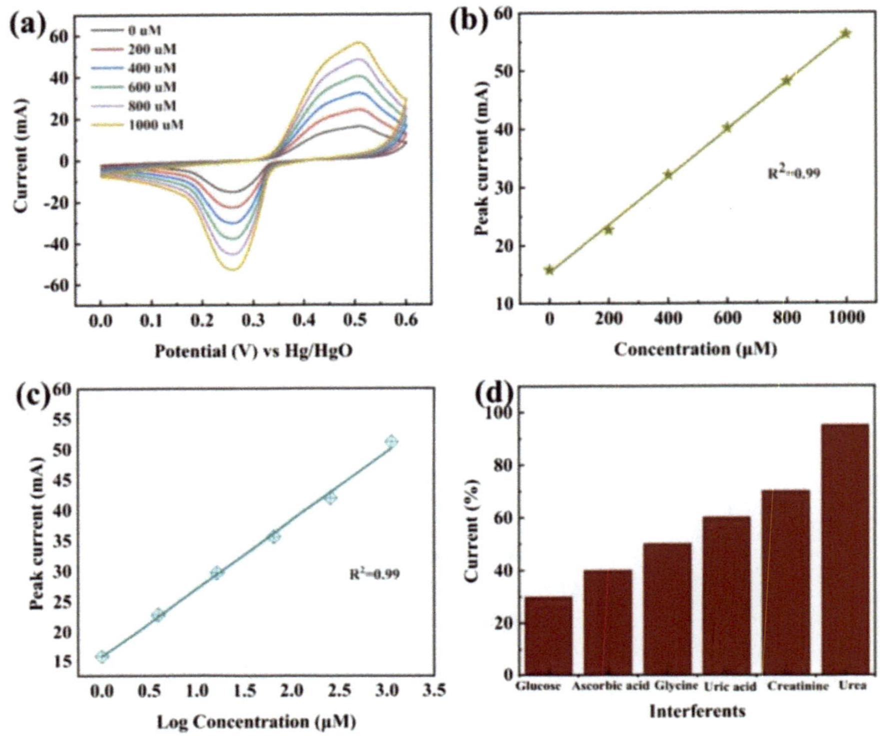

Fig. 12.3 Electrochemical performance of the La@$NiCoFe_2O_4$@Ti_3C_2-based sensor for creatinine detection: **a** Cyclic voltammograms recorded across a concentration range of 0–1000 μM creatinine, **b** plot of peak current versus creatinine concentration, **c** logarithmic plot of creatinine concentration against peak current response, (d) comparative current response of creatinine and common interfering species, demonstrating sensor selectivity [58]

contamination of water and food chains. To address this, a novel electrochemical sensor was developed using gadolinium-doped zinc oxide (Gd@ZnO) nanorods embedded in sulfur-doped reduced graphene oxide (S-rGO) [59]. The sensor witnessed low limits of detection (0.0071 μM and 0.0023 μM) and high sensitivity. The sensor also showed strong selectivity and was effectively applied to real biological and environmental samples, achieving recovery rates between 97 and 101%.

A simple hydrothermal approach was employed to synthesize rare-earth vanadates, TVO_4 (T=Ho, Y, Dy), which were evaluated as effective electrocatalysts for the simultaneous detection of nitrofurazone (NF) and roxarsone (RX) [60]. The substitution of different rare earths at the T-site of the vanadates has impacted the morphology and electrochemical behavior. Among various rare-earth dopants used in TVO_4, $DyVO_4$ demonstrated faster charge transfer capabilities with enhanced electrical conductivity. $DyVO_4$-based electrodes exhibited improved selectivity and sensitivity for detecting NF and RX across a wide range of concentrations (NF: 0.01–264 μM; RX: 0.01–21 μM and 36–264 μM). Even the detection limits were very low, at 0.002 μM for NF and 0.0009 μM for RX.

The hydrothermally synthesized nanoparticles of cerium niobium oxide (CNO) were investigated for their potential in electrochemical sensing. The main objective of this study was to optimize the pH and scan rate for the electrochemical oxidation of 4-nitrophenol. The sensor had a strong linear response over a concentration range of 0.01 to 192.01 μM, with a detection limit of 0.042 μM, which is very good. When it came to detecting 4-nitrophenol, a glassy carbon electrode modified with CNO nanoparticles showed outstanding repeatability, reproducibility, and long-term stability [61].

3.3 *Pesticide and Environmental Pollutant Detection Sensor*

A benzimidazole-based fungicide named carbendazim (CDM) is extensively employed to preserve crops from pests and diseases. The samarium-based metal (SmM) nanosheets have been fabricated through a sonochemical method that witnessed swift electron transfer and enhanced electrochemical activity. When samarium-based metal nanosheets were cast on a screen-printed carbon electrode (SPCE), this modified electrode exhibited a significant improvement in the performance of detecting CDM. The current response of the modified electrode was nearly double (1.97-fold) in comparison with the unmodified electrode, analyzed by cyclic voltammetry curves. The modified electrode fabricated as a sensor displayed a broad linear detection range (0.01–22 μM), high sensitivity (13.34 $\mu A\ \mu M^{-1}\ cm^{-2}$), with an impressive limit of detection of 1.0 nM [62].

The hydrothermally synthesized porous neodymium molybdate nanostructures (pf-NdM NSs) demonstrated superior electrocatalytic performance for the reduction of methyl parathion (MP) because of their high surface area and porous nature [63]. The sensor made up of pf-NdM NSs showed a low detection limit of 5.7 nM, a broad linear detection range from 0.5 to 300 μM, and high sensitivity (1.88 $\mu A\ \mu M^{-1}\ cm^{-2}$), along with good selectivity toward the target analyte.

A nanocomposite of neodymium molybdate ($Nd_2Mo_3O_{12}$) incorporated with reduced graphene oxide (RGO) was synthesized utilizing co-precipitation method as an electrochemical sensor for detecting metol (ML) [64]. The electrical conductivity of the metal molybdate was increased with the addition of RGO. The NdMO/RGO-modified sensor demonstrated better sensitivity for ML detection, because of its large electroactive surface area, efficient electron transfer, and superior electrocatalytic properties, achieving a low detection limit of 0.005 μM over a wide linear range (0.01–1770 μM).

3.4 Humidity Sensor

The performance and durability of gas sensors composed of metal–oxide–semiconductor (MOS) materials are adversely affected by air humidity. To overcome this limitation, a humidity-resistant design was developed by encapsulating rare-earth-doped tin oxide nanofibers with a mesoporous silica molecular sieve layer (SBA-15). The hydrophobic SBA-15 coating acts as a barrier against water vapor while allowing efficient hydrogen diffusion, and the incorporation of rare earth elements enhances sensor responsiveness and reduces the required operating temperature. Specifically, Er-SnO_2/SBA-15 and Tb-SnO_2/SBA-15 sensors demonstrated high sensitivity to 10 ppm hydrogen at 280 °C, with response values of 27.71 and 33.68—significantly higher than undoped SnO_2. These sensors also exhibited rapid response and recovery times (< 1 s), a low detection limit (200 ppb), and excellent selectivity. Notably, the presence of the SBA-15 layer improved signal retention across varying humidity levels (from 25 to 85% RH), increasing from 38.8% to 60.0% for Er-doped and from 25.6% to 57.8% for Tb-doped configurations [65].

Highly porous green pigments of $CoCr_2O_4$ doped with rare earth elements (Ce, Gd, Ho, and Sm) were synthesized through a rapid and cost-effective solution combustion method. The humidity sensing behavior of these materials was evaluated by measuring their electrical resistance across a wide range of relative humidity (0–98%) and frequencies (20 Hz to 20 MHz) at room temperature. A consistent decrease in resistance with increasing humidity was observed. Among the doped samples, Sm-doped $CoCr_2O_4$ exhibited the most pronounced response to humidity variations [66]. Overall, the rare earth-doped pigments demonstrated a linear response, high sensitivity, and minimal hysteresis, highlighting their potential for effective humidity sensing applications.

4 Conclusion

The precise tailoring of material properties at the nanoscale is becoming more and more crucial for making high-performance electrochemical sensors. Due to structural stability, superior redox activity, and tunable electronic properties, metal oxides showed better sensor performance. But their practical efficacy is limited owing to their less conductive behavior and semiconductive nature. The introduction of rare earth elements with metal oxides not only resolves this issue but also enhances the sensor performance. The rare-earth dopants enhanced the catalytic behavior, analyte interactions, and charge transfer properties by controlling oxygen vacancies, lattice defects, and additional active sites. The chapter discusses improved sensitivity, long-term stability and selectivity via these defects in a variety of sensor applications like gas sensors, bio sensor, humidity sensor, and pesticide and environmental pollution detection sensor. The integration of defect engineering, nanotechnology, and ideal doping opens up plenty of new possibilities for the field of electrochemical sensing.

Future research should focus on finding new dopant materials, scalable fabrication methods, and ways to connect these technologies with digital platforms so that we can solve new analytical problems in real time, with low power, and at high efficiency.

References

1. Saputra, H.A.: Electrochemical sensors: basic principles, engineering, and state of the art. Mon.Shefte Für Chem.-Chem. Mon. **154**, 1083 (2023). https://doi.org/10.1007/s00706-023-03113-z
2. Goshisht, M.K., Patra, G.K., Mahal, A., Singh, A.K., Shobha, N., Parshad, M.: Recent advances in electrochemical biosensors for the detection of pathogens, diseases biomarkers, and heavy metal ions. Inorg. Chim. Acta **574**, 122403 (2024). https://doi.org/10.1016/j.ica.2024.122403
3. Madadelahi, M., Romero-Soto, F.O., Kumar, R., Tlaxcala, U.B., Madou, M.J.: Electrochemical sensors: types, applications, and the novel impacts of vibration and fluid flow for microfluidic integration. Biosens. Bioelectron. **272**, 117099 (2025). https://doi.org/10.1016/j.bios.2024.117099
4. Zabitler, D., Ülker, E., Turan, K., Erdoğan, N.Ö., Tığ, G.A.: Electrochemical sensor for biological samples monitoring. Top. Catal. (2025).https://doi.org/10.1007/s11244-025-02080-5
5. He, Q., Wang, B., Liang, J., Liu, J., Liang, B., Li, G., Long, Y., Zhang, G., Liu, H.: Research on the construction of portable electrochemical sensors for environmental compounds quality monitoring. Mater. Today Adv. **17**, 100340 (2022). https://doi.org/10.1016/j.mtadv.2022.100340
6. Ferdiana, N.A., Bahti, H.H., Kurnia, D., Wyantuti, S.: Synthesis, characterization, and electrochemical properties of rare earth element nanoparticles and its application in electrochemical nanosensor for the detection of various biomolecules and hazardous compounds: a review. Sens. Bio-Sens. Res. **41**, 100573 (2023). https://doi.org/10.1016/j.sbsr.2023.100573
7. Fergus, J.W.: Sensing mechanism of non-equilibrium solid-electrolyte-based chemical sensors. J. Solid State Electrochem. **15**, 971 (2010). https://doi.org/10.1007/s10008-010-1046-4
8. Baranwal, J., Barse, B., Gatto, G., Broncova, G., Kumar, A.: Electrochemical sensors and their applications: a review. Chemosensors **10**, 363 (2022). https://doi.org/10.3390/chemosensors10090363
9. Zhang, S., Liu, S., Yang, D., Li, L.: High-entropy catalysts for electrocatalytic reduction reaction. J. Alloy. Compd. **1009**, 176970 (2024). https://doi.org/10.1016/j.jallcom.2024.176970
10. Tovar-Lopez, F.J.: Recent progress in micro- and nanotechnology-enabled sensors for biomedical and environmental challenges. Sensors **23**, 5406 (2023). https://doi.org/10.3390/s23125406
11. Jalalvand, A.R., Karami, M.M.: Roles of nanotechnology in electrochemical sensors for medical diagnostic purposes: a review. Sens. Bio-Sens. Res. **47**, 100733 (2025). https://doi.org/10.1016/j.sbsr.2024.100733
12. Ramzan, S., Rahim, A, Ahmad, A., Alsaiari, M.: Metal oxide and their sensing applications. In: Micro and Nano Technologies, Nanomaterials-Based Electrochemical Sensors: Properties, Applications and Recent Advances, pp. 155–176. Elsevier (2024). https://doi.org/10.1016/b978-0-12-822512-7.00013-2.
13. Thiruppathi, M., Thiyagarajan, N., Ho, J.A.A.: Applications of metals, metal oxides, and metal sulfides in electrochemical sensing and biosensing. In: Environmental Chemistry for a Sustainable World, Metal, Metal-Oxides and Metal Sulfides for Batteries, Fuel Cells, Solar Cells, Photocatalysis and Health Sensors, vol. 62, pp. 209–244. Springer (2021).https://doi.org/10.1007/978-3-030-63791-0_7

14. Maheshwaran, S., Chen, W.H., Lin, S.-L., Ghorbani, M., Hoang, A.T.: Metal oxide-based electrochemical sensors for pesticide detection in water and food samples: a review. Environmental Science Advances **3**, 154 (2023). https://doi.org/10.1039/d3va00313b
15. Mashhadian, A., Jian, R., Tian, S., Wu, S., Xiong, G.: An overview of electrochemical sensors based on transition metal carbides and oxides: synthesis and applications. Micromachines **15**, 42 (2023). https://doi.org/10.3390/mi15010042
16. Deng, Y.: Semiconducting metal oxides: composition and sensing performance. In: Semiconducting Metal Oxides for Gas Sensing, pp. 105–147. Springer (2023). https://doi.org/10.1007/978-981-99-2621-3_4
17. Yoon, Y., Truong, P.L., Lee, D., Ko, S.H.: Metal-Oxide nanomaterials synthesis and applications in flexible and wearable sensors. ACS Nanoscience Au **2**, 64 (2021). https://doi.org/10.1021/acsnanoscienceau.1c00029
18. Meenakshi, S., Pandian, K., Gopinath, S.: Quantitative simultaneous determination of pentoxifylline and paracetamol in drug and biological samples at graphene nanoflakes modified electrode. J. Taiwan Inst. Chem. Eng. **107**, 15 (2020). https://doi.org/10.1016/j.jtice.2019.11.011
19. Alhalili, Z.: Metal oxides nanoparticles: general structural description, chemical, physical, and biological synthesis methods, role in pesticides and heavy metal removal through wastewater treatment. Molecules **28**, 3086 (2023). https://doi.org/10.3390/molecules28073086
20. Baig, N., Sajid, M., Saleh, T.A.: Recent trends in nanomaterial-modified electrodes for electroanalytical applications. TrAC, Trends Anal. Chem. **111**, 47 (2019). https://doi.org/10.1016/j.trac.2018.11.044
21. Lee, Y.Y., Sriram, B., Wang, S.F., Kogularasu, S., Chang-Chien, G.P.: A comprehensive review on emerging role of rare earth oxides in electrochemical biosensors. Microchem. J. **193**, 109140 (2023). https://doi.org/10.1016/j.microc.2023.109140
22. Arris, F.A., Benoudjit, A.M., Sanober, F., Salim, W.W.A.W.: Characterization of electrochemical transducers for biosensor applications. In: Multifaceted Protocol in Biotechnology, pp. 119–137. Springer (2018). https://doi.org/10.1007/978-981-13-2257-0_11
23. Tong, L., Wu, L., Su, E., Li, Y., Gu, N.: Recent advances in the application of nanozymes in amperometric sensors: a review. Chemosensors **11**, 233 (2023). https://doi.org/10.3390/chemosensors11040233
24. Fathy, M.A., Bühlmann, P.: Next-generation potentiometric sensors: a review of flexible and wearable technologies. Biosensors **15**, 51 (2025). https://doi.org/10.3390/bios15010051
25. He, S., Yuan, Y., Nag, A., Feng, S., Afsarimanesh, N., Han, T., Mukhopadhyay, S.C., Organ, D.R.: A review on the use of impedimetric sensors for the inspection of food quality. Int. J. Environ. Res. Public Health **17**, 5220 (2020). https://doi.org/10.3390/ijerph17145220
26. Brinda, K.N., Yhobu, Z., Nagaraju, D.H., Budagumpi, S.: Working principle and sensing mechanism of electrochemical sensors. In: 2D Materials-Based Electrochemical Sensors, pp. 9–44. Elsevier (2023). https://doi.org/10.1016/b978-0-443-15293-1.00009-4
27. Chavali, M.S., Nikolova, M.P.: Metal oxide nanoparticles and their applications in nanotechnology. SN Appl. Sci. **1**, 607 (2019). https://doi.org/10.1007/s42452-019-0592-3
28. Agnihotri, A.S., Varghese, A., Nidhin, M.: Transition metal oxides in electrochemical and bio sensing: a state-of-art review. Appl. Surf. Sci. Adv. **4**, 100072 (2021). https://doi.org/10.1016/j.apsadv.2021.100072
29. Ojeery, A.A., Farea, M.A.: Enhancing sensitivity and selectivity in gas sensors: A novel approach using metal oxide-conducting polymer composites. Inorg. Chem. Commun. **172**, 113726 (2024). https://doi.org/10.1016/j.inoche.2024.113726
30. Chen, S., Liu, P., Li, Y., Zhu, M., Dang, Y., Zhou, Y.: Engineering the doping amount of rare earth element erbium in $CdWO_4$: influence on the electrochemical performance and the application to the electrochemical detection of bisphenol A. J. Electroanal. Chem. **904**, 115867 (2021). https://doi.org/10.1016/j.jelechem.2021.115867
31. Rahman, M.H., Rahaman, M.Z., Chowdhury, E.H., Motalab, M., Hossain, A.K.M.A., Roknuzzaman, M.: Understanding the role of rare-earth metal doping on the electronic structure and optical characteristics of ZnO. Mol. Syst. Des. Eng. **7**, 1516 (2022). https://doi.org/10.1039/d2me00093h

32. Sebastian, S., Raj, C.S.A., Diana, P., Ganesh, V., Awwad, N.S., Yahia, I.S., AlAbdulaal, T.H.: Rare earth elements terbium (Tb), erbium (Er), ytterbium (Yb) doped zinc oxide nanoparticles: synthesis structural optical and gas sensing studies. Physica B **694**, 416384 (2024). https://doi.org/10.1016/j.physb.2024.416384
33. Marikutsa, A., Rumyantseva, M., Konstantinova, E.A., Gaskov, A.: The key role of active sites in the development of selective metal oxide sensor materials. Sensors **21**, 2554 (2021). https://doi.org/10.3390/s21072554
34. Hossain, M.K., Ahmed, M.H., Khan, M.I., Miah, M.S., Hossain, S.: Recent progress of rare earth oxides for sensor, detector, and electronic device applications: a review. ACS Appl. Electron. Mater. **3**, 4255 (2021). https://doi.org/10.1021/acsaelm.1c00703
35. George, J.M., Antony, A., Mathew, B.: Metal oxide nanoparticles in electrochemical sensing and biosensing: a review. Microchim. Acta **185**, 358 (2018). https://doi.org/10.1007/s00604-018-2894-3
36. Arumugam, B., Palanisamy, S., Janagaraj, G., Ramaraj, S.K., Chiesa, M., Subramaianan, C.: Real-time detection of 2, 4, 6-trichlorophenol in environmental samples using Fe_3O_4 nanospheres decorated graphitic carbon nitride nanosheets. Surfaces and Interfaces **39**, 102930 (2023). https://doi.org/10.1016/j.surfin.2023.102930
37. Yu, C., Gou, L., Zhou, X., Bao, N., Gu, H.: Chitosan–Fe_3O_4 nanocomposite based electrochemical sensors for the determination of bisphenol A. Electrochim. Acta **56**, 9056 (2011). https://doi.org/10.1016/j.electacta.2011.05.135
38. Daizy, M., Ali, M., Bacchu, M., Aly, M.A.S., Khan, M.: ZnO hollow spheres arrayed molecularly-printed-polymer based selective electrochemical sensor for methyl-parathion pesticide detection. Environ. Technol. Innov. **24**, 101847 (2021). https://doi.org/10.1016/j.eti.2021.101847
39. Li, R., Zhe, T., Bai, F., Xu, Z., Li, M., Bu, T., Li, F., Fang, H., Wang, L., Lü, X.: Hierarchical SnO_2 nanoparticles designed based on in situ derivatization strategy for rapid and sensitive imidacloprid detection. Microchem. J. **190**, 108691 (2023). https://doi.org/10.1016/j.microc.2023.108691
40. Arham, Z., Ramli, A.K., Natsir, M., Nurdin, M.: Photoelectrode properties based on TiO_2 photocatalyst doped non-metallic agent as initial model for detection and degradation of pesticide pollutants. Biointerface Res. Appl. Chem. **13**, 563 (2023). https://doi.org/10.33263/briac136.563
41. Gao, W., Wen, D., Ho, J.C., Qu, Y.: Incorporation of rare earth elements with transition metal–based materials for electrocatalysis: a review for recent progress. Mater. Today Chem. **12**, 266 (2019). https://doi.org/10.1016/j.mtchem.2019.02.002
42. Trinadh, T., Khuntia, H., Anusha, T., Bhavani, K., Kumar, J., Brahman, P.: Synthesis and characterization of nanocomposite material based on graphene quantum dots and lanthanum doped zirconia nanoparticles: an electrochemical sensing application towards flutamide in urine samples. Diam. Relat. Mater. **110**, 108143 (2020). https://doi.org/10.1016/j.diamond.2020.108143
43. Monsef, R., Salavati-Niasari, M.: Electrochemical sensor based on a chitosan- molybdenum vanadate nanocomposite for detection of hydroxychloroquine in biological samples. J. Colloid Interface Sci. **613**, 1 (2022). https://doi.org/10.1016/j.jcis.2022.01.039
44. Tagaya, M., Ikoma, T., Yoshioka, T., Motozuka, S., Xu, Z., Minami, F., Tanaka, J.: Synthesis and luminescence properties of Eu(III)-doped nanoporous silica spheres. J. Colloid Interface Sci. **363**, 456 (2011). https://doi.org/10.1016/j.jcis.2011.07.066
45. Li, Y., Chen, K., Liu, Y., Ma, J., Liao, Y., Yang, H., Cheng, J., Yue, Q., Yuan, K., Ren, Y., Zou, Y., Deng, Y.: Gadolinium-doped mesoporous tungsten oxides: rational synthesis, gas sensing performance, and mechanism investigation. Nano Res. **16**, 7527 (2022). https://doi.org/10.1007/s12274-022-5274-6
46. An, D., Wang, Y., Dai, J., Guo, L., Liu, N.: Synthesis of Pr doped In_2O_3 nanoparticles and their enhanced ethanol gas-sensing performance. Ceram. Int. **50**, 4945 (2023). https://doi.org/10.1016/j.ceramint.2023.11.237

47. Bai, Y., Fu, H., Ran, X., Chen, Y., Wang, N., Yang, X., Xiong, S., Li, S., An, X.: La, Ce co-doped In_2O_3 hierarchical microstructure with high sensing performance towards n-butanol. Sens. Actuators, B Chem. **397**, 134642 (2023). https://doi.org/10.1016/j.snb.2023.134642
48. Xu, H., Li, J., Li, P., Shi, J., Gao, X.: Effect of rare earth doping on electronic and gas-sensing properties of SnO_2 nanostructures. J. Alloy. Compd. **909**, 164687 (2022). https://doi.org/10.1016/j.jallcom.2022.164687
49. Bhardwaj, A., Bae, H., Mathur, L., Mathur, S., Song, S.J.: Cubic Bi_2O_3-based electrochemical nitric oxide sensor using double perovskite oxide electrodes. J. Electrochem. Soc. **169**, 117510 (2022). https://doi.org/10.1149/1945-7111/aca2e0
50. Murugesh, P., Udaiyar, P.S., Mani, N.: Design of Ce incorporated tricobalt tetroxide Co_3O_4 sensor with boosted gas sensing performance toward nitrogen dioxide. J. Mater. Sci. Mater. Electron. **35** (2024). https://doi.org/10.1007/s10854-024-12681-7
51. Wu, G., Sun, J., Zhang, Z., Guo, D., Liu, J., Liu, L.: Recent advances in biological applications of nanomaterials through defect engineering. Sci. Total. Environ. **816**, 151647 (2021). https://doi.org/10.1016/j.scitotenv.2021.151647
52. Zhou, M., Liu, Y., Su, Y., Su, Q.: Plasmonic oxygen defects in MO_{3-x} ($M = W$ or Mo) nanomaterials: synthesis, modifications, and biomedical applications. Adv. Healthcare Mater. **10**, 2101331 (2021). https://doi.org/10.1002/adhm.202101331
53. Ran, Y., Li, Y., Cui, X., Lai, T., Yao, L., Zhao, R., Wang, L., Wang, Y.: Sm-doped SnO_2 nanoparticles synthesized via solvothermal method as a high-performance formaldehyde sensing material for gas sensors. J. Mater. Sci.: Mater. Electron. **32**, 8249 (2021). https://doi.org/10.1007/s10854-020-05216-3
54. Guan, Y., Wang, D., Zhou, X., Sun, P., Wang, H., Ma, J., Lu, G.: Hydrothermal preparation and gas sensing properties of Zn-doped SnO_2 hierarchical architectures. Sens. Actuators, B Chem. **191**, 45 (2013). https://doi.org/10.1016/j.snb.2013.09.002
55. Rani, N., Kumar, V., Kumar, S., Yadav, S., Babu, S.: Sm-doped SnO_2 nanostructures for aqueous ammonia sensing application. J. Mater. Sci.: Mater. Electron. **35**, 1961 (2024). https://doi.org/10.1007/s10854-024-13721-y
56. Gayathri, K., Ravichandran, K., Sridharan, M., Shonali, N.G., Mohan, R., Ravikumar, P.: Influence of Sm incorporation into the SnO_2 lattice for enhanced NH_3 gas sensing ability. Ceram. Int. **51**, 26681 (2025). https://doi.org/10.1016/j.ceramint.2025.03.350
57. Saadh, M.J., Kaur, M., Yadav, A., Ibrahim, S.M., Mahdi, M.S., Hussein, S.A., Hamoody, A.M., Shakir, M.N., Alasheqi, M.Q., Ahmad, I., Abosaoda, M.K.: Neodymium doped nanosheet-assembled hierarchical Co_3O_4 nanoflowers modified electrode as an electrochemical sensor for the voltammetric determination of linezolid in the presence of theophylline. Microchem. J. **212**, 113311 (2025). https://doi.org/10.1016/j.microc.2025.113311
58. Waris, M.H., Muzaffar, N., Mumtaz, M.A., Afzal, A.M., Iqbal, M.W., Mumtaz, S., Ali, M., Alaraidh, I.A., Okla, M.K., Munnaf, S.A.: High performance lanthanum-doped nickel cobalt ferrites on titanium carbide MXene electrode material for superior hybrid device and precision creatinine sensing. Appl. Phys. A **131**, 220 (2025). https://doi.org/10.1007/s00339-025-08306-3
59. Selvam, S.P., Alagarsamy, S., Chen, T., Chen, S., Ramachandran, R., Al-Mohaimeed, A.M., Ali, M.A., Yu, J., Elshikh, M.S.: Gadolinium doped ZnO nanorods on chalcogenide-modified graphene for enhanced voltammetric detection of methyl carbamate. Compos. B Eng. **291**, 112061 (2025). https://doi.org/10.1016/j.compositesb.2024.112061
60. Sriram, B., Baby, J.N., Hsu, Y.F., Wang, S.F., George, M.: Scheelite-type rare earth vanadates TVO_4 ($T = $ Ho, Y, Dy) electrocatalysts: Investigation and comparison of T site variations towards bifunctional electrochemical sensing application. Chem. Eng. J. **451**, 138694 (2022). https://doi.org/10.1016/j.cej.2022.138694
61. Manjula, N., Pulikkutty, S., Chen, T.W., Chen, S.M., Fan, C.H., Ali, M.A., Al-Hemaid, F.M.: Electrochemical sensor based on cerium niobium oxide nanoparticles modified electrode for sensing of environmental toxicity in water samples. Colloids Surf., A **637**, 128277 (2022). https://doi.org/10.1016/j.colsurfa.2022.128277

62. Kokulnathan, T., Kumar, E.A., Wang, T.J.: Synthesis of two-dimensional nanosheet like samarium molybdate with abundant active sites: real-time carbendazimin analysis in environmental samples. Microchem. J. **158**, 105227 (2020). https://doi.org/10.1016/j.microc.2020.105227
63. Ganesan, M., Devi, R.K., Liao, A.H., Lee, K.Y., Gopalakrishnan, G., Chuang, H.C.: 3D-flower-like porous neodymium molybdate nanostructure for trace level detection of organophosphorus pesticide in food samples. Food Chem. **396**, 133722 (2022). https://doi.org/10.1016/j.foodchem.2022.133722
64. Shanlee, S.S.R., Sundaresan, R., Chen, S.M., Balaji, R., Jeyapragasam, T., Peng, J.Y., Jothi, A.I.: High-performance electrochemical sensor based on neodymium molybdate/reduced graphene oxide ($Nd_2Mo_3O_{12}$/RGO) for rapid detection of carcinogenic organic pollutants in water samples. Surfaces Interfaces **40**, 103020 (2023). https://doi.org/10.1016/j.surfin.2023.103020
65. Yang, X.Y., Pan, J.L., Wang, S.S., Chen, D., Wang, Y.C., Wang, S.K., Cairang, L., Zhang, Z.X., Sun, G.Z., Pan, X.J., Zhou, J.Y.: Humidity-resistant hydrogen sensors based on rare-earth-doped tin dioxide nanofibers with hydrophobic mesoporous silica sieve encapsulation. Sens. Actuators, B Chem. **412**, 135770 (2024). https://doi.org/10.1016/j.snb.2024.135770
66. Angadi, V.J., Tudorache, F., Kumar, A., Manjunatha, S.O., Prakash, C., Wang, S., Sehgal, S.S., Ubaidullah, M., Ayachit, N.H.: Fabrication of rare earth(R= Ce, Gd, Ho and Sm) doped $CoCr_{2-x}R_xO_4$ pigments for capacitive and resistive humidity sensor applications. Ceram. Int. **49**, 38691 (2023). https://doi.org/10.1016/j.ceramint.2023.09.204

Recent Progress and Future Outlook of REE-Doped Metal Oxides

Komaljeet Kour, Nupur Aggarwal, Naveen Kumar, and Anu Kapoor

1 Introduction

Rare-earth metal oxide nanostructures have emerged as a new innovative material, offering exceptional and highly efficient performance meeting the growing world demand for materials in various technological applications [1]. Due to the increasing global demand of these high performance materials in various technological applications, the current research is concentrated on producing these highly efficient and cost-effective materials by optimizing the key factors related to these nanomaterials, including functional diversity, structural adaptability, and integration capabilities [2, 3]. Recent research has unveiled the extent of co-doping, which facilitates the simultaneous incorporation of several dopants into the metal oxide matrix such that the fine tailoring of various electrical, optical, and catalytic properties of these rare earth-doped metal oxides is possible [4]. Optimization of some other different parameters such as shape, size, temperature, and some other parameters, we can regulate different properties such as increase the surface area, conductivity, and adsorption properties, turning these rare earth-doped metal oxides an ideal candidate in different applications such as catalysis, sensing, and energy storage. Nanoscale engineering makes new avenues to apply these materials in a variety of technologies. Recent studies are also leaning toward the integration of these materials into the flexible electronic devices, expanding their applications in the future-generation electronic and optoelectronic devices [5]. Further, the use of sophisticated characterization methods such as electron microscopy and X-ray diffraction gives an important insight into their structural and electronic characteristics, allowing researchers to maximize their performance. In this chapter, we will discuss the advances in the

K. Kour · N. Aggarwal · A. Kapoor
Department of Physics, UIS, Chandigarh University, Mohali 140413, Punjab, India

N. Kumar (✉)
Department of Physics, Galgotias University, Greater Noida 203201, Uttar Pradesh, India
e-mail: naveensethi99@gmail.com

© The Author(s), under exclusive license to Springer Nature Singapore Pte Ltd. 2026
N. Kumar et al. (eds.), *Rare Earth-Doped Metal Oxide Nanostructures*, Engineering Materials, https://doi.org/10.1007/978-981-95-6192-6_13

use of rare earth-doped metal oxides, focusing on themes like co-doping techniques, nano-architectural engineering, device integration, and advanced characterization methods. These aspects have synergistically pushed the performance envelope and broadened the areas of applicability of rare earth-doped systems [6]. Table 13.1 provides a detailed summary of novel developments in rare earth-doped metal oxides. In this chapter, we aim to showcase how recent research is significantly expanding the capabilities and applications of rare earth-doped metal oxides in a diverse range of fields.

2 Enhanced Functionalities Through Co-doping Strategies

Co-doping, whereby two or more dopants are introduced simultaneously to a host metal oxide matrix, has become an effective approach toward tailoring the properties of material. Co-doping can be used to tune the electrical, optical, structural, and magnetic properties of a material. In doped rare earth systems, co-doping by other rare earth ions (for example, Eu^{3+}/Gd^{3+}) or by transition metals (for example, Fe^{3+}, Mn^{2+}) can also induce synergies that amplify optical, magnetic, or electronic conductivities. For example, co-doping ZnO with Gd^{3+} and Mn^{2+} can enhance room-temperature ferromagnetism and photocatalytic performance. Likewise, the co-doping of Ce^{4+} with other rare earth ions can enable mixed valence states, enhancing redox activity and oxygen vacancy creation key factors in catalysis and sensing [12]. Co-doping also has the ability to circumvent typical setbacks like concentration quenching or lattice distortion, resulting in more stable and efficient materials.

2.1 Nano-Architectures and Surface Engineering

Nano architecture is a new evolving fundamental concept of nano architecture, in which new materials are produced using interdisciplinary research. In this technique, the production of functional materials is done via using principles of nanotechnology with a fusion of other disciplines under controlled environment [13]. In this growing world there is a continuous need of such materials which fulfill the social needs including energy storage and conversion, food packaging, environmentally friendly, biomedical and integrated good production [14]. Nanofabrication has made it possible to have tight control over the architecture and surface chemistry of doped metal oxides. Rational nano-architectures like core–shell nanostructures, hollow nano-spheres, hierarchical structures, and mesoporous networks provide increased surface area, light absorption, and tunable charge transport routes. Surface modification with the incorporation of functional groups, passivation layers, or interface modification enhances material stability, charge separation, and environmental compatibility. In rare earth-doped systems, the engineering of the nanoscale interface is the key for maximizing emission efficiency, magnetic coupling, and catalytic

Table 13.1 Emerging trends and functional enhancements in rare earth-doped metal oxide systems

Advancement area	Description	Key examples	Statistical highlights	Impacts and benefits	Reference
Co-doping strategies	Simultaneous doping with multiple elements to synergistically enhance properties	ZnO co-doped with Eu and Tb	100% MB was degraded in just 15 min using EU and Tb co-doping	High-performing catalysts used in environmental applications	[7]
		ZnO co-doped with Gd and La	The maximum degradation efficiency with Gd/La@ZnO nanoflowers was 91% for Rh B and 74% for TCN	High catalytic efficiency under solar-light irradiation	[8]
Nanoarchitectures and surface engineering	Design of advanced nanostructures and surface modifications	Rare earth (RE)-doped Co_3O_4 nanostructures with abundant oxygen vacancies	Eu-doped Co_3O_4 NPs showed the specific capacitance of 1021.3 F g^{-1} at 2 Ag^{-1}	Different dopants can be used to change the structure from large nanosheets to tiny nanosheets used for high-efficiency energy storage	[9]
Flexible and wearable device integration	Incorporation into soft, stretchable platforms	Deposition of $Y_2O_3:Eu^{3+}$ thin film onto quartz fabric substrate with EBE	Improved photoluminescence by 124.73% with 15 min of deposition	Viable for smart textiles Having applications in security warnings, anti-counterfeiting and decoration purposes	[10]
In-situ and operando characterization	Real-time monitoring of structural, optical, and electronic changes	Ce-doped TiO_2 studied via XANES and EXAFS	Real-time observation of $Ce^{3+} \rightarrow Ce^{4+}$ transitions with UV-light irradiation	Insight into mechanical and optical properties Enables predictive tuning of performance	[11]

selectivity. Such design principles are highly effective in energy-related applications like batteries, supercapacitors, and solar-driven processes. Munawar et al. used the rare earth elements to develop a high porous nanostructure by dual-doping them with TiO_2 and found that supercapacitors based on Ce/Ho co-doped TiO_2 retained 99.28% of their capacity after 5000 cycles. Thus porous co-doped TiO_2 can be a potential candidate in electrochemical applications [15]. The Nd_2O_3-MoS_2 composite synthesized by Rosaiah et al. showed a higher catalytic and energy storing performance. It also exhibited hydrogen evolution reactions making it ideal for hydrogen production due to improved active site [16].

2.2 Integration into Flexible and Wearable Devices

With the fast development of wearable electronics and flexible sensors, introducing rare earth-doped metal oxides to flexible platforms has drawn significant interest. Their potential to exhibit special luminescence, magnetism, and sensing properties makes them appealing for real-time health monitoring, environment sensing, and energy harvesting devices. New research has shown the effective introduction of doped oxides to flexible substrates like polymers, fibers, and biodegradable films. Methods such as inkjet printing, electrospinning, and solution casting are being modified to allow low-temperature and scalable processing, which is important for preserving substrate integrity [17]. This transition to flexible integration not only increases the applicability of rare earth-doped materials but also benefits from trends toward soft and sustainable electronics. Dutta et al. successfully synthesized the erbium-doped ZnO nanoparticles and integrated it in the PVDF membranes to form EZ-PVDF nanocomposite films which are a potential candidate for the application of piezoelectric nanogenerators and HMI applications [18]. Fan et al. successfully synthesized a flexible piezoelectric nanogenerator by co-doping rare earth element ceramics embedded in the PVDF and PDMS composite film which generated an enhanced output and showed a potential application in biomedical applications [19]. Wang et al. successfully synthesized a fluorescence-recovered wearable hydrogel patch using RENPs of $NaGdF_4$ doped with Yb, Er, and Ce for monitoring the composition of sweat [20].

2.3 Real-Time Monitoring of Materials

Advanced characterization techniques have greatly improved our knowledge of rare earth-doped metal oxides, especially under actual operating conditions. Operando and in-situ methods, e.g., in-situ X-ray diffraction (XRD), Raman spectroscopy, transmission electron microscopy (TEM), and X-ray absorption spectroscopy (XAS) enable scientists to track changes in structure, chemistry, and electronics during synthesis, reaction, or device operation. Such approaches offer valuable information about dopant behavior, phase changes, defect creation, and catalytic processes. For instance, operando XAS may ascertain oxidation state dynamics of Ce in doped TiO_2 redox cycling, whereas in-situ TEM allows direct observation of nanoparticle growth and transformation. The use of these advanced tools has brought deeper mechanistic insight and allowed rational design of materials with controllable and predictable properties. Nehra et al. synthesized lanthanum-doped tungsten trioxide and characterized the samples using XRD, Raman spectroscopy, FTIR, PL, TEM, EDX, and UV–visible for the optical, structural properties of the material. Kinetic studies were done to check the degradation rate and it was found out that the synthesized La-doped WO3 follows first order kinetics [21]. Saeed et al. used machine learning tools to random tree, linear regression, instance-based K-nearest neighbors, multilayer perceptron, random forest, support vector machine-regression, and additive regression were used to predict the bandgap of CeO_2 and Nd_2O_3 nanoparticles. Mikolajczyk et al. used both experimental and simulation methods to characterize the synthesized RE-TiO_2 [22].

This table (Table 13.2) presents a comparative study of important properties in rare earth-doped metal oxides compared to their undoped equivalents, with respect to variations in electrical conductivity, optical absorption, magnetic properties, and catalytic activity. It illustrates how doping with rare earth can tremendously increase or alter these characteristics for different applications in energy, catalysis, and electronics.

3 Challenges and Limitations

3.1 Synthesis Scalability

Commercialization of large-scale production of rare earth-doped metal oxide nanostructures is a significant challenge. Traditional methods, such as sol–gel and hydrothermal synthesis, are multistep, which can slow down the production process and render it difficult to maintain material quality as batch size increases. Equipment used, e.g., high-temperature furnaces and pressure reactors, is costly and hard to scale up for commercial-level production. Older batch processes are less efficient, but newer continuous processes are more scalable but less common.

The price of rare earth elements is also an issue, as they are costly and not always readily available. Additionally, the synthesis routes consume a great deal of energy, which adds to operational expenses [29]. Environmental concerns, such as waste production, can be present during mass production. Additionally, it is challenging to keep material quality and consistency at higher scales, and the process of material testing becomes more labor-intensive.

Table 13.2 Comparative analysis of properties in rare earth-dope versus undoped metal oxides

Metal oxide system	Undoped metal properties	Rare earth-doped properties	Benefits of doping	References
ZnO (Dy, Fe, Co-doped)	Diamagnetic behavior at high magnetic field	For 2% Dy-doped ZnO: paramagnetic with weak ferromagnetism and For 2% Fe CO-doped 2% DY doped ZnO: antiferromagnetic with weak ferromagnetism behavior	Enhances spintronic performance via combined doping of Fe and DY	[23]
TiO_2 (Ce-doped)	Lower photocurrent density, hydrogen yield is 12.64 μmol.cm^{-2}.h^{-1}	Improved photocurrent density ~ 2.6 times more, more hydrogen yield ~ 37.03 μmol.cm^{-2}.h^{-1}	Boosts hydrogen production efficiency, photoelectrolysis of water	[24]
TiO_2 (Ce-doped)	Bandgap: ~ 3.2 eV	Bandgap narrowed to 2.75 eV, 88% UV & 77% visible degradation of Spiramycin antibiotic	Better pollutant removal under both UV and visible light	[25]
ZrO_2 (Ce-doped)	Bandgap is 5.1 eV	Bandgap reduced to ~ 3.50 eV, better conductivity ~ 4.23×10^{-5} S.m^{-1}	Enhances dielectric properties; suitable for high-k materials and electronics	[26]
ZnO (Nd-doped)	Bandgap 3.29 eV, high resistivity 1.54×10^{-3} Ω.cm	Reduced bandgap (3.21 eV), resistivity dropped to 0.26×10^{-3} Ω.cm	Improves optoelectronic properties; good for LEDs and solar cells	[27]
TiO_2 (Mn/Ce co-doped)	Bandgap ~ 3.23 eV, weak visible light activity	Reduced bandgap, improved UV–visible absorption, modified lattice	Broad spectrum light absorption, potential in photovoltaic cells	[28]

3.2 *Material Stability*

For the applicability of rare earth-doped metal oxide nanostructures for a large span of time, the stability factor is very important. But these materials face issues with the mechanical, chemical, and thermal stability. Most oxides undergo phase transformation at high temperatures, which can disrupt their structure and functionality. These get easily contaminated by atmospheric chemicals, moisture, and humidity, which leads to the corrosion and loss of functionality. Nanostructures with small size and high surface area can make them brittle, and serving doping can introduce defects and reduce the overall strength. Long-term mechanical stress, chemical exposure, and thermal cycling are some factors which can cause aging, reducing the functionality.

The use of composite materials, protective coatings and improved synthesis conditions to increase the crystallinity are some strategies being investigated to improve the stability factor in these materials. To make these nanostructures economically feasible for real-world applications the stability issues need to be resolved.

3.3 Economic Considerations

For the rare earth-doped metal oxide nanostructures to be widely used, their financial sustainability is important. Despite their great potential, there are some financial barriers. The cost for the extraction and processing of these rare earth elements (REEs) is very high, and the varying market prices impact the total costs. The cost is further increased by the fact that these REEs are limited in nature. The synthesis techniques used to synthesize these materials, like sol–gel and hydrothermal methods are also very expensive and energy consuming posing challenges to their large scale production. For small-scale businesses, scaling up laboratory scale to industrial scale production requires a significant equipment investigation. The success of these materials will depend on demand in sectors like energy storage, catalysis, and optoelectronics, but competition from other technologies may have an impact on prices. Because producers must follow sustainability guidelines, environmental regulations further raise costs. Finally, when evaluating the long-term economic viability of these materials, it is critical to consider the total life cycle costs of production, use, and disposal. The economic landscape of rare earth-doped metal oxides is anticipated to change in the direction of more creative applications as research advances and new approaches to cost containment and sustainable practices are created [30].

4 Future Work

4.1 Advancements in Characterization Techniques

Improving our understanding and analysis of rare earth-doped metal oxide nanostructures is crucial for their future. To reveal their structure, functionality and behavior, different characterization techniques are utilized. Today scientists are able to follow the real-time evolution during synthesis with in-situ methods, making it possible to study dynamic processes such as charge transport or catalytic activity. Precise details of rare earth-doping effect are given by spectroscopic methods such as Raman spectroscopy and IR spectroscopy, and synchrotron based X-ray absorption. High-resolution imaging equipment like TEM, AFM, and STM give a detailed view of the surface morphology and interior architecture at the nanoscale. Machine learning and computer simulation are becoming high-powered tools for recognizing patterns in large datasets and predicting material response. Finally, having guidelines and standard procedures for testing these materials is necessary in providing consistency, reliability, and innovation in the field. Overall, all of these innovations will not only cause us to know more but will also cause us to move faster application and implementation on the ground in different sectors.

4.2 Sustainable Synthesis Approaches

Demand for doped-rare earth metal oxide nanostructures is increasing, and hence it is important to emphasize environmentally friendly and long-term sustainable synthesis approaches. This includes the use of green chemistry principles, including the use of non-toxic, environmentally benign precursors and replacing greener solvents in place of toxic solvents. Manufacturing can also be made more sustainable by energy efficient synthesis, including low temperature synthesis and sustainable power sources. Using green catalysts and recycling byproducts to minimize waste serves to support the concepts of the circular economy. Microreactors and continuous flow systems are just some of the new technologies that enable greater efficiency, scalability, and less waste. Researchers are able to assess the environmental implications of each method through using sustainability indicators and lifecycle analysis. By combining innovation with sustainability, the development of such materials can go on in a way that is both good business and good for the planet.

4.3 Expanding Applications in Emerging Technologies

Rare earth-doped metal oxide nanostructures are being used increasingly in advanced technologies with new solutions for industries. They have the ability to improve the efficiency of solar cells and improve energy storage devices like batteries and supercapacitors by offering greater charge capacity and lifespan in renewable energy. In sophisticated catalysis, the materials reduce toxic emissions in catalytic converters and enable eco-friendly reactions in environmental processes like CO_2 reduction and wastewater treatment. In electronics, they improve the brightness and color accuracy of LEDs and hold potential in quantum dots and photonic devices that can transform communication and data technologies. They offer improved imaging for diagnostic purposes in biomedicine and can be utilized in drug delivery systems for targeted delivery to maximize efficacy and reduce side effects. In the environmental cleanup process, they can filter out contaminants from water and air and be utilized as sensitive materials in sensors for real-time environmental monitoring. With such a wide range of applications, these nanostructures are at the forefront of developing more sustainable, efficient, and effective technologies that respond to global challenges and enhance everyday life.

5 Long-Term Research Directions

Rare earth-doped metal oxide nanostructures are increasingly finding applications in novel technologies that provide new industrial solutions. Through the enhancement of renewable energy source lifespan and charge capacity, they can enhance solar cell efficiency as well as energy storage devices such as batteries and supercapacitors. Advanced catalysis materials minimize toxic emissions in catalytic converters and ensure environmentally benign reactions in applications such as wastewater treatment and CO_2 reduction. They enhance LEDs light intensity and color fidelity in electronics and have the potential for application in quantum dots and photonic devices, which may revolutionize data and communication technologies. They can be employed in drug delivery systems in targeted delivery to have enhanced efficacy and less side effects, and they provide better imaging for diagnosis in biomedicine. They can remove impurities from water and air for environmental cleanup and can be employed as sensitive materials in sensors for real-time environment monitoring. With all these varied uses, these nanostructures form a major part of developing more sustainable, efficient, and effective technologies to address global challenges and enhance everyday life.

6 Conclusion

Rare earth-doped metal oxide nanostructures represent the frontier developments in material science, with unique optical, electrical, and magnetic properties. This chapter has discussed the fundamental characteristics of such materials, new advancements in synthesis techniques, and promises for applications in emerging technologies. Incorporation of rare earth elements into metal oxides unlocks new functionalities for enhancing the performance of devices in optoelectronics, catalysis, energy storage, and environmental remediation. Going ahead, in order to ensure increased adoption, overcoming the challenges of scalability, material stability, and economic feasibility will be quintessential in these material applications.

Interdisciplinary collaborations are going to be vital in transferring theoretical research to practical applications. Researchers with diverse expertise could address a lot of complicated issues and encourage innovation. With respect to their development and synthesis, sustainable approaches must accompany synthesis along with comprehensive lifecycle assessments to ensure that the materials meet their intended role and objectives with respect to environmental and societal improvement. Future work should focus on developing a deeper understanding of properties, new synthesis methods, and new applications which could address some global problems. Commitment to these long-term research directions will allow science to better exploit the potential of rare earth-doped metal oxide nanostructures for transformative benefits to society and in steps toward a sustainable future.

References

1. Patil, A.S., Patil, A.V., Dighavkar, C.G., Adole, V.A., Tupe, U.J.: Synthesis techniques and applications of rare earth metal oxides semiconductors: a review. Chem. Phys. Lett. **796**, 139555 (2022)
2. Zhou, Y.: Controllable design, synthesis and characterization of nanostructured rare earth metal oxides. Phys. Sci. Rev. **5**(3), 20180084 (2020)
3. He, Y., Zhou, W., Xu, J.: Rare earth-based nanomaterials for supercapacitors: preparation, structure engineering and application. Chemsuschem **15**(12), e202200469 (2022)
4. Zhang, J., et al.: A brief review of co-doping. Front. Phys. **11**(6), 117405 (2016)
5. Hossain, M.K., Ahmed, M.H., Khan, M.I., Miah, M.S., Hossain, S.: Recent progress of rare earth oxides for sensor, detector, and electronic device applications: a review. ACS Appl. Electron. Mater. **3**(10), 4255–4283 (2021)
6. Sharma, N., Choudhury, S.P.: Gas sensing using metal oxide semiconductor doped with rare earth elements: a review. Mater. Sci. Eng., B **307**, 117505 (2024)
7. Ahmad, I., Akhtar, M.S., Ahmed, E., Ahmad, M., Keller, V., Khan, W.Q., Khalid, N.R.: Rare earth co-doped ZnO photocatalysts: solution combustion synthesis and environmental applications. Sep. Purif. Technol. **237**, 116328 (2020)
8. Palanivel, B., Macadangdang, R.R., Jr., Hossain, M.S., Alharthi, F.A., Kumar, M., Chang, J.H., Gedi, S.: Rare earth (Gd, La) co-doped ZnO nanoflowers for direct sunlight driven photocatalytic activity. J. Rare Earths **41**(1), 77–84 (2023)
9. He, Y., Zhou, W., Li, D., Liang, Y., Chao, S., Zhao, X., Zhang, M., Xu, J.: Rare earth doping engineering tailoring advanced oxygen-vacancy Co3O4 with tunable structures for high-efficiency energy storage. Small **19**(8), 2206956 (2023)
10. Chen, Z.: Rare earth activated phosphors coating for photoluminescent textile (2016)
11. Wu, T.S., Syu, L.Y., Lin, B.H., Weng, S.C., Jeng, H.T., Huang, Y.S., Soo, Y.L.: Reduction of dopant ions and enhancement of magnetic properties by UV irradiation in Ce-doped TiO2. Sci. Rep. **11**(1), 7668 (2021)
12. Raza, T., Arshad, M., Batool, Z., Ahmad, M., Murtaza, I., Ali, A., Raza, M.A., Zaheer, H., Kousar, R.: High capacitive rare-earth co-doped transition metal/graphene oxide composites as effective electrode material for supercapacitors. J. Rare Earths (2025)
13. Ariga, K.: Nanoarchitectonics: what's coming next after nanotechnology? Nanoscale Horizons **6**(5), 364–378 (2021)
14. Gupta, D., Varghese, B.S., Suresh, M., Panwar, C., Gupta, T.K.: Nanoarchitectonics: functional nanomaterials and nanostructures—a review. J. Nanopart. Res. **24**(10), 196 (2022)
15. Munawar, T., Manzoor, S., Jabbour, K., Nisa, M.U., Sardar, S., Mukhtar, F., Osman, S.M., Esan, M.F., Ashiq, M.N., Iqbal, F.: Nanostructural engineered titanium dioxide by rare earth metals dual doping for electrochemical supercapacitor applications. J. Korean Ceram. Soc. **61**(4), 545–557 (2024)
16. Rosaiah, P., Rao, K.M., Vattikuti, S.V.P., Dhanalakshmi, R., Al-Zahrani, F.A.M.: Efficient electrochemical energy conversion and storage performance of combined rare-earth metal oxide-Mos2 nanocomposite. Available at SSRN 5274950
17. Park, S.J., Jeon, J.Y., Kang, B.C., Ha, T.J.: Wearable temperature sensors based on lanthanum-doped aluminum-oxide dielectrics operating at low-voltage and high-frequency for healthcare monitoring systems. Ceram. Int. **47**(4), 4579–4586 (2021)
18. Dutta, S., Chakraborty, T., Sharma, S., Mondal, D., Saha, A., Pradhan, A.K., Chakraborty, C., Das, S., Sutradhar, S.: Fabrication of rare earth-doped ZnO-PVDF flexible nanocomposite films for ferroelectric response and their application in piezo-responsive bending sensors. Dalton Trans. **53**(34), 14347–14363 (2024)
19. Fan, Y., Jia, Z., Zhang, Z., Gu, S., Du, W., Lin, D.: Flexible composites with rare-earth element doped polycrystalline particles for piezoelectric nanogenerators. Micromachines **15**(11), 1280 (2024)

20. Wang, M., Lin, B., Chen, Y., Liu, H., Ju, Z., Lv, R.: Fluorescence-recovered wearable hydrogel patch for in vitro detection of glucose based on rare-earth nanoparticles. ACS Biomater. Sci. Eng. **10**(2), 1128–1138 (2024)
21. Nehra, P., Khatri, A., Kumar, A., Singh, S., Rana, P.S.: Synthesis and characterization of rare earth metal doped tungsten trioxide photocatalyst for degradation of Rhodamine B dye. J. Rare Earths **41**(9), 1302–1310 (2023)
22. Nadolna, J., Mikolajczyk, A., Wyrzykowska, E., Mazierski, P., Grzyb, T., Wei, Z., Kowalska E, Caicedo PN, Zaleska-Medynska A, Puzyn T., Puzyn, T.: Visible-light photocatalytic activity of rare-earth-metal-doped Tio2: experimental analysis and machine learning for virtual design. Available at SSRN 4633102
23. Malhotra, J.S., Singh, A.K., Khosla, R., Sharma, S.K., Sharma, G., Kumar, S.: Investigations on structural, optical and magnetic properties of Fe and Dy co-doped ZnO nanoparticles. J. Mater. Sci.: Mater. Electron. **29**(5), 3850–3855 (2018)
24. Lin, B.L., Chen, R., Zhu, M.L., She, A.S., Chen, W., Niu, B.T., Chen, Y.X., Lin, X.M.: Enhanced photoelectrochemical water splitting performance of Ce-Doped TiO2 Nanorod array Photoanodes for efficient hydrogen production. Catalysts **14**(9), 639 (2024)
25. Khan, H.: Cerium-doped titanium dioxide (CeT) hybrid material, characterization and Spiramycin antibiotic photocatalytic activity. Catalysts **15**(6), 512 (2025)
26. Sravanthi, M.N., Sudagar, J.: High temperature studies on cerium dioxide-doped zirconium oxide for structural, optical and dielectric properties. ES Mater. Manuf. **28**, 1493 (2025)
27. Nowsherwan, G.A., Azhar, M., Anwar, N., Ahmed, M., Ali, Q., Nowsherwan, N., Riaz, S., Naseem, S., Lai, W.C.: Exploring the advanced optoelectronic properties of Nd-Doped ZnO thin films for next-generation perovskite solar cells. ACS Omega **10**(18), 18251–18269 (2025)
28. Elafia, Z., Messous, M.Y., Hammouti, B., Cherkaoui, M.: Synthesis and characterization of Mn/Ce-Doped TiO2: investigation of structural and optical properties. Moroc. J. Chem. **11**(2), J-Chem. (2023)
29. Qian, F., Hu, C., Jiang, W., Zhang, J., Peng, L., Song, L., Chen, Q.: General and scalable strategy for synthesis of Pt-rare earth alloys as highly durable oxygen reduction electrocatalysts. Chem. Eng. J. **468**, 143665 (2023)
30. Mehtab, A., Ahmed, J., Alshehri, S.M., Mao, Y., Ahmad, T.: Rare earth doped metal oxide nanoparticles for photocatalysis: a perspective. Nanotechnology **33**(14), 142001 (2022)

MIX
Papier aus verantwortungsvollen Quellen
Paper from responsible sources
FSC® C105338
FSC
www.fsc.org

If you have any concerns about our products,
you can contact us on
ProductSafety@springernature.com

In case Publisher is established outside the EU,
the EU authorized representative is:
Springer Nature Customer Service Center GmbH
Europaplatz 3, 69115 Heidelberg, Germany

Printed by Libri Plureos GmbH
in Hamburg, Germany